Advances in Intelligent and Soft Computing

145

T0137899

Editor-in-Chief

Prof. Janusz Kacprzyk
Systems Research Institute
Polish Academy of Sciences
ul. Newelska 6
01-447 Warsaw
Poland
E-mail: kacprzyk@ibspan.waw.pl

For further volumes:
http://www.springer.com/series/4240

Ford Lumban Gaol and Quang Vinh Nguyen (Eds.)

Proceedings of the 2011 2nd International Congress on Computer Applications and Computational Science

Volume 2

 Springer

Editors
Ford Lumban Gaol
Bina Nusantara University
Perumahan Menteng
Caking
Indonesia

Quang Vinh Nguyen
School of Computing and Mathematics
University of Western Sydney
Penrith
New South Wales
Australia

ISSN 1867-5662   e-ISSN 1867-5670
ISBN 978-3-642-28307-9   e-ISBN 978-3-642-28308-6
DOI 10.1007/978-3-642-28308-6
Springer Heidelberg New York Dordrecht London

Library of Congress Control Number: 2012932484

Printed on acid-free paper

Springer is part of Springer Science+Business Media (www.springer.com)

Foreword

The Proceedings of the 2011 2nd International Congress on Computer Applications and Computational Science (CACS 2011) are a compilation of current research findings in computer application and computational science. CACS 2011 comprises a keynote speech, plenary speech, and parallel sessions where all of the papers have been reviewed thoroughly by at least three reviewers and the program chairs.

This book provides state-of-the-art research papers in Computer Control and Robotics, Computers in Education and Learning Technologies, Computer Networks and Data Communications, Data Mining and Data Engineering, Energy and Power Systems, Intelligent Systems and Autonomous Agents, Internet and Web Systems, Scientific Computing and Modeling, Signal, Image and Multimedia Processing, and Software Engineering.

The book provides an opportunity for researchers to present an extended exposition of such new works in all aspects of industrial and academic development for wide and rapid dissemination. At all times, the authors have shown an impressive range of knowledge on the subject and an ability to communicate this clearly.

This book explores the rapid development of computer technology that has an impact on all areas of this human-oriented discipline. In the recent years, we are faced with new theories, new technologies and computer methods, new applications, new philosophies, and even new challenges. All these works reside in industrial best practices, research papers and reports on advanced inter-discipline projects.

The book makes fascinating reading and should be of great interest to researchers involved in all aspects of computational science and computer applications, and will help attract more scholars to study in these areas.

Ford Lumban Gaol
Quang Vinh Nguyen

Contents

A Novel Approach for Extracting Nociceptive-Related Time-Frequency Features in Event-Related Potentials

Li Hu, Weiwei Peng, and Yong Hu

Abstract. In the present study, we are mainly objective to develop a novel approach to extract nociceptive-related features in the time-frequency domain from the event-related potentials (ERPs) that were recorded using high-density EEG. First, the independent component analysis (ICA) was used to separate single-trial ERPs into a set of independent components (ICs), which were then clustered into three groups (symmetrically distributed ICs, non-symmetrically distributed ICs, and noise-related ICs). Second, the time-frequency distributions of each clustered group were calculated using continuous wavelet transform (CWT). Third, the principal component analysis (PCA) with varimax rotation was used to extract time-frequency features from all single-trial time-frequency distributions across all channels. Altogether, the developed approach would help effectively extracting nociceptive-related time-frequency features, thus yielding to an important contribution to the study of nociceptive-specific neural activities.

Li Hu
Department of Orthopaedics and Traumatology,
The University of Hong Kong, Hong Kong,
China Key Laboratory of Cognition and Personality
and School of Psychology, Southwest University,
Chongqing, China
e-mail: hultju@gmail.com

Weiwei Peng · Yong Hu
Department of Orthopaedics and Traumatology,
The University of Hong Kong, Hong Kong, China
e-mail: ww.peng@gmail.com, yhud@hkucc.hku.hk

F.L. Gaol et al. (Eds.): Proc. of the 2011 2nd International Congress CACS, AISC 145, pp. 1–6.
springerlink.com   © Springer-Verlag Berlin Heidelberg 2012

1   Introduction

Pain sensation related components existed and merged in the multi-channel elec-troencephalography (EEG). However, it is usually sank in a lot of EEG back-ground signals. Pain sensation related components existed and merged in the multi-channel electroencephalography (EEG). However, it is usually sank in a lot of EEG background signals. The feature of pain related EEG is a problem in current studies.

The pain related features could be extracted from transient increase (event-related synchronization [ERS]) or a transient decrease (event-related desynchroni-zation [ERD]) in synchrony of the underling neuronal populations [1-2]. These transient modulations (ERS and ERD) can be explained as the reflection of changes in parameters that control oscillations in neuronal networks [1], and may represent the neuronal mechanisms involved in cortical activation, inhibition and binding [1, 3-4]. In addition, the magnitude of ERS and ERD may be modulated by changes in stimulus intensity and repetition [5-7], and fluctuations in vigilance, expectation, task complexity and emotional status [3, 8-9].

As a type of event-related modulations, laser-evoked potentials (LEPs) are cor-tical brain responses elicited by laser stimuli. LEPs are considered the best tool for assessing function of nociceptive pathways in physiological and clinical studies [10-12] as laser heat pulses excite Aδ and C fibre free nerve endings in the super-ficial skin layers selectively (i.e. without coactivating Aβ mechanoreceptors) [13-14], and thus provide a purely nociceptive input.

In the present study we develop a novel approach to extract nociceptive-related features in the time-frequency domain from the event-related potentials (ERPs) that were recorded using high-density EEG.

2   Methods

2.1   Subject

EEG data were collected from 37 healthy volunteers (18 females) aged 27 ±4.5 (mean ±SD, range= 22 to 41 years). All participants gave their written informed consent and were paid for their participation. The local ethics committee approved the procedures.

2.2   Nociceptive Stimulation and Experimental Design

Radiant-heat stimuli were generated by an infrared neodymium yttrium aluminium perovskite (Nd:YAP) laser with a wavelength of 1.34 μm (Electronical Engineer-ing, Italy). Laser pulses activate directly nociceptive terminals in the most superfi-cial skin layers. Laser pulses were directed at the dorsum of both left and right hand and foot, on a squared area (5x5 cm) defined prior to the beginning of the experimental session. A He-Ne laser pointed to the area to be stimulated. The laser pulse was transmitted via an optic fibre and its diameter was set at approximately 6 mm (28 mm2) by focusing lenses. The pulse duration was 4 ms. One energy of stimulation was used in each of the four conditions. The average energies was 2.2±0.3 J (right hand). At these energies laser pulses elicited a clear pinprick pain, related to the activation of Aδ fibres. After each stimulus, the laser beam target was shifted by approximately 1 cm in a random direction, to avoid nociceptor fatigue or sensitization.

Before the recording session the energy of the laser stimulus was individually adjusted using the method of limits (laser step size: 0.25 J). In the recording session, laser-evoked EEG responses were obtained following the stimulation of the dorsum of the right hand. In total, we delivered 30 laser pulses, using an inter-stimulus interval (ISI) ranging between 5 and 15 s.

2.3   EEG Recording and Analysis

Participants were seated in a comfortable chair in a silent, temperature-controlled room. They wore protective goggles and were asked to focus their attention on the stimuli and relax their muscles. The electroencephalogram (EEG) was recorded using 64 Ag-AgCl scalp electrodes placed according to the International 10-20 system, referenced against the nose. Electro-oculographic (EOG) signals were simultaneously recorded using surface electrodes. Signals were amplified and digitized at a sampling rate of 1,000 Hz.

EEG data were processed using EEGLAB, an open source toolbox running in the MATLAB environment. Continuous EEG data were band-pass filtered between 1 and 30 Hz. EEG epochs were extracted using a window analysis time of 1500 ms (500 ms pre-stimulus and 1000 ms post-stimulus) and baseline corrected using the pre-stimulus interval. Trials contaminated by eye-blinks and movements were corrected using an Independent Component Analysis (ICA) algorithm. In all datasets, individual removed independent components (ICs) had a large EOG channel contribution and a frontal scalp distribution.

First, epoched EEG data were decomposed using ICA, which is the method to perform blind source separation of scalp EEG. When applied to multi-channel EEG recordings, ICA unmixes the signal recorded on the scalp into a single linear combination of ICs, each having a maximally independent time course and a fixed scalp distribution. The separated ICs were then clustered into three groups (symmetrical cluster, non-symmetrical cluster, and noise).

Second, the separated EEG signal (both symmetrical cluster and non-symmetrical cluster) across all recording electrodes were then translated into the time-frequency domain using continuous wavelet transform (CWT). It offers an optimal compromise for time-frequency resolution and is therefore well suited to explore event-related modulations of the EEG spectrum in a wide range of frequencies. In addition, the time-frequency analysis is optimal to explore not only the phase-locked responses (ERP) but also the non-phase-locked responses (ERS/ERD).

At last, both phase-locked and non-phase-locked brain responses were mixed on the time-frequency plane, and these joint time-frequency components (ERP/ERS/ERD) from the time-frequency representation should be separated correctly before extracting nociceptive-related time-frequency features. Recent studies indicated that principal component analysis (PCA) with varimax rotation was a suitable method to separate ERP, ERS and ERD from a set of time-frequency representations of single-trial ERPs. PCA is a way of orthogonal transformation to convert a set of observations of possibly correlated variables into a set of values of uncorrelated variables (PCs), and varimax rotation is a change of coordinates used in PCA that maximizes the sum of the variances of the squared loadings.

3 Results

The time-frequency features of each cluster were extracted using PCA with varimax rotation. For the symmetrical cluster, four main PCs were displayed in Figure 1. The first PC (low frequency ERP), which explained more than 50% of the total variance and showed a maximal power at Cz, represented the N2-P2 complex in the time domain. The second PC, which displayed a maximal power at parietal and occipital regions, represented the non-phase-locked Alpha ERD. The third and fourth PCs, which demonstrated a maximal power at central and frontal regions, represented the non-phase-locked Beta band ERS.

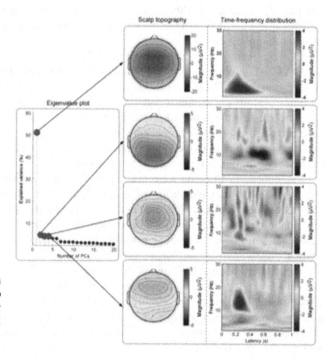

Fig. 1 PCA with varimax rotation to isolate time-frequency features of symmetrical cluster.

For the non-symmetrical cluster, PCA with varimax rotation provided three main PCs (Figure 2). The first PC, which was bilaterally distributed but displayed a clear maximum on the centro-parietal electrodes overlying the hemisphere contralateral to the stimulated side, represented the Mu band ERD. The second PC, which maximal at the contralateral centro-parietal region, represented the N1 wave in the time domain. The third PC, which maximal at bilaterally occipital regions, represented a typical Alpha band ERD.

Among all these time-frequency features, both Mu band ERD and N1-related ERP were maximal at the contralateral centro-parietal regions, which is compatible with the somatotopical representation of the body in the postcentral gyrus, thus possibly serving as the "body marker" of LEP.

Fig. 2 PCA with varimax rotation to isolate time-frequency features of non-symmetrical cluter. Note that the scalp topographies of mu-ERD (top part) and N1-related ERP (middle part) are maximal at the controlateral central regions, which are compatible with the somatotopic representation of the body in the postcentral gyrus.

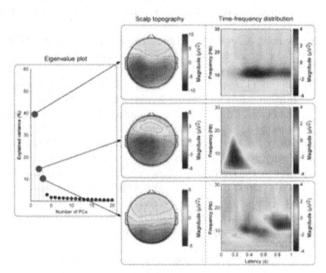

4   Discussion

In the present study, we developed a novel approach based on ICA, time-frequency analysis, and PCA with varimax rotation, which is able to separate time-frequency features effectively and reliably. This approach would be easily applied in various ERP studies across different modalities. Among several time-frequency features that were obtained using the proposed approach, both Mu band ERD and N1-related ERP were possibly represented as the "body marker" of LEPs, which may related to the nociceptive-specific neural activities.

Acknowledgment. This work was partially supported by S.K. Yee Medical Foundation (207210), and a grant from the Research Grants Council of the Hong Kong (GRF 712408E).

References

[1] Pfurtscheller, G., Lopes da Silva, F.H.: Event-related EEG/MEG synchronization and desynchronization: basic principles. Clin. Neurophysiol. 110, 1842–1857 (1999)
[2] Mouraux, A., Iannetti, G.D.: Across-trial averaging of event-related EEG responses and beyond. Magn. Reson. Imaging 26, 1041–1054 (2008)
[3] Ploner, M., et al.: Pain suppresses spontaneous brain rhythms. Cereb. Cortex 16, 537–540 (2006)
[4] Mouraux, A., et al.: Non-phase locked electroencephalogram (EEG) responses to CO_2 laser skin stimulations reflect central interactions between A partial partial differential- and C-fibre afferent volleys. Clin. Neurophysiol. 114, 710–722 (2003)
[5] Ohara, S., et al.: Attention to a painful cutaneous laser stimulus modulates electrocorticographic event-related desynchronization in humans. Clin. Neurophysiol. 115, 1641–1652 (2004)
[6] Stancak, A., et al.: Desynchronization of cortical rhythms following cutaneous stimulation: effects of stimulus repetition and intensity, and of the size of corpus callosum. Clin. Neurophysiol. 114, 1936–1947 (2003)
[7] Iannetti, G.D., et al.: Determinants of laser-evoked EEG responses: pain perception or stimulus saliency? J. Neurophysiol. 100, 815–828 (2008)
[8] Del Percio, C., et al.: Distraction affects frontal alpha rhythms related to expectancy of pain: an EEG study. Neuroimage 31, 1268–1277 (2006)
[9] Mu, Y., et al.: Event-related theta and alpha oscillations mediate empathy for pain. Brain Res. 1234, 128–136 (2008)
[10] Bromm, B., Treede, R.D.: Laser-evoked cerebral potentials in the assessment of cutaneous pain sensitivity in normal subjects and patients. Rev. Neurol (Paris) 147, 625–643 (1991)
[11] Iannetti, G.D., et al.: Usefulness of dorsal laser evoked potentials in patients with spinal cord damage: report of two cases. J. Neurol Neurosurg. Psychiatry 71, 792–794 (2001)
[12] Cruccu, G., et al.: Recommendations for the clinical use of somatosensory-evoked potentials. Clin. Neurophysiol. 119, 1705–1719 (2008)
[13] Bromm, B., Treede, R.D.: Nerve fibre discharges, cerebral potentials and sensations induced by CO_2 laser stimulation. Hum. Neurobiol. 3, 33–40 (1984)
[14] Carmon, A., et al.: Evoked cerebral responses to noxious thermal stimuli in humans. Exp. Brain. Res. 25, 103–107 (1976)

Numerical Simulation of Turbulent Flow through a Francis Turbine Runner

Hu Ying and Hu Ji

Abstract. Three-dimensional turbulent viscous flow analyses for a Francis turbine runner are performed by solving Reynolds averaged Navier-Stokes equations closed with the realizable k-ε turbulence model. The discretization is carried out by the finite element method based FVM on a patched block-structured grid system. The boundary conditions for the turbulent properties are treated with a particular attention. In a new research view, CFX-TASCflow software are used to calculate the 3D unsteady turbulent flow in a model Francis turbine runner and to simulate the 3D turbulent flow field in it. The results at the design operating condition are presented in this paper. They show that the flow field structures and the characteristics of swirling flow and its developing process in the Francis turbine can be well predicted. It can be concluded that the results are able to provide important guidance for the hydraulic design of a Francis turbine or its optimization.

I   Introduction

As one of the most important components in turbine, runner can change the water potential energy into the rotating mechanical energy directly. Many characteristic parameters in turbine such as the flow capacity, the hydraulic efficiency, the efficiency of cavitation, the systematic stability of turbine during working and the adaptability for case, are closely related with the geometric model of the runner. In addition, the global layout of the turbine is also closely bound up with the geometric model of the runner. So, it is significant to design an optimized geometric model to a runner for a high performance turbine.

Hu Ying · Hu Ji
School of Mechanical and Electrical Engineering,
Kunming University of Science and Technology, Kunming, Yunnan, China
e-mail: huyingtennis@hotmail.com, 371836147@qq.com

F.L. Gaol et al. (Eds.): Proc. of the 2011 2nd International Congress CACS, AISC 145, pp. 7–12.
springerlink.com   © Springer-Verlag Berlin Heidelberg 2012

Traditionally, there are two methods to design a runner. One method is that by comprehensively considering the design parameters provided by the hydropower station, designers determine a rough runner model firstly, relying on their design experience. Then repeatedly, designers will not stop the model experiencing and modifying the geometric model of the runner until the performance of turbine meet the demand given by the hydropower station. Another method is to modify an existing turbine to satisfy some new demand, in which the selected runner's characteristic parameters are similar to the desired parameters given by the hydropower station. Since both methods need to conduct the model experiment dozens of times, they are time consuming and costly. In addition, it is not easy to measure the interior flow distribution visually in the model experiment. With the development of the Computational Fluid Dynamics (CFD), the technology of computer simulation is widely used as one of the powerful way to research the interior turbulent flow distribution of the runner [1, 2]. By analyzing the results of CFD software, we can modify or optimize the geometric model of the runner properly [3, 4].

In this study, a three-dimensional unsteady flow distribution in a Francis Turbine is simulated based on Reynolds time-averaged governing equations, using the Realizable k-ε turbulent mode in CFX-TASCflow software. And the refined three-dimensional interior turbulent flow distribution of the runner is obtained in the meantime.

2 Governing Equation

Since it requires fairly high computational cost to solve the Navier-Stokes equation in the case of turbulent flow, in practice, it is probably impossible to solve the three-dimensional Navier-Stokes equation directly. A time average method is introduced by engineers to solve the transient Navier-Stokes equation. Generally speaking, this method regards the turbulent flow as a composition of two simple flows—the time average flow and the transient pulsating flow. The Reynolds averaged variable is defined as below.

$$\bar{\phi} = \frac{1}{\Delta t} \int^{+\Delta t} \phi(t) dt$$

Where ϕ is the original variable, $\bar{\phi}$ is the Reynolds averaged variable. Let ϕ' denotes the transient pulsating variable; the relationship among these three variables is expressed as below.

$$\phi = \bar{\phi} + \phi' \ (\phi = V, u, v, w, p)$$

Combining the Reynolds averaged equation, we can derive both the time averaged continuity equation and the Reynolds averaged Navier-Stokes equation, RANS for short. For simplicity, we omit the overline label, which distinguishes the time averaged variable from the original variable. Only the time averaged transient pulsating variable make an exception.

$$\frac{\partial}{\partial x_i}(\rho u_i) = 0 \tag{1}$$

$$\frac{\partial}{\partial x_j}(\rho u_i u_j) = -\frac{\partial p}{\partial x_i} + \frac{\partial}{\partial x_j}(\mu \frac{\partial u_i}{\partial x_j} - \rho \overline{u_i' u_j'}) + s_i \tag{2}$$

Where $-\rho \overline{u_i' u_j'}$ is the Reynolds stress.

Because (1) and (2) are not closed, a turbulence model must be applied to close the equations.

3   Numerical Computation

To solve the equation group (1) (2), some hypothesis about Reynolds stress must be brought in [5]. In this study, we use the Realizable k-ε turbulent mode as the hypothesis to help us solve the equation group (1) (2). The relationship between Reynolds stress and gradient of average velocity, namely, the relationship among viscosity of the turbulent fluid, the kinetic energy of turbulent fluid and the dissipation rating , is obtained by utilizing the hypothesis of Realizable k-ε turbulent model [6].

In this study, the geometric model is divided into several sub-geometries. The Body-Fitted Coordinates is used when we mesh these sub-geometries with the structured network. The central finite difference method is adopted at the diffusion grid area, and the second upwind difference method is used for at convective grid area. The algorithm of implicit time-integration method is used during we solved the equation set based on the Finite Volume Method. This method ensures that the Kinematic Equation and the Continuity Equation could be solved directly and simultaneously, preventing solving process from the repeat of assuming initial pressure, solving and modifying the pressure. By utilizing this method, the calculation process is faster and more stable than the traditional.

The boundary condition is defined as the boundary of inlet, outlet, the wall and periodical boundary condition.

4   Case Study

In this study, a three-dimensional unsteady flow distribution of the runner in a Francis Turbine is simulated by using the CFX-TASCflow software. In this working condition, the flux is 762.8l/s, the rotate speed is 72.85r/min, water head is 30m, and the diameter of the runner is 0.36952m. This Francis Turbine has 15 blades, and 15 computational areas are contained congruously. 2970 nodes in total are generated in this model.

The geometric model of this runner as shown in fig.1 is designed in the interactive turbine software, named BladeGen. And the grid, as shown in fig.2, is generated in TurboGrid software. TurboGrid has the most efficient, innovative meshing algorithm, and the high-quality grid can be automatically generated by this program for the most of the geometric model of the runner.

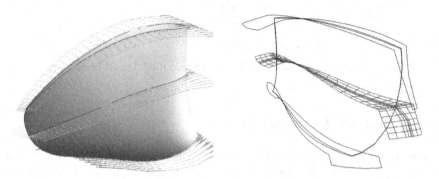

Fig. 1 Computer visualization of a model runner blade

Fig. 2 Grids on mid-section in a flow passage

Velocity field of the cross section of the blade is displayed in fig.3(a), fig.3(b) and fig.3(c). To be specific, the velocity field at the top part of the blade is displayed in fig.3(a). And fig.3(b) and fig.3(c) shows the velocity field at the middle part and the bottom part of the blade respectively. As shown in these pictures, the relative velocity angle of the flow is in accord with the angle of the positioned blade. There is no impact at the regions near the blade, and also, there is no reversed flow or vortex etc. around the blade. So, a reasonable velocity field is obtained in this simulation. In addition, we can see from these pictures that velocity of the flow accelerates at the end of the blade section, because the flow channel gradually shrinks from the inlet to the outlet.

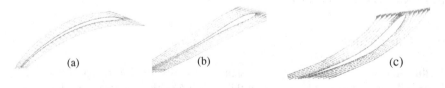

Fig. 3 Velocity vectors: (a)   near crown   (b)mid-section (c)near shroud

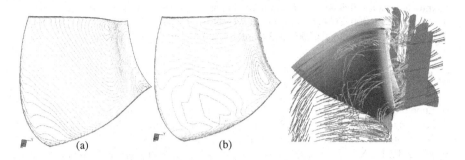

Fig. 4 Pressure contour plot (a)on face of blade   **Fig. 5** Trace plot through flow passage
(b)on back of blade

Surface pressure on the blade of the runner is displayed in fig.4(a) and fig.4(b). The contour plot of the pressure on the front surface of the blade is displayed in fig.4(a), and the pressure on the back surface of the blade is displayed in fig.4(b). From these contour plots, we can see the pressure is evenly distributed on the both surfaces of the blade, and the pressure on the front surface of the blade is little bigger than it on the back. Surface pressure gradually decreased from the inlet of the blade to the outlet. This tendency is accord with the reality. There is a possibility for cavitation to be generated at the back surface of the blade because low pressure appeared at a large area of the outlet of the blade.

The result of the three-dimensional flow trace of the Francis Turbine from the inlet of the guide vane to the outlet of the runner is shown in fig.5. Almost all the detail of the flow lines are shown visually and explicitly in this picture. As it is shown, the flow distribution in the runner's flow channel is changed gradually and it is rather ideal for this flow distribution. A high hydraulic efficiency can be obtained by this runner.

5   Conclusion

By utilizing the Realizable k-ε turbulent mode and the Navier-Stokes equation, the three-dimensional flow distribution in the flow channel of the runner can be simulated with the proper boundary condition in the CFD-TASCflow software. The flow distribution at the cross section of the blade and the surface pressure on the blade also can be figured out by calculation based on the solution of three-dimensional flow distribution. The modification of the geometric model of the runner can be instructed by these simulated results, and a better hydraulic efficiency runner can be found by repeat this numerical method. Facts have shown that an optimal geometric model of the runner can be obtained by simulating the three-dimensional flow distribution in the flow channel of the runner, using the CFD software. This method can reduce the number of the trails of the experiment and the cost of design, also, it can shorten the design circle. It lays a foundation for a high efficient, reasonable and reliable way of a runner design.

Acknowledgment. Science Foundation of Yunnan Provincial Science and Technology Department (No. 2009ZC029) and Research Foundation for Doctors, Kunming University of Science and Technology（2009-023）are thanked for their financial support.

References

1. Enomoto, Y.: Design optimization of a Francis turbine runner using multi-objective genetic algorithm. In: 22nd IAHR Symposium on Hydraulic Machinery and Systems, Stockholm, Sweden, June 29-July 2, vol. A01-2, pp. 1–10 (2004)
2. Wang, F., Li, X., Ma, J.: Experimental investigation of characteristic frequency in unsteady hydraulic behaviour of a large hydraulic turbine. Journal of Hydrodynamics 21, 12–19 (2009)
3. Hu, Y., Cheng, H., Hu, J., Li, X.: Numerical simulation of unsteady turbulent flow through a Francis turbine. Wuhan University Journal of Natural Sciences 16, 179–184 (2011)
4. Shao, Q., Liu, S., et al.: Three-Dimensional simulation of unsteady flow in a model Francis hydraulic turbine. Tsinghua Science and Technology 9, 694–699 (2004)
5. Shih, T.-H., Liou, W.W., Shabbir, A.: A New k-ε Eddy-Viscosity Model for High Reynolds Number Turbulent Flows-Model Development and Validation. Computers & Fluids 24, 227–238 (1995)
6. Tao, W.: Numerical Heat Transfer, pp. 432–473. Jiaotong University Press (2001)
7. TASCflow Theory Documentation. Advanced Scientific Computing Ltd., Canada (1995)

Learning Is a Team Sport … Analysis of a Model for Interactions in Computer Science Education

Leila Goosen

Abstract. This paper introduces Computer Science students' need to learn team interaction skills. A model that emphasizes the importance of structuring team interactions for effectiveness in Computer Science education by ensuring that key variables are present, is then offered. Findings regarding an analysis of this model in terms of positive interdependence, individual accountability, interpersonal and small-team skills, promotive interaction and team processing is further detailed. This is done by indicating to what extent, and how, educators ensured that the key variables were implemented. The paper finally offers conclusions, including a summary and an indication of the importance of this research.

1 Introduction

Computer Science "projects are growing in complexity, and require teams to solve problems and develop solutions" [1 p. 460]. Such "team projects provide students with an opportunity to" develop shared mental models and clarify their ideas. [2] believes that when team members engage in such coordinated efforts to share in the interactive creation of a joint solution to the problem at hand, as they learn new concepts during team interactions, their actions "necessarily involve... discussion between the team members" [2 p. 40]. Students feel that such discussions provide "a method that significantly fosters the development of a wider breadth of knowledge" [3 p. 17], and enables them to "construct and extend conceptual understanding of what is being learned" [4 p. 28]. Many benefits of team-based learning also correspond to transferable skills that "change the ways in which students learn" [2 p. 33] - not only while interacting in teams, but in such a way "that they can subsequently perform higher as individuals" [4 p. 23].

Leila Goosen
University of South Africa, PO Box 392 UNISA 0003
e-mail: GooseL@UNISA.ac.za

F.L. Gaol et al. (Eds.): Proc. of the 2011 2nd International Congress CACS, AISC 145, pp. 13–18.
springerlink.com

The examples described above only represent some of the uses "of teams in education (as) supported by the literature. Studies show that" industry increasingly "highly values team skills in graduates" [5 p. 465], as various forms of team interactions have been applied in learning, and are accepted as "part of the 21st century educational lexicon" [2 p. 34]. In order for Computer Science students to be responsive to these changing requirements, while learning programming, they must also be provided "with team experiences that will allow them to" develop the necessary skills to "become effective and productive team players in society and the workplace" [1 p. 461]. However, these same authors refer to students who "graduate with technical skills, but lack team project experiences" [1 p. 460], as educators "often refrain from using teams because of the effort involved with ... team assessment" or administration. [4 p. 26-27] believe that educators "tend to have a very limited understanding of", and "need considerably more sophistication about the ways in which", they should "structure and monitor meaningful learning experiences for students" [6 p. 132] in terms of key variables to realize effective team interactions. This paper will therefore focus on the analysis of a model for team interactions in Computer Science education.

2 Key Variables of Effective Team Interactions

Many educators think they use effective team interactions for learning, "when in fact they are simply seating students together" [4 p. 26]. According to [7 p. 366], the following key variables mediate the effectiveness of team interactions:

1. *Positive interdependence* exists when individual team members "perceive that they can attain their goals if and only if the other individuals with whom they are ... linked (in the team) attain their goals" [7 p. 366].
2. *Individual accountability* is present "when the performance of individual students is assessed, the results are given back to the individual and the (team), and the member is held responsible by (other team members) for contributing his or her fair share to the group's success" [4 p. 23].
3. "*Promotive interaction* occurs as individuals encourage and facilitate each other's efforts to accomplish the group's goals" [7 p. 368].
4. The appropriate use of *social skills* requires that *interpersonal and small team skills* such as trust building, "communication, and conflict-management skills have to be taught" [4 p. 24].
5. *Team "processing* occurs when group members ... reflect on which member actions were helpful and unhelpful and ... make decisions about which actions to continue or change" [7 p. 369].

These key variables, together with the allocation of specific roles in teams and assessment in a team context, formed part of a model for team interactions in Computer Science education accessible in a previous paper [8]. The research presented in this paper consisted of an analysis of the ways in which team interactions were being used, specifically in terms of the key variables. This was done to determine the effectiveness of the model, as well as identify possible shortcomings. The analysis compared the key variables as described in the model with the experiences of Computer Science educators, in terms of how they implemented these - their final year students completed team projects as part of their assessment requirements. Semi-structured interviews with educators and document analysis were used as sources of data.

3   Analysis of a Model for Team Interactions

3.1   Positive Interdependence

Educators strengthened *positive interdependence* during team interactions by ensuring that the project was "divided so that each group member (was) responsible for doing one aspect of the assignment" [7 p. 367]. In the projects investigated, students decided between themselves how to divide their projects in terms of the allocated sections they would each complete, with 5% of each student's final mark allocated based on fair division of tasks between team members.

Results in a study by [9 p. 178] indicate that "the majority of the students prefer that a small portion of the" incentives received from achieving the common goal (in this case marks for the project) "be allocated based on individual contributions, while the majority ... is divided equally among the team members". In contrast, in the team projects investigated, more than half of students' marks were allocated "according to the personal contributions of each team member".

3.2   Individual Accountability

Individual accountability is important for team effectiveness, since some members tend to be more than willing to let especially the academically stronger members of the team do all or most of the work [4], unless mechanisms are in place "so that students could be held responsible for their tasks" [1 p. 460]. Individual accountability was initially established by requiring teams to submit the task allocation to each team member to educators at the first due date. Each team member was held individually accountable for those parts of the project that the team had agreed was her/his responsibility, in that 5% of the final team marks allocated were based on whether the teams completed their projects as initially planned in terms of the original division of labor indicated.

At each cyclic meeting, team members briefly stated what they had "been working on and what progress has been made" [5 p. 467]. These regular meetings, together with the staggered due dates used, "also motivated the students to make meaningful contributions so that they could have something significant to report" or submit. This is in line with the suggestion by [4 p. 27], "that students should be required to report on their own contributions to their project".

Individual assessment served "as the ultimate source of each member's personal accountability", with each one individually responsible for displaying their own knowledge and skills in programming by applying these to their parts of the project. Individual accountability was finally also structured by having each student write their class tests and examinations with regard to programming knowledge and skills individually [4 p. 23].

3.3 Interpersonal and Small-Team Skills

Unskilled team members cannot interact effectively. "Studies show that students require instruction on", and should "receive concrete definitions about *team skills*" team skills" [5 p. 465, 468]. Despite similar advise from [1 p. 462], students in the team projects investigated did not receive training to "identify the key elements of an effective (team) and learn the need for planning in order to effectively accomplish the designated task(s)". However, as indicated previously, by the first due date each team needed to provide the educator with an account of how they had planned their work, with details written down of, for example, the division of tasks between various members.

In order for them to be able to achieve their mutual goals, students have to be taught how to "communicate accurately and unambiguously (and) ... resolve conflicts constructively" [7 p. 369]. Students in the teams investigated did not receive any training in "resolving conflicts that arose within the team" [1 p. 460], which could provide direction if there were problems that needed to be solved.

The educators interviewed did not specifically mention anything about students creating ground rules that governed students' behavior in the teams. However, documentation did lay the bulk of responsibility with regard to regulating students' contributions to the team at the feet of the students themselves. Students were required to document all attempts to resolve issues around getting a team member to contribute, as well as the results of these attempts - only after many such attempts where they did not succeed, were students allowed to have the educator acknowledge that the team had made every attempt to sort the problem out.

3.4 Promotive Interaction

Promotive interaction, which "occurs as individuals encourage and facilitate each other's efforts to accomplish the group's goals", was supposed to take place each time the teams met [7 p. 368]. These meetings presented students with platforms for "exchanging needed resources, such as information and materials, and processing information more efficiently and effectively".

Educators tried to foster an environment in which students realized that "promotive interaction is characterized by individuals ... acting in trusting and trustworthy ways" within the teams, and team interactions where they provided each other with mutual support [4]. As the students sat in their teams to carry on their discussions at their cyclic meetings, they needed to provide "explanations, elaboration, and guidance to help their peers understand … key principles related to" their responsibilities within the project [6 p. 125]. This also included them encouraging each other to exert the effort needed to achieve their mutual goal [7]: the successful completion of the project. They needed to inspire each other to such an extent that team members' possible competing agendas were overcome when "faced with a conflict between their urge to express personal skills, and the unavoidable need to cooperate with their teammates" [9 p. 178].

3.5 Team Processing

Educators were encouraged to have as much of the team interactions as possible happen in class. Observations by the educator could thus be "implemented to effectively monitor the progress of" [5 p. 469] "each group and documenting the contributions of each member" [4 p. 24], which could help students to "reflect on how well they (were) functioning and how they may improve their learning processes". In this way, team "processing may result in … (continuously) improving students' skills in working as part of a team".

Educators were also able to produce an assessment "of the work produced by students in teams" [2 p. 40] "based on their … responsibilities within the project" [1 p. 460], and to provide "appropriate feedback to students reflective of" their own observations in terms of students' participation in, and mastery and application of, team interaction skills [5 p. 467]. The results of team processing could finally be used to inform the actions to "be taken to correct situations where students (were) … persistently disrespectful or uncooperative" - these included that students could eventually "be required to complete the assignments on their own" [1 p. 462] or to work "among themselves rather than be an undue burden to other students' team experiences" [5 p. 468].

4 Conclusions

Computer Science students need to gain the "knowledge and experience … required to work effectively in teams" [1 p. 460], so that they can operationalize these interaction procedures "to make a difference in business" and industry, government, education and other organizational settings [7]. There is a significant body of literature available in Computer Science education that discusses team interactions in learning to program and computer team projects for students.

After thus introducing the paper, the journey towards the analysis of a model in Computer Science education was started by offering a review from literature of some of the key variables for effective team interactions in the second section. Section 3 "provides the basis for … understanding the internal dynamics that" made team interactions work – it shows how educators had structured the key variables for team interactions to be effective; so that students were "positively interdependent, individually accountable to do their fair share of the work, promote each other's success, appropriately use social skills, and periodically process how they can improve the effectiveness of their efforts" [4 p. 27].

In this way, the paper provides some examples of the ways in which educators "should be able to put these into practice successfully while" students are working together in teams [8 p. 51], and the contribution that these "could make to effective learning and teaching in" Computer Science education. Educators who ensure that the key elements for team interactions in the model as described in this paper were effectively implemented in their students' teams should enable most teams to interact effectively to bring in the required team projects according to specifications and initial planning. The study reported on in this paper contributes uniquely to the literature by analyzing the theoretical predictions made in the model, regarding how these key variables can be effectively and practically structured for application in team interactions, so that students successfully achieve their team learning outcomes.

References

1. Smarkusky, D., Dempsey, R., Ludka, J., de Quillettes, F.: Enhanching team knowledge: Instruction vs. experience. In: Proceedings of the 36th SIGCSE Technical Symposium on Computer Science Education, pp. 460–464. ACM Press, New York (2005)
2. Whatley, J., Bell, F., Shaylor, J., Zaitseva, E., Zakrzewska, D.: Collaboration across Borders: Peer Evaluation for Collaborative Learning. In: Proc. Informing Science and Information Technology Education Joint Conf., pp. 33–48 (2005)
3. Hassanien, A.: Student Experience of Group Work and Assessment in Higher Education. Journal of Teaching in Travel & Tourism 6(1), 17–39 (2006)
4. Johnson, D., Johnson, R., Smith, K.: The State of Cooperative Learning in Postsecondary and Professional Settings. Educ. Psychol. Rev. 19, 15–29 (2007)
5. McKinney, D., Denton, L.F.: Affective Assessment of Team Skills in Agile CS1 Labs: The Good, the Bad, and the Ugly. In: Proceedings of the 36th SIGCSE Technical Symposium on Computer Science Education, pp. 465–469. ACM Press, New York (2005)
6. Peterson, S.E., Miller, J.A.: Comparing the Quality of Students' Experiences during Cooperative Learning and Large-Group Instruction. Journal of Educational Research 97(3), 123–133 (2004)
7. Johnson, D.W., Johnson, R.T.: An educational psychology success story: Social interdependence theory and cooperative learning. Educational Researcher 38(5), 365–379 (2009)
8. Goosen, L., Mentz, E.: Groups can do IT: A Model for Cooperative Work in Information Technology. In: Kendall, M., Samways, B. (eds.) Learning to Live in the Knowledge Society, pp. 45–52. Springer, Heidelberg (2008)
9. Hazzan, O.: Computer Science Students Conception of the Relationship between Reward (Grade) and Cooperation. In: Proceedings of the 8th Annual Conference on Innovation and Technology in Computer Science Education, pp. 178–182. ACM Press, New York (2003)

Static Power Optimization for Homogeneous Multiple GPUs Based on Task Partition

Yisong Lin, Tao Tang, and Guibin Wang

Abstract. Recently, GPU has been widely used in High Performance Computing (HPC). In order to improve computational performance, several GPUs are integrated into one computer node in practical system. However, power consumption of GPUs is very high and becomes as bottleneck to its further development. In doing so, optimizing power consumption have been draw broad attention in the research area and industry community. In this paper, we present an energy optimization model considering performance constraint for homogeneous multi-GPUs, and propose a performance prediction model when task partitioning policy is specified. Experiment results validate that the model can accurately predict the execution of program for single or multiple GPUs, and thus reduce static power consumption by the guide of task partition.

1   Introduction

As the evolution of programming model of GPU, supercomputer integrated with several homogeneous GPUs, draw broad attention in the research area of high performance computing [1] and industry community. TH-1A [2], ranked 1st in the Top500 [3] list of Nov. 2010, is a typical supercomputer. Being one of the core component of TH-1A, GPUs provide the main computing resource in processor.

However, power consumption is the main bottleneck to high performance computing systems integrated multiple GPUs as accelerators. In one hand, the GPUs have very high performance and performance power ratio. In the other hand, the power consumption of a single GPU is much higher than a single CPU. When GPU is a basic component of supercomputer system, its high energy consumption not only increases computational cost, but also reduces the system reliability because of

Yisong Lin · Tao Tang · Guibin Wang
National Laboratory of Parallel and Distributed Processing,
National University of Defense Technology
e-mail: linyisong@live.cn

F.L. Gaol et al. (Eds.): Proc. of the 2011 2nd International Congress CACS, AISC 145, pp. 19–29.
springerlink.com   © Springer-Verlag Berlin Heidelberg 2012

heat [4]. Therefore, it is very important to reduce power consumption for practical systems.

As the development of the CMOS technology, the percentage of static power consumption is increased steadily [5]. So reducing static power is an important technique for energy optimization. The basic idea of static power optimization is to turn off the idle devices. Therefore, for heterogeneous supercomputer systems, the basic idea of power optimization is to shut down necessary GPUs based on the schedule of static task partition of parallel program. In this paper, we present an energy optimization model for multiple heterogeneous GPUs systems and evaluate it by detailed experiments.

Our paper is organized as follows: In section 2, we analyze the problems definition and target architecture. Then in section 3, we present an energy optimization model considering performance constraint for homogeneous multi-GPUs, and present a performance prediction model when task partitioning policy is specified. In section 4, we evaluate our model by several important experiments. Finally, we conclude our paper in a summary and an outlook.

2 Problems Definition and Target Architecture

In this section, we state the problem definition and target architecture. Figure.1 shows the basic system architecture with GPUs, including a host processor and multiple GPUs. Every process unit, including cores of CPUs and processing elements of GPU, has its local memory. Host CPU can issue DMA command to GPUs, when exchanging data between main memory of CPU and local memory of GPU.

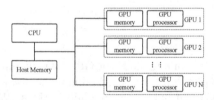

Fig. 1 System architecture with GPUs

There is only one access channel for the GPUs in one computing node. And all of the GPUs need to access the main memory, so there is only one GPU can access the main memory by DMA engine. If multiple GPUs want to exchange data at the same time, they must serialize their DMA requests. In order to enable parallelism among multiple GPUs, we can partition the thread space of GPU kernel function, and different sub-partition of thread space can be executed by different GPU. As we have emphasized above, every sub-kernel need host processor to provide input data, and they must serialize the DMA commands. Hence, we get a typical example of communication and computation pipeline of time-space graph in Figure.2.

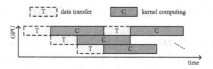

Fig. 2 Example of communication and computation pipeline of time-space graph

In Figure.2, vertical direction represents different GPU and horizontal direction represents the time passed. The dark shadow in solid rectangle represents the time when GPU executes a kernel function, while the light shadow in dot rectangle represents the time when GPU exchange data between main memory and local memory. Clearly, the computing rectangles in one single line of a time-space graph cannot overlap, while the computing rectangles in different lines can overlap. The communication rectangles in deferent lines cannot overlap, because there is only one GPU can access the main memory at the same time.

The problem of static task partition based power optimization can be defined as follows. Assume that there are M GPUs in the system; the static power of one single GPU is P. By task partitioning $\mathscr{C}$, there are $N(N \leq M)$ GPUs computing, and the other GPUs standing by in low power or shutting down states. When a GPU consume no energy in standing by state and consume P energy in computing state, for a parallel program executing T seconds, the energy the program consuming is

$$E = N \cdot P \cdot T. \tag{1}$$

As P is a constant value, we can rewrite the equation as follow,$E \propto N \cdot T$.

In summary, the total energy consumed by a parallel program proportion the area of the rectangular in time-space graph.

Several factors can affect time-space graph. Firstly, the complexity of computation and communication of the parallel program determine the execution time of kernels and arrangement of time-space graph. Secondly, the task partitioning of parallel program determines the data size of a single kernel. Finally, if we want to get an optimal task partition model, we must combine the characteristics of the parallel program and task partition model.

At last, we get the formal definition of static task partition based power optimization problem. Given a parallel program K and performance restriction S, find a task partition $\mathscr{C}$, which partition the thread space of K in multiple sub-spaces and distribute them onto multiple GPUs, so as to minimize the area of rectangular of the time-space graph.

3 Energy Optimization Model for Homogeneous Multi-GPUs

In this section, we present an energy optimization model considering performance constraint for homogeneous multi-GPUs system. Firstly, let's consider the special case without performance constraint. Then, we will extend the special case to a more general case, with performance constraint, and give task partition algorithm.

3.1   Static Energy Optimization without Performance Constraint

If no performance constraint exists, the best partition plan is to let one GPU compute the kernel and the other GPU idle to save energy. The reason is that the data transferring between main memory and local of GPU must be serialized when multi-GPUs are used to compute, no matter how many GPUs are used. It means that it is imposable for multi-GPUs to get linearity or super linearity speedup, that is to say the horizontal direction reduction of time-space graph is not proportional with vertical direction increasing. In conclusion, static power consumption of multi-GPUs is higher than that of single one.

Next, we discuss the task partition problem of single GPU to minimize static power consumption. For single GPU, the lowest static power consumption means the shortest execution time and the highest performance. So, this problem is also a performance optimization problem.

Kernel function is consisted of many parallel threads, which can be partitioned into sets of threads (named thread block). Thread blocks are mapped to Stream Multiprocessor (SM) when the kernel function is executing. Usually, there are several SM in one GPU. In order to use GPU effectively, the parallelism of tasks should be appropriate in order to ensure that every SM in the GPU get a thread block. On one hand, if the number of thread blocks is lower than the number of SMs in GPU, some SM will idle which means wastes of energy. On the other hand, if the number of thread blocks is higher than the number of SMs in GPU, the computation workload on top of SMs in GPU will be ensured and the GPU will be used more effectively. In a word, we assume that, given a kernel function, the minimal task size of task partition is e.

Assuming that the size of transmitted data of task e is $Data(e)$ and the computation time of task e is $Comp(e)$. $Data(e)$ is the size of transmitted data from host processor as the input of task e. $Comp(e)$ is the time during which the GPU executes the task e. $Data(e)$ and $Comp(e)$ determine the shape of time-space graph. The relationship between $Data(e), Comp(e)$ and the plan of task partition is as following.

When $Data(e)$ is given, $T_{(datatransfer)}(Data(e))$, the time to transfer data, can be get from the following equation.

$$T_{data\ transfer}(Data(e)) = T_{startup} + Data(e)T_{unit}. \qquad (2)$$

where $T_{startup}$ is the start up time when transferring data, T_{unit} is the time transferring a unit of data. Usually, $T_{startup}$ is not varying with the size of transferred data and is very short, so the default value of $T_{startup}$ is zero in order to simplify our model. The simplified model is as following, $T_{data\ transfer} \propto Data(e)$.

As for $Comp(e)$, the computation time of task e, it is clearly that $Comp(e)$ is not growing with value of e. If we assign more thread blocks to a SM in a GPU, it will take the SM more time to compute the thread blocks and the pipeline will be used more effectively, meanwhile, the GPU access latency of host memory will be hidden too. In conclusion, the computation time of GPU is glowing slower than the data size of tasks. In other words, $Comp(k \cdot e) \leq k \cdot Comp(e)$ $(k \geq 1)$.

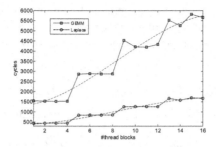

Fig. 3 Relationships between execution times of GEMM and Laplace and their thread block number per SM

To illustrate the relationships between execution times of applications and their thread block number per SM, Figure.3 shows an example. Using GPGPUSim, a GPU simulator with 4 SMs, we get the execution time of GEMM and Laplace. Horizontal direction is the number of thread block (The size of a single thread block is 16×16). From Figure.3, the execution time of applications is growing with thread block number like ladders. The reason is that there are 4 SMs in a GPU, when the number of thread block assigned to GPU is not an integer multiple of the number of thread block, some SM will be in idle state because of load imbalance. So when performing task partition, the minimal unit of task is the integer multiple of e.

It is hard to analysis the relationship between computing time and size of tasks formally, so we get the relationship by profiling. In Figure.3, the dotted line is the 3 polynomial flitting curves of experimental results. For the rest of the paper, we use the flitting curves instead of $Comp(k \cdot e)$.

At last, let us show the relationship between data transfer time and number of thread block. As shown above, the unit of task partition is thread block. However, computation is used as standard for task partition. Meanwhile, there are many different relationship computation and data required by the computation. Take matrix multiplication, $C = A \times B$, as an example, where matrix B and C are allocated distributed among GPUs and matrix A is allocated in GPU local memory fully. Assume that the sizes of A, B and C are all $N \times N$, and the basic unit of partition is a column of matrix. So we can get the data size required to transfer is $Data(e) = N \times N + N$. If the computation is doubled in some GPU, meaning that the GPU compute 2 column of C, the data size required to transfer is $Data(2e) = N \times N + 2N$. Some programs, like matrix multiplication, have more computational complexity than memory complexity, so computation increases faster than the amount of data transfer. However, some programs, like matrix addition, computation and memory complexity are the same. We define the function G, to illustrate the relationship between computation and data required by the computation.

$$Data(k \cdot e) = G(k, Data(e)). \tag{3}$$

G can be get from formal analysis of program.

Now we can discuss the time-space graph of computation and communication, when there is no performance requirement. The ratio between computation and communication R is the most property of time-space graph. When e is given, $R(e)$ is defined as following,

$$R(e) = Comp(e)/T_{data\ transfer}(Data(e)). \tag{4}$$

If the tasks assigned to a GPU is $k \cdot e$ $(k > 1)$, $R(k \times e)$ is defined as following,

$$R(k \cdot e) = Comp(k \cdot e)/T_{data\ transfer}(G(k, Data(e))). \tag{5}$$

Programs, like matrix addition, with the same computation and memory complexity, have $G(k, Data(e)) = k \cdot Data(e)$. So we can infer 2 properties $R(k \cdot e) < R(e)$, $R(i \cdot e) < R(j \cdot e)$.

Programs, with more computation complexity than memory complexity, have $G(k, Data(e)) < k \cdot Data(e)$, so the relationship between $R(k \cdot e)$ and $R(e)$ is not constant, depending on the property of programs.

When the total computation of kernel function is F and the unit of task partition is $k \cdot e$, based on the value of $R(k \cdot e)$, we can get the optimum task partition. Figure.4 shows the computation and communication time-space graph when $R(k \cdot e) > 1$, Figure.4(a), and $R(k \cdot e) < 1$, Figure.4(b), respectively.

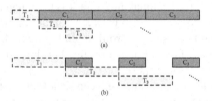

Fig. 4 Computation and communication time-space graph of a single GPU

It should be clear that $T1$, $T2$ and $T3$ are data transferring operation for the same GPU. To illustrate more effectively, we place $T1$, $T2$ and $T3$ in different line of time-space graph. Figure.4(a) is the case of $R(k \cdot e) > 1$. We can get the execution time of kernel function as following,

$$T_a(k) = T_{data\ transfer}(G(k, Data(e))) + \frac{F}{k \cdot e}Comp(k \cdot e). \tag{6}$$

Figure.4(b) is the case of $R(k \cdot e) < 1$. We can get the execution time of kernel function as following,

$$T_b(k) = Comp(k \cdot e) + \frac{F}{k \cdot e}T_{data\ transfer}(G(k, Data(e))). \tag{7}$$

Obviously, $T_a(k) = T_b(k)$ when $R(k \cdot e) = 1$.

As mentioned above, for programs with more computation complexity than memory complexity, it is hard to find the relationship between $R(k \cdot e)$ and $R(e)$, because it depends on the characteristics of program. However, for programs with the same computation and communication complexity, $R(k \cdot e)$ decreases with increase of k monotonicly. If we make sure that $R(e) < 1$, the execution of kernel function is $T_b(k)$. Generally, the execution time of kernel function can be calculated by $T(k)$.

$$T(k) = \begin{cases} T_a(k) & R(k \cdot e) \geq 1 \\ T_b(k) & R(k \cdot e) < 1 \end{cases} \tag{8}$$

So the problem of static power optimization can be formally defined as follows, selecting k from $[1, \frac{F}{e}]$ to minimize $T(k)$.

3.2 Static Energy Optimization with Performance Constraint for Multi-GPUs

When there is no performance constraint, the static energy optimization problem for multi-GPUs can be reduced to the problem of performance optimization for single-GPU. In this section, we discuss the problem of static energy optimization with performance constraint for multi-GPUs.

Assume that F is the total tasks, and computation communication ratio is $R(k \cdot e)$ if the unit of task partition is $k \cdot e$. Figure.5 gives two types of pipelines.

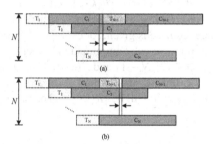

Fig. 5 Computation and communication time-space graph of multiple GPUs

In Figure.5(a), $Comp(k \cdot e)$, the execution of task $C_i(k \cdot e threadblocks)$, is longer than N times of T_i, the data transferring time of C_i. The left dotted line in Figure.5(a) is the time when T_N completes, and the right dotted line is the time when $T(N+1)$ starts so as to $C(N+1)$ can execute after C_1 immediately. Figure.5(b) shows a different case, when C_1 completes, $T(N+1)$ is still running. So $C(N+1)$ must wait for C_1. The model of static energy optimization with performance constraint for multi-GPUs is as follows.

If $R(k \cdot e) \geq N$, every GPU is busy except for startup and drain process of pipeline. So we can conclude that the latency of data transferring can be hidden by computation when it is not in startup process. As the number of tasks one GPU gets is $\frac{F}{N \cdot k \cdot e}$, the execution time of the program is as follows,

$$T_a(k) = N \cdot T_{data\ transfer}(G(k, Data(e))) + \frac{F}{N \cdot k \cdot e} Comp(k \cdot e). \tag{9}$$

If $R(k \cdot e) < N$, stall will occur in pipelines due to data waiting. So we can assume that the computation latency of GPU can be hidden by data transferring. The execution time of program is as follows,

$$T_b(k) = \frac{F}{k \cdot e} T_{data\ transfer}(G(k, Data(e))) + Comp(k \cdot e). \tag{10}$$

Generally, the execution time of kernel function can be calculated by $T(k)$.

$$T(k) = \begin{cases} T_a(k) & R(k \cdot e) \geq N \\ T_b(k) & R(k \cdot e) < N \end{cases} \tag{11}$$

From the function of $T(k)$, $T(k)$ of single GPU is a special case of multiple GPUs. After getting $T(k)$ of multiple GPUs, we analyze the task partitioning algorithm for multiple GPUs to minimize static power.

Based on $T(k)$, we can get minimal execution time curve as Figure.6. Minimal execution time curve describe the minimal execution time when the number of GPUs used is given. Minimal execution time curve is a non-increasing curve. When the number of GPUs used is given, for example i, the minimal execution time, assuming S_i, and optimum unit of task partition, assuming k_i, can be calculated from $T(k)$. It should be clear that, when the performance is optimal, if $R(k \cdot e) \geq i$, all the i GPUs will be used; if $R(k \cdot e) < i$, $\lceil R(k \cdot e) \rceil$ GPUs are sufficient to hide the computation latency and $min(i, \lceil R(k \cdot e) \rceil) = \lceil R(k \cdot e) \rceil$ is the number of GPUs used.

When the performance required is given, for example, S_c(dotted line in Figure.6) is the minimal execution time; we can compute the number of GPUs used for computing. If $S_C \in [S_{i+1}, S_i)$ $(0 \leq i \leq N-1)$ and $R(k \cdot e) \geq i+1$, the minimal number of GPUs used will be $i+1$. If $S_C \in [S_{i+1}, S_i)$ $(0 \leq i \leq N-1)$ and $R(k \cdot e) < i+1$, the minimal number of GPUs used will be $\lceil R(k \cdot e) \rceil$. The other unused GPU will be closed to save energy.

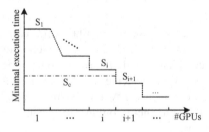

Fig. 6 Minimal execution time curve

4 Experiment Results and Analysis

In order to evaluate our energy optimization model for homogeneous multi-GPUs, we extend GPGPUSim [6] to fulfill our requirements, by adding cycle accurate data transferring function and adding host processor support to drive multiple GPUs.

4.1 Experiment Benchmarks

We select 8 typical benchmarks from signal processing, financial simulation and scientific computing. 5 of the 8 benchmarks, *BlackScholes, dwtHaar1D, MersenneTwister, MonteCarlo* AND *SobolQRNG*, are from CUDA SDK 3.0. *BlackScholes(BS)* and *MonteCarlo(MC)* are from financial areas, the former implements *BlackScholes* model[7], a mathematical model of a financial market containing certain derivative investment instruments, while the later is a type of computational algorithms that rely on repeated random sampling to compute their results. *dwtHaar1D(DW)* implements wavelet transformation of signals. *MersenneTwister(MT)* is the kernel of *Mersenne Twister*[8], which generate pseudo-random numbers. *SobolQRNG(SQ)* compute quasi-random numbers based on *Sobol*[9] algorithm. The other 3 benchmarks are from scientific area, including *matrix multiplication(MM), matrix vector multiplication(MV)* and *Laplace transformation(LP)*.

Table 1 shows the benchmarks description. Top 5 of the benchmarks have common characteristics. The data size is linearly growing with thread block size, so $G(k, data(e)) = k \cdot data(e)$. For other 3 benchmarks, $G(k, data(e)) < k \cdot data(e)$.

Table 1 Benchmark description

Benchmark	Data Size	Thread Block Size	Number of thread block	Data/ Thread Block	Data/k thread block
BS	2MB	256	256	8KB	$8k$ KB
MC	1MB	256	128	8KB	$8k$ KB
DH	2MB	256	1024	2KB	$2k$ KB
MT	4MB	256	256	16KB	$16k$ KB
SQ	8×2MB	256	512	4KB	$4k$ KB
MM	2×4MB	16×16	4096	128 KB	$64(1+k)$ KB
MV	8MB	128	2048	64.5KB	$64k + 0.5$ KB
LP	2MB	16×16	2048	1.3KB	$1152k + 144$ B

4.2 Experiment Results and Analysis

Figure.7 shows the model predicted and simulated normalized execution time of single GPU, assuming $k = 1$ and $\#SM = 4$. We set the number SM in a GPU 4 and 8 respectively, getting two group of results. All the other configurations are the same as Quadro FX5600[6] except number of SMs in a GPU. As there are huge gap among the run time of different benchmarks, we give normalized results when $k = 1$ and $\#SM = 4$. In addition, we profile $Comp(k, e)$ when the unit of task partition is

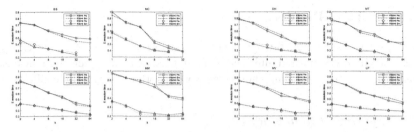

Fig. 7 Model predicted and simulated normalized execution time of single GPU, assuming $k = 1$ and $\#SM = 4$

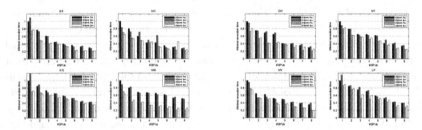

Fig. 8 Model predicted and simulated normalized minimal execution time of multiple GPUs, assuming $k = 1$ and $\#SM = 4$

14. From the experiment results, the model can predict the execution of program accurately for single GPU. Meanwhile, we can see the relationship between execution time and the unit of task partitioning. So the model can be used to guide energy optimization when there is no performance constraint.

Next, we take the experiments of homogeneous multi-GPUs. The configurations of experiments are the same as the above experiment except that the SMs of a GPU are set to 4. Figure.8 shows model predicted and simulated normalized execution time of single GPU, assuming $k = 1$ and $\#SM = 4$. We can see from Figure.8 that our model can predict the minimal execution of program accurately, which can be used for minimizing the power consumption for multiple GPUs. From Figure.8, we can see that the optimum task partition is usually 5-10 units of thread blocks. The reason is that small tasks are hard to exploit the parallelism inside SMs and CUDA programming model requires that there are 8 thread blocks in one SM simultaneously. Large tasks are hard to hide the main memory access latency and overhead.

5 Conclusions

In this paper, we have analyzed how to optimize pow consumption for homogeneous multi-GPU computer system based on the static task partition. For the long running kernel, we can assign it to multiple GPUs by task partitioning. However, power consumption is an important issue to computers, so optimizing performance of long running kernels and reducing power consumption are the problems we concerned.

We present an energy optimization model considering performance constraint for homogeneous multi-GPUs, and give the performance prediction model when task partitioning policy is specified. We attribute the problem of static power optimization for homogeneous multi-GPUs to the problem of task partitioning problems. Based on GPGPUSim, we build simulation environments for multiple GPUs. Experiment results validate that the model can predict the execution of program accurately for single and multiple GPUs, and thus can be used to guide the task partition for static power optimization.

Acknowledgements. This work is supported by the National Natural Science Foundation of China (NSFC) No.60921062 and 60873016.

References

1. Luebke, D., Harris, M., Govindaraju, N., Lefohn, A., Houston, M., Owens, J., Segal, M., Papakipos, M., Buck, I.: In: Proceedings of the 2006 ACM/IEEE Conference on Supercomputing, SC 2006. ACM, New York (2006)
2. Yang, X.J., Liao, X.K., Lu, K., Hu, Q.F., Song, J.Q., Su, J.S.: Journal of Computer Science and Technology 26(3), 344 (2011)
3. Top500, http://www.top500.org
4. Yang, X., Liao, X., Xu, W., Song, J., Hu, Q., Su, J., Xiao, L., Lu, K., Dou, Q., Jiang, J., Yang, C.: Frontiers of Computer Science in China 4(4), 445 (2010)
5. Schaller, R.: IEEE Spectrum 34(6), 52 (1997)
6. Bakhoda, A., Yuan, G., Fung, W., Wong, H., Aamodt, T.: In: IEEE International Symposium on Performance Analysis of Systems and Software, ISPASS 2009, pp. 163–174 (2009)
7. Black, F., Scholes, M.: The pricing of options and corporate liabilities. Political Economy 81 (1973)
8. Matsumoto, M., Nishimura, T.: ACM Trans. Model. Comput. Simul. 8, 3 (1998)
9. Bratley, P., Fox, B.L.: ACM Trans. Math. Softw. 14, 88 (1988)

A Dynamic Structure of Counting Bloom Filter

Jianhua Gu and Xingshe Zhou

Abstract. Counting Bloom Filter based on the counter-array structure has the shortcoming of counter overflow and less space-efficient. To address these shortcomings, we propose a dynamic structure for Counting Bloom Filter which dynamically changes the counter size according to the number of inserted elements. Hence it not only makes a better use of memory space but also eliminates counter overflow. We put up with the methods of addition and subtraction bit by bit while inserting and deleting elements to effectively reduce the times of memory access. In this way, an effective tradeoff can be achieved between counter access speed and space efficiency. Besides, to reduce excessive memory allocation/deallocation cost caused by consecutively changing counter size, we propose a configurable delayed shrinking algorithm which can appropriately delay the counter size shrinking based on user's configuration. The experiment results show that our dynamic structure could meet the needs of most application scenarios.

1   Introduction

Bloom Filter is a data structure with extremely high space efficiency. Since 1970 when Bloom Filter[1] was proposed by B. Bloom, it has been widely applied to database field. Counting Bloom Filter (CBF) as one of the variations, adds the operation of element deletion to the existing insertion and query operation so that it is able to signify the set of dynamic variation. CBF has been used in broad fields such as Web cache memory sharing, longest-prefix matching[7], and stream data processing[5].

Although [2] believes that 4-bit counter is enough for most applications, this memory method with fixed length is not flexible enough because it has two

Jianhua Gu · Xingshe Zhou
School of Computer Science and Technology,
Northwestern Polytechnical University, Xi'an China, 710072
e-mail: {gujh,zhouxs}@nwpu.edu.cn

F.L. Gaol et al. (Eds.): Proc. of the 2011 2nd International Congress CACS, AISC 145, pp. 31–37.
springerlink.com   © Springer-Verlag Berlin Heidelberg 2012

defects. First, though there is very low possibility of counter overflow, certain measures and strategies are still needed for a robust application system when overflow occurs. The overflow solution in [3] may cause false negative which is not accepted by Bloom Filter. Second, fixed length determines that we must choose the counter size based on the worst situation, which will lead to a waste of storage space when there are only few set elements.

This paper presents a CBF-based dynamic structure which allows counter size to dynamically change according to the number of inserted elements. This not only improves space efficiency but also eliminates overflow. We allow users to find a tradeoff between counter access speed and flexibility through configurable initial counter size. In particular, when the initial counter length is set as 4, our structure becomes a CBF with the function of overflow treatment. In order to bring down overhead, we put up with the methods of addition and subtraction bit by bit during element insertion and deletion to effectively reduce the frequency of memory access. In this way, an effective balance is achieved between counter access speed and space efficiency. What is more, to reduce excessive memory allocation/deallocation overhead caused by consecutively changing counter size within a short time, we propose a configurable delayed shrinking algorithm which can appropriately delay the counter size shrinking based on user's configuration.

2 Related Work

There are already some studies[4] improving CBF based on its shortcomings of low space efficiency and inflexible structure caused by fixed counter length. d-Left Counting Bloom Filter tries to store the "fingerprint" of set elements by d-Left Hashing. It can provide the same functions with CBF with half or even smaller storage space. However, d-Left Counting Bloom Filter does not follow the basic structure of Bloom Filter and thus cannot be converted into Bloom Filter as easily as CBF. What is more, similar to the counter overflow of CBF, d-Left Counting Bloom Filter also has the possibility of bucket overflow[8].

Despite the fact that Spectral Bloom Filter[5] is designed for multi-set, it keeps the logical structure of CBF and adopts the counter storage method with higher space efficiency. It arranges the least number of bits required by every counter to a bit string and then access counters through index structure. Such physic structure is evidently advantageous in space efficiency when indicating multi-sets with highly different multiplicity; whereas it has no advantages at all in either storage space or counter access speed when indicating ordinary sets due to its complicated index structure. To avoid the overhead by complicated index structure, Dynamic Count Filter[6] simplifies Spectral Bloom Filter. It stores all the counters with a basic array and an overflow array; and when the counter length changes the basic array does not change and overflow array is reconstructed. This structure is a bit similar to our common dynamic structure, but Dynamic Count Filter is designed for the application of multi-set and data flow. [9] also introduces a data structure with the same function with CBF to denote multi-set more effectively in the place of CBF, but it does not provide relevant experiment data to prove its analysis. All of the above solutions are designed for multi-set, yet their structures are much more complicated than CBF; Dynamic Count Filter has comparatively simpler structure, it is not common though.

The above works have made a certain progress in improving space efficiency and structural flexibility. However, they either expand the application of CBF to certain fields or lead to certain function loss while improving one aspect, and thus cannot really substitute CBF. Our objective is to propose a CBF with relatively simple but flexible structure and high space efficiency which allows users to configure suitable dynamic structure according to specific application scenarios, and in the end substitutes the original counter-array structured CBF and supports the operation of elements deletion.

3   Dynamic Structure of Counting Bloom Filter

3.1   Motivation

As is stated above, the basic structure of CBF is a counter array, so counter overflow may happen. For a robust application system, certain measures need to be taken to keep it work well during overflow. Some solutions may cause false negative.

And then, the fixed length of counter may lead to the inflexibility and the waste of storage because the length is set according to the worst situation.

3.2   Basic Dynamic Structure

The fundamental idea to deal with the above problem is to have the counter length dynamically change based on the number of inserted elements. For this, we design a dynamic structure of CBF—DCBF (Dynamic Counting Bloom Filter).

> **Definnition1** DCBF can be put into a six tuple:
> DCBF = <BF[i], b, m, Hk, δ, λ>.
> Here,
> BF[i] refers to standard Bloom Filter No.i (i=1......b);
> b stands for the length of DCBF;
> m refers to the length of a standard BF (it has m bits);
> H_K refers to k independent hash functions (h_1, ..., h_k);
> δ stands for the initial length of DCBF;
> λ stands for the factor for delayed shrinking.

It takes several times of memory access for the above dynamic structure to access the counter and read relevant counter data. Suppose the number of bit array structured BF (counter length) is b, and they have fixed sequence of low bit to high bit in the counter. We define the value of counter No.i as Ci, the No.i bit of BF No. j is BFj[i] (i=1...m and j=1...b), then the binary form of Ci can be written as $BF_b[i]$ $BF_{b-1}[i]...BF_1[i]$. Therefore, to read the value of counters, the structure only needs to read the corresponding bit in each bit array.

δ is used to balance counter access speed and space efficiency. λ is used in delayed shrinking algorithm which will be discussed in section 3.4.

For a better understanding, we first introduce a special example of DCBF — BDCBF (Basic DCBF) when $\delta = 1$, $\lambda = 1$. In BDCBF, the counter length is decided by the maximum value among m counters. Suppose the value of counter No.i is Ci, then the counter length b = $\lceil \log_2 \max(C_i) \rceil$.

3.3   Operations on DCBF

DCBF supports the operations of element insertion, query and deletion, and their algorithms are shown by algorithm 1, algorithm 2 and algorithm 3 respectively.

```
Algorithm 1 Insert(element)
For(i=1 to k)
{  bit=H_k(element); counter=0;
   for(j=1 to b)
   {   if(0==BF_j[bit]) { BF_j[bit]=1; N[j]++; break }
       else { BF_j[bit]=0; N[j]--; counter++;  }        }
     if(counter>=b)
     {    b++;
       Allocation a BF_b;
       BF_b[bit]=1;        }
}
```

When inserting an element into DCBF, we need k Hash functions to carry out computation on the inserted element and to count the number of corresponding counters in DCBF according to Hash results. The common method is to add 1 to the corresponding counter value read from DCBF and then write it into the counter. Therefore altogether it needs 2b times to read and write the memory. The insertion algorithm in this paper adopts the addition bit by bit so as to greatly reduce the times of reading and writing memory. Likewise, when deleting elements from DCBF we adopt the subtraction bit by bit to reduce the number of reading and writing memory.

```
Algorithm 2   Query(element)
   counter1=0;
     for(i=1 to k)
     {   bit=H_k(element);counter2=0;
       for(j=1 to b)
       {  if(1==BF_j[bit]) { counter1++;break }
          else counter2++;
        }
     if(counter2>=b)break;
     }
   if(counter1>=k) return true;
return false;
```

When indicating an integer with a binary bit string, generally speaking the probability of bit No. 0 (the lowest bit) of the binary bit string being zero is 50%. To generalize, there's 1/4 probability to read and write twice and 1/8 probability to read and write three times. In this situation, the average times to read and write the memory for an element insertion is

$$2*\sum i*(1/2^i) \ , \quad i=1..b$$

Usually application only wants to know whether there exist relevant elements in the set instead of the specific number of the elements, therefore in query operation, we only need to make sure that the corresponding counter is not zero so as to reduce the times of reading and writing memory.

```
Algorithm 3  Delete(element)
For(i=1 to k)
{ bit=Hₖ(element);
  for(j=1 to b)
  { if(1==BFⱼ[bit]) { BFⱼ[bit]=0;N[j]--; break }
    else { BFᵢ[bit]=1; N[j]++; counter++; }    }
  DelayShrink(level); }
```

3.4   Delayed Shrinking

When the BF's top bit becomes vacant after element deletion, it is not necessary to withdraw this BF immediately to shrink counter length; in contrast, it can be deallocated after a delay of a certain period of time so as to avoid the overhead by too much allocation and deallocation of bit array. We makes improvement on[6] so that it becomes simpler and brings down the overhead of storage space.

It[6] costs a lot to judge whether all the counter values are smaller than the threshold value T especially when m is very big. Therefore, we propose an optimization algorithm. We do not record the number of counters that are bigger than the threshold value; instead, we record the number of 1 in every BF_i.

We define N[i] as the number of 1(i=1..b) in BF_i. When there is a new BF, its N[] is 0. The maximum value of N[i] is m, which means that all the bits in BF_i are set as 1.

We use *level* to indicate the degree that withdraws BF when the top BF_b's all bits are 0. The smaller level value, the higher degree (possibility) of withdrawing BF_b; otherwise, the bigger level value, the lower possibility of withdrawing BF_b. We modulate the possibility of withdrawing BF_b by level value.

$$level=(int)b*\lambda$$

In this formula, $0 \le \lambda \le 1$, then *level*=0..b. We can control *level* through λ and in the end control the withdrawal of BF (see algorithm 4).

```
Algorithm 4   DelayShrink(level)
If( (all N[i]==0)
for(i=b to b-level){  free(BF_b);b--; }
```

4   Experiment and Results Analysis

Insertion and deletion operation is the basic operation of bloom filter. We test the speed of insertion algorithm bit by bit we put up with. The test1 is that to read every corresponding bit of each array, and to add by 1, and then write these bits back array. The test2 is our addition bit by bit. The test3 is the integer addition that put all m-bit into one integer. In theory, the speed of integer addition is the fastest among the three methods. So we make the integer addition as the reference through the three methods. We add by to the m-bit DCBF from 0 to maximum and compute the time it takes. The computer's CPU is Intel(R) Pentium(R) 2.80GHz with 40G disk and 512M memory. The experience results show as Table.1. The Table.1 shows that the time of test2 is close to that of test3 and verifies the DCBF's efficiency in both time and space.

Table 1 Comparison of three methods

Time (ms) \ Method \ m	2	3	4	5	6	7	8	9	10	11	12	13	14	15	16
Test1	0.21	0.45	1.03	1.99	4.31	9.59	21	45	100	222	475	1033	2202	4726	10182
Test2	0.14	0.14	0.34	0.62	1.09	2.16	4.2	8.2	16.	32.9	65	133	265	529	1057
Test3	0.03	0.03	0.07	0.14	0.17	0.31	0.5	1.1	2.3	4.17	8.7	17.5	34.5	63.7	142

5　Conclusion

This paper presents a dynamic structure of Counting Bloom Filter which can dynamically adjust the counter length according to the number of inserted elements and allow users to find a better balance between counter access speed and space efficiency through configuration of parameters. The theoretical analysis and experimental results provide strong evidence for us to believe that this dynamic structure can substitute the original counter-array structure so that Counting Bloom Filter has no overflow, higher flexibility and higher space efficiency.

References

[1] Bloom, B.: Space/Time Tradeoffs in Hash Coding with Allowable Errors. Communications of the ACM 13(7), 422–426 (1970)
[2] Fan, L., Cao, P., Almeida, J., Broder, A.Z.: Summary Cache: A Scalable Wide-Area Web Cache Sharing Protocol. IEEE/ACM Transactions on Networking 8(3), 281–293 (2000)
[3] Mitzenmacher, M.: Compressed Bloom Filters. IEEE/ACM Transactions on Networking 10(5), 604–612 (2002)
[4] Broder, A., Mitzenmacher, M.: Network applications of Bloom filters: A survey. Internet Mathematics 1(4), 485–509 (2004)
[5] Cohen, S., Matias, Y.: Spectral Bloom Filters. In: Proceedings of the 2003 ACM SIGMOD Conference, pp. 241–252 (2003)
[6] Aguilar-Saborit, J., Trancoso, P., Muntes-Mulero, V., Larriba-Pey, J.L.: Dynamic Count Filters. SIGMOD Record 35(1), 26–32 (2006)

[7] Dharmapurikar, S., Krishnamurthy, P., Taylor, D.: Longest Prefix Matching using Bloom Filters. In: Proceedings of the ACM SIGCOMM, pp. 201–212 (2003)

[8] Bonomi, F., Mitzenmacher, M., Panigrahy, R., Singh, S., Varghese, G.: An Improved Construction for Counting Bloom Filters. In: Azar, Y., Erlebach, T. (eds.) ESA 2006. LNCS, vol. 4168, pp. 684–695. Springer, Heidelberg (2006)

[9] Pagh, A., Pagh, R., Rao, S.: An Optimal Bloom Filter Replacement. In: Proc. of the Sixteenth Annual ACM-SIAM Symp. on Discrete Algorithms, pp. 823–829 (2005)

Draft Line Detection Based on Image Processing for Ship Draft Survey

Xin Ran, Chaojian Shi, Jinbiao Chen, Shijun Ying, and Keping Guan

Abstract. Draft line detection is the first and significant step for ship draft survey, usually determined by visual observation manually. In order to overcome the man-made error, an automated draft line detection method based on image process is proposed in this paper. Firstly, the image containing draft line is obtained and pre-processed, then the contours including draft line are extracted from the image by Canny edge detection algorithm, finally the draft line is detected by Hough transform. The experimental results show that the proposed method is effective to detect the draft line and helpful to improve the accuracy of ship survey.

1 Introduction

Water-borne vessels, with the ability to carry large amounts of cargo economically, are particularly suitable for transporting dry bulk cargo, liquid cargo and other similar type cargo that can be loaded by clamshell buckets, conveyors or pumped into the vessel holds. However, it is important to obtain accurate readings of the vessel draft to determine the amount of cargo that has been loaded onto the vessel. Burness Corlett & Ptns (IOM) Ltd. presented a technical report to analyze the various factors that may cause measurement errors of ship draft survey, and concluded that the accumulated errors will result in 0.5%~1% deviation of the final ship survey [1].

The ship draft marks are located at 6 specific positions around the freeboard, the marine surveyors will observe the draft lines and read the numbers before and after unloading cargoes, then use them to calculate the weight of cargoes. However, this is simply a visual estimate and can vary a number of centimeters depending on the surveyor and wave conditions. Conditions on oceans and rivers can

Xin Ran · Chaojian Shi · Jinbiao Chen · Shijun Ying · Keping Guan
Merchant Marine College, Shanghai Maritime University,
Shanghai, P.R. China
e-mail: xinran@shmtu.edu.cn

F.L. Gaol et al. (Eds.): Proc. of the 2011 2nd International Congress CACS, AISC 145, pp. 39–44.
springerlink.com   © Springer-Verlag Berlin Heidelberg 2012

drastically affect the draft line measurements when compared to readings taken within an area such as a protected port. Furthermore the exact point along the curve of the wave being measured is currently determined by the surveyor which leads to differences amongst surveyor practices.

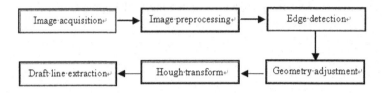

Fig. 1 The flowchart of the proposed method

Fig. 2 The original ship draft line image

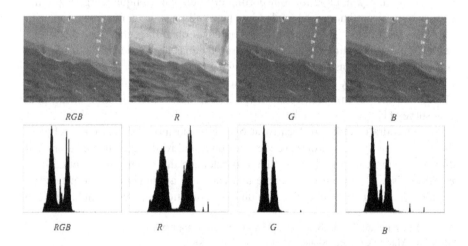

Fig. 3 The original image and its red, green and blue channel with their histogram

In order to overcome the disadvantages of manual observation, many researchers proposed new measurement devices or method to replace human eyes. David Ray et al. designed a portable draft measurement device to obtain the distance from the top of the ship deck to the level of the water in which the ship is floating [2]. Huayao Zheng et al. proposed a new level sensor system for ship stability analysis and monitor, which positioned double pressure sensors below the water surface of the ship [3]. J. Wu and R. Cai studied the factors in vessel draft survey, and analyzed the reasons of errors, thus provided the countermeasure to increase precious of draft survey [4].

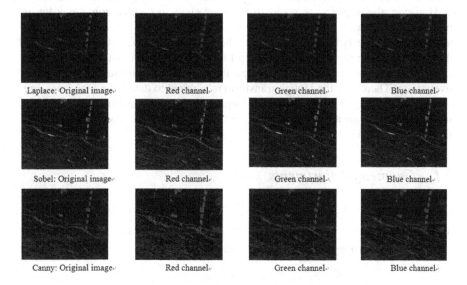

Fig. 4 The experimental results to detect the draft line with Laplace, Sobel and Canny detector

2   Draft Line Detection Method

The proposed method is composed of a series of image processing algorithm, from image preprocessing, geometry transform to edge detection. The flowchart is shown as Fig.1. The original images are taken by surveyor around the ship, which are usually not suitable for direct detection of draft line due to inappropriate position or view angle of surveyor, and also due to the influence of sunshine or wave conditions. Thus the preprocessing should be adopted so that the draft line signal in the images can be enhanced.

Fig.2 shows one of the draft line images the surveyor obtained. It can be seen that the image is sideling and a fake water line appears on the real draft line due to the shipboard flushing of sea wave. An affine transform algorithm is used to adjust the image making the draft line horizontal, which will be discussed later. At preprocessing stage, the red, green and blue channel are divided from the original image, each channel and its histogram are shown in Fig.3. It is noticed that the draft line is more distinct in red channel than in other channels. The peak distribution of pixel grey scales in each channel's histogram also gives the same judgment. So the red channel will be split from the original image and used at the subsequently step.

Because the ship draft line is exactly the consecutive contour or edge in the image, the edge detection algorithm is used to extract the line feature in the image. The effect of several edge detection methods is verified through the practical experiments for the original image and its three channels, including Laplace, Sobel and Canny detection operator. The results shown in Fig.4 illustrate that the best way to extracting draft line is Canny operator adopted in red image channel.

However, the edge image result from edge detection method contains a lot of lines, including draft mark, watermark on the shipboard, wave crests and other noises, and obviously the draft line. Then the mission at the next step is to pick out the draft line from the edge image. From Fig.4 it can be seen that most of the detected edge is short, but the draft line and the upper fake watermark line are relatively long. Therefore, the Hough transform is adopted to detect the long lines and exclude other shot lines in the image. Depending on the common sense that the watermark line is always at upper position than draft line, the lower and true draft line will be picked out at the final step.

2.1 Geometry Adjustments

Let's define the proper edge image as f, the coordinates of pixel in the image as (x, y), and the distorted image of f as f', the coordinates of pixel as (x', y'), and then the distortion transform can be defined as:

$$x' = r(x, y) \qquad y' = s(x, y) \tag{1}$$

where $r(x, y)$ and $s(x, y)$ are the spatial transform. In the situation discussed here, the image taken by surveyor should be adjusted by rotation and translation, and the affine transform will be a good choice. The rotation degree can be defined by computing the inclination of the draft line or watermark line; the translation can be defined by the position of draft mark. The affine transform can be defined as:

$$\begin{pmatrix} \tilde{r} \\ \tilde{s} \\ 1 \end{pmatrix} = \begin{pmatrix} a_{11} & a_{12} & a_{13} \\ a_{21} & a_{22} & a_{23} \\ a_{31} & a_{32} & a_{33} \end{pmatrix} \begin{pmatrix} r \\ s \\ 1 \end{pmatrix} \tag{2}$$

2.2 Hough Transform

The Hough transform is a technique that can be used to isolate features of a particular shape within an image. The classical Hough transform is most commonly used for the detection of lines because it can be defined in a specific parametric form [5]. In the image space, the straight line can be described as $y = ax + b$, where a is the slope of the line and b is the y-intercept and can be graphically plotted for each pair of image points (x, y). However, a convenient equation for describing a set of lines uses parametric or normal notion:

$$x\cos\theta + y\sin\theta = \rho \qquad (3)$$

where ρ is the length of a normal from the origin to this line and θ is the orientation of the line with respect to the X-axis. Therefore, a set of points that form a straight line will produce sinusoids which cross at the parameters for that line. Thus, the problem of detecting collinear points can be converted to the problem of finding concurrent curves.

Before adjustment   After adjustment

Fig. 5 The result of geometry adjustment by affine transform

3   Experiments and Discussion

The practical experiment and results in different steps are presented to illustrate the effect of the proposed method. Fig.2 is the image of ship draft line obtained by surveyor from the port. The image is sideling due to the inappropriate position or view angle of surveyor and a fake water line appears on the real draft line because of the shipboard flushing of sea wave. Fig.3 shows the red, green and blue channel of the original image and their histogram. The different experimental results to detect the draft line with Laplace, Sobel and Canny detector are shown in Fig.4, from the results it can be seen that Canny detector with red channel of the original image can achieve the best result. Fig.5 shows the result of geometry adjustment by affine transform, the draft line is horizontal and the draft mark is adjusted. Fig.6 (a) shows the draft line detection result, the two longer lines, the draft line and the upper waterline, are detected and illustrated in green. The final result is shown in Fig.6 (b); the draft line is extracted exactly according to the position relationship between the draft line and waterline.

4 Conclusions

A ship draft line detection method is presented in this paper, which can extract the draft line exactly through a serious of image processing techniques. The experimental results show that the proposed method is effect and can be used instead of visual observation. After detecting the draft line successfully, an automatic ship draft mark reading method can be developed, then the ship draft survey system can be developed, which will be our future work.

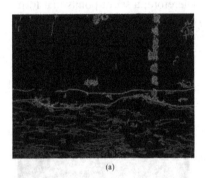

(a)   (b)

Fig. 6 The draft line detection result, (a) the draft line and waterline, (b) the final result

Acknowledgment. The article is sponsored by Science & Technology Program of Shanghai Maritime University (No.20110006), Shanghai Leading Academic Discipline Project (No. S30602), National key technology R&D program (No.2009BAG18B04) and Major National Science and Technology Projects (No.2008ZX05027-005-003).

References

1. Burness Corlett & Ptns(IOM) Ltd., Measurement of cargo loaded by draft survey. Technical Report. BCP/J/5616 (May 1995)
2. Ray, D., Wallace, R., Eugene, W., Michael, J.: Portable draft measurement device and method of use therefore. US Patent 6347461 (2002)
3. Zheng, H., Huang, Y., Ye, Y.: New level sensor system for ship stability analysis and monitor. IEEE Transactions on Instrumentation and Measurement 48(6), 1014–1017 (1999)
4. Wu, J., Cai, R.: Problem In Vessl'S Draft Survey And Countmeature To Increase Its Precision. Journal of Inspection and Quarantine 20(1), 79–80 (2010)
5. Fernandes, L.A.F., Oliveira, M.M.: Real-time line detection through an improved Hough transform voting scheme. Pattern Recognition 41(1), 299–314 (2008)

High Throughput Computing Application to Transport Modeling

Mahmoud Mesbah, Majid Sarvi, Jefferson Tan, and Fateme Karimirad

Abstract. Optimization of road space allocation (RSA) from a network perspective is computationally challenging. Analogues to the Network Design Problem (NDP), RSA can be classified as a NP-hard problem. In large scale networks when the number of alternatives increases exponentially, there is a need for an efficient method to reduce the number of alternatives as well as a computational approach to reduce the computer execution time of the analysis. A heuristic algorithm based on Genetic Algorithm (GA) is proposed to efficiently select Transit Priority Alternatives (TPAs). In order to reduce the execution time, the GA is modified to implement two parallel processing techniques: A High Performance Computing (HPC) technique using Multi-threading (MT) and a High Throughput Computing (HTC) technique. The advantages and limitations of the MT and HTC techniques are discussed. Moreover, the proposed framework allows for a TPA to be analyzed by a commercial package which is a significant provision for large scale networks in practice.

1 Introduction

Mesbah et al. (2010) proposed a bi-level optimization program for road space allocation (RSA). The main objective was to identify the roads on which a bus lane should be introduced. The authors showed that this Mixed Integer Non-Linear (MINL) formulation is an NP-hard problem and therefore, is computationally challenging (Mesbah et al., 2010). For large scale networks, a heuristic approach is adopted to find reasonable solutions for this bi-level structure. This type of

Mahmoud Mesbah
The University of Queensland University
e-mail: mahmoud.mesbah@uq.edu.au

Majid Sarvi
Department of Civil Engineering, Monash University, Vic 3800, Clayton
e-mail: Majid.sarvi@monash.edu

F.L. Gaol et al. (Eds.): Proc. of the 2011 2nd International Congress CACS, AISC 145, pp. 45–51.
springerlink.com

problem can be classified under the umbrella of Network Design Problems (NDP) which has a wide range of applications in Engineering. The goal is then to find the optimal combination of links to be added/modified in the network in order to minimize a certain objective function. Since the RSA problem is NP-hard, application of the proposed analytical and heuristic optimization methods to large-scale problems requires extensive computational power. Such an application is only feasible by deployment of advanced computational techniques. One of the effective responses of computer scientists to such demands was the development of High Performance Computing (HPC) (Strohmaier et al., 2005), which aims to provide fast computations at the processor level and/or at the throughput level in terms of applications completed. This supports the simultaneous execution of many jobs, but within the same multiprocessor computer. This is particularly important where executing jobs must communicate amongst themselves and/or share the same memory. However, there are alternative computing paradigms such as High Throughput Computing (HTC). HTC is distinctly aimed at providing large amounts of processing capacity taken together over a long period of time. Yet another notion is Many Task Computing (MTC), which simply implies that many tasks are supported, whether or not they are long-duration tasks, and regardless of the number of processors per computer. MTC bridges the gap between HPC and HTC systems. Despite the terminological complexities, the goal is clear. In this paper, although HPC taken broadly may likewise apply, we focus particularly on the HTC approach, to distinguish the use of several independent computers on a network as against our previous work using a single multiprocessor. The aim of this paper is to demonstrate the application of HTC in solving a large-scale optimization problem in Transportation Engineering. In this study, a package called VSIUM is employed. VISUM is developed for Windows only and it is available for purchase with software and hardware locks. The broad aim of this paper is to adopt the application of a general-purpose HTC resource to a practical problem in transportation engineering.

2   Transit Priority Optimization

The transit priority optimization is formulated in this paper as a bi-level optimization program (Shimizu et al., 1997, Bard, 1998). Let the upper level be the objective function and constraints from the system manager's point of view. The upper level determines the TPA or the links on which priority would be provided for transit vehicles (decision variables). The aim of the upper level is to achieve System Optimal (SO), thus the objective function includes a combination of network performance measures.

The upper level can be formulated as follows (Mesbah et al., 2010):

$$MinZ = \alpha \sum_{a \in A} x_a^c t_a^c(x) + \beta(\sum_{a \in B} x_a^b t_a^b(x) + \sum_{i \in I} w_i^b) + \gamma \sum_{a \in A} \frac{x_a^c}{Occ^c} l_a Imp^c + \eta \sum_{a \in B} f_a s_a Imp^b \quad (1)$$

S.t.

$$\sum_{a \in A_2} Exc_a \phi_a \leq Bdg \tag{2}$$

$$\phi_a = 0 \text{ or } 1 \quad \forall a \in A_2 \tag{3}$$

Note that $f_a = \sum_{p \in L} f_p \xi_{p,a}$ where $\xi_{p,a}$ is the bus line-link incident matrix and $t_a^b(x)$ is the in-vehicle travel time.

The first two terms in the objective function are the total travel time by car and bus. The next two terms represent the various other impacts of these two modes including emission, noise, accident, and reliability of travel time. The factors $\alpha, \beta, \gamma, \eta$ not only convert the units, but also enable the formulation to attribute different relative weights to the components of the objective function. Equation (2) states that the cost of the implementation should be less than or equal to the budget. The decision variable is ϕ_a by which the system managers try to minimize their objective function (Z). If $\phi = 1$ buses can speed up to free flow speed, while the capacity of the link for cars is reduced to $Cpc_{1,a}^c$.

At the lower level, it is users turn to maximize their benefit. Based on the decision variables determined at the upper level, users make their trips. In this paper, the traditional four step method is adapted for transport modeling. The lower level consists of three models: Modal split model, Traffic assignment model (car demand), and Transit assignment model (bus demand).

Once the demand is determined, users choose their travel mode. Then, the car demand segment of the total demand is assigned to the network. The last step at the lower level formulation is the assignment of transit demand.

The bi-level structure, even in the simplest form, with linear objective function and constraints is a NP-hard problem and is difficult to solve (Ben-Ayed and Blair, 1990). Furthermore, in this problem, the upper level objective function and the UE traffic assignment are non-linear. This adds to the complexity of the problem. This paper employs a Genetic Algorithm approach to find an approximate solution to the optimal answer. The output of the model is the combination of transit exclusive lanes which minimizes the proposed objective function.

3 Genetic Algorithm Solution

A Genetic Algorithm (GA) is an iterative search method in which the new answers are produced by combining two predecessor answers. A number of studies applied GA to transit networks. Two recent examples are a transit network design problem considering variable demand (Fan and Machemehl, 2006) and minimization of transfer time by shifting time tables (Cevallos and Zhao, 2006). In this work, GA is applied to the RSA problem. A gene is defined to represent the binary variable ϕ and a chromosome is the vector of genes (Φ) which represents a TPA.

A chromosome (or TPA) contains a feasible combination of links on which an exclusive lane may be introduced (set A_2). Therefore, the length of the chromosome is equal to the size of A_2.

3.1 Parallel GA - Multi-Threading (MT) Approach

The operating system (OS) creates process *threads* to run computer applications. If no provision is made in the coding, the OS creates only one thread by which all the required processes of the code are run successively. In the presence of multiple running applications, there could be more than one thread processed by a Central Processing Unit (CPU) at a time. Execution of multiple threads can result in parallel processing of an application if the machine is capable of handling it, such as when a *multi-core* machine is in use. Either way, this technique of executing multiple threads in parallel is called *Multi-threading* in computer science.

The speed up achieved by the multi-threading approach depends on the number of cores on a machine and the efficiency of the OS on running parallel processes. In this study, the implementation is in the Microsoft Windows OS since the TPA are evaluated by VISUM and this package is only available in the Windows platform.

However, given that the TPA evaluations are carried out using a commercial package, and all commercial packages need licenses, multi-threading approach can save considerably on *license* costs for some packages. Since all instances of a commercial package are triggered on one machine, only a single VISUM license is sufficient to be provided, but at the expense of performance.

3.2 Parallel GA – the HTC Approach

The multi-threading approach can run a certain number of TPA evaluations in parallel. Hence, multi-threading is an approach that delivers speed-ups due to the parallelism afforded by many computations that can run simultaneously on many cores. However, as mentioned earlier, the number of threads on a machine is limited, or at least, there is a limit to the speed up expected. A distributed computing approach such as HTC schedules TPA evaluations as computational jobs to several computers on a network, each being independent with its own set of cores and local memory. Therefore, there is virtually no limit on the number of processing tasks that can be done simultaneously as the number of computers in a network is not necessarily bounded. The trade-off is a need to distribute the task to available computers in the network, manage the queue, data transfers, provide an inter-process message-passing system in some cases, then collect and integrate the results. There are a number of existing systems that provide for these things. In this study, Condor is used.

3.3 Adaptation of the RSA Problem

Although the issues emerging in adopting a general HPC tool to the RSA problem are tool specific, important lessons can be learned from them for application of any commercial package on the Windows platform. As VISUM is a commercial package, it is protected from being used beyond the maximum number of running instances that is allowed by the purchased license. Additionally, VISUM comes with a hardware lock, which contains the license data.

To evaluate a function, a user submits a *job* on the *submission machine*, and Condor will copy the application program to the *worker* node as a job or *task*. When the job is completed, any output data file identified in the job's *command file* for copying back will be downloaded back to the submission machine, and then the application copy is removed on the worker node. Windows differentiates between the local or remote launch of an application. Windows also consults the user permissions to run an application either locally or remotely. Condor runs VISUM remotely as a 'system' user. The VISUM 'COM server' was set to grant suitable permissions to launch Condor jobs from a 'remote system' user.

4 Numerical Example

In this section, three GA implementation approaches (SGA, PGA-MT, and PGA-HTC) are applied to an example network. The layout of the network used is a grid network consists of 86 nodes and 306 links. All the circumferential nodes together with Centroid 22, 26, 43, 45, 62, and 66 are origin and destination nodes. A 'flat' demand matrix of 30 Person/hr is traveling from all origins to all destinations. The total demand for all the 36 origin destination is 37,800 Person/hr. There are 10 bus lines covering transit demand in the network. The frequency of service for all the bus lines is 10 mins. There are a total number of 120 links (one directional) in the network on which an exclusive lane can be introduced.

Table 1shows the seven computers used in this study. The last column in Table 1 is the time spent for evaluation of one TPA on each machine.

Three types of SGA, PGA by Multi-threading (MT) approach, and PGA by HTC approach are run in this section. The First experiment (datum) for the MT approach is performed on machine 4 with 4 threads, and for the HTC approach on machines 1, 2, 3, and 6 with a total of 22 threads.

In this section first it is shown how GA moves towards the optimal answer. Then it will be demonstrated that the parallelization approaches, including MT and HTC, do not affect the number of evaluations in the GA or the rate of improvement in the objective function. The parallelization approaches merely reduce the time required for evaluations by performing a number of evaluations in parallel. The minimum objective function value achieved in a run with 400 generations was equal to -4.757.

Table 1 Computers used in the experiments

Machine	CPU	Cores	Threads	Windows	VISUM	Evaluation Time (s)
1	Intel Core i7 CPU 860 @ 2.8 GHz	4	8	7 64-bit	11.03 64-bit	65
2	Intel Core i7 CPU Q820 @ 1.73 GHz	4	8	7 64-bit	11.03 64-bit	147
3	Intel Core 2 Quad CPU Q6600 @ 2.4 GHz	4	4	7 64-bit	11.03 64-bit	101
4	Intel Core 2 Quad CPU Q6600 @ 2.4 GHz	4	4	XP 64-bit	11.01 32-bit	122
5	Intel Core 2 Quad CPU Q6600 @ 2.4 GHz	4	4	XP 64-bit	11.01 32-bit	121
6	Intel Core 2 Duo CPU E8500 @ 3.16 GHz	2	2	XP 64-bit	11.01 32-bit	88
7	Intel Pentium 4 CPU @ 3.2 GHz	2	2	XP 32-bit	11.01 32-bit	226

Table 2 presents the effects of the number of available threads on the execution time in the HTC approach. In all of the experiments/runs population size is 40, cp=0.98, and mp=0.01. In general, the Average Time per Evaluation (ATE) decreases when the number of threads increases which is an expectable outcome.

Table 2 Comparison of the HTC speed up by varying cores

Experiment Code	Number of Cores	Number of Threads	Computers	Number of Evaluations on Generation 50	Execution Time (sec)	Average time per evaluation (sec)	Speed up	Efficiency
E210	1	1	4	1680	241475	143.7	1	1
E225	6	10	2, 6	1724	37296	21.6	6.643	0.664
E226	10	14	2, 4, 6	1481	28013	18.9	7.597	0.543
E227	10	18	1, 2, 6	1626	24918	15.3	9.378	0.521
E228	16	22	1, 2, 4, 6	1659	19335	11.7	12.331	0.560
E230	20	26	1, 2, 4, 5, 6	1614	19053	11.8	12.173	0.468
E231	22	28	1, 2, 4, 5, 6, 7	1645	26235	15.9	9.011	0.322

mp =0.01

The lowest ATE belonged to experiment E228 with 11.7sec. However, two important observations can also be made from Table 2. First is that the ATE has remained constant when the number of threads increased from 22 to 26. This means although the resources to perform the evaluation job were increased, the ATE was not improved. This technique saves repeating evaluations which considerably reduces the GA execution time. Therefore, although the average number of evaluations per generation is 33, the first generations evaluate close to 40 (which is the population size) evaluations, while the last generations evaluate just over 20

TPAs. When close to 40 TPAs are being evaluated, both experiments of E228 and E230 may allocate 2 or less evaluations to a thread.

According to the above logic, the ATE in experiment E231 should also be similar to E228 and E230. However, as mentioned in Table 2, ATE has considerably increased in E231. In this experiment, a very slow CPU of machine 7 is added to the pool.

5 Concluding Remarks and Future Work

This paper presented the development and implementation of High Throughput Computing (HTC) to a transport optimization problem. Road Space Allocation (RSA) optimization problem was tackled using 3 variants of a Genetic Algorithm, namely Serial Genetic Algorithm (SGA), Parallel Genetic Algorithm (PGA) using Multi-threading (MT) technique and PGA using HTC. The final answer of the optimization problem was independent from the type of GA variant.

The performance of the GA variants was compared in a numerical example. The PGA-MT approach with 4 'threads' available could reduce the execution time by 3.2 to 3.7 times compared to SGA. The PGA-HTC approach with 18 'threads' could decrease the execution time by 9.3-9.8 times in different examples. Although the 'efficiency' of the MT approach was higher than the HTC, MT approach cannot be used to solve large scale (real world network) examples since the total number of threads on a computer is limited. In contrast, there is no limit on the number of cores/threads which can be employed in the HTC approach. One of the novel aspects of this study was the successful implementation of the HTC approach to the road space application study using commercial software on Windows platform.

References

Bard, J.F.: Practical bilevel optimization: algorithms and applications. Kluwer Academic Publishers, Boston (1998)

Ben-Ayed, O., Blar, C.E.: Computational difficulties of Bilevel Linear Programming. Operations Research 38, 556–560 (1990)

Fan, W., Machemehl, R.B.: Optimal Transit Route Network Design Problem with Variable Transit Demand: Genetic Algorithm Approach. Journal of Transportation Engineering 132, 40–51 (2006)

Foster, I.T., Zhao, Y., Raicu, I., Lu, S.: Cloud Computing and Grid Computing 360-Degree Compared. In: Grid Computing Environments Workshop, GCE 2008 (2008)

Mesbah, M., Sarvi, M., Ouveysi, I., Currie, G.: Optimization of Transit Priority in the Transportation Network Using a Decomposition Methodology. Transportation Research Part C: Emerging Technologies (2010)

Shimizu, K., Ishizuka, Y., Bard, J.F.: Nondifferentiable and two-level mathematical programming. Kluwer Academic Publishers, Boston (1997)

Strohmaier, E., Dongarra, J.J., Meuer, H.W., Simon, H.D.: Recent trends in the marketplace of high performance computing. Parallel Computing 31, 261–273 (2005)

Towards an Organic PLM Approach Providing a Semantic Engineering Desktop

Detlef Gerhard

The paper describes research efforts to address three core problems of the complete process of bringing a new product or service to market from an IT perspective and especially focusing on Product Lifecycle Management systems (PLMS). Today's major challenges are the support of globally distributed and multi-disciplinary teams, complexity, and enterprise knowledge management. Essential building block is the use of semantic web technologies for IT systems integration and also for semantic enrichment of data. The aim is to give a strategic view of usage and integration of supportive technologies and tools. This contribution introduces an approach to future IT working environment for engineers in the sense of a „Semantic Engineering Desktop" and answers the question how mashup technologies can contribute to more individual and flexible engineering application integration and knowledge management without losing data and process security.

1 Introduction

The term "Organic Computing" (OC) has been around for a few years now [1] and describes an architectural approach for biologically-inspired IT systems with "organic" properties. OC is based on the insight that every technical system is surrounded by large collections of other more or less autonomous systems, which communicate freely and organize themselves in order to perform the actions and services that are required. Main characteristic properties of an OC system besides self-organization are self-configuration self-optimization, and context-awareness, all with the aim to master complexity and establish human machine interfaces oriented accordingly to user needs.

The complexity of today's business and engineering IT environments embedding PLMS in enterprise spanning configurations which are spread over different

Detlef Gerhard
Mechanical Engineering Informatics and Virtual Product Development Division
Vienna University of Technology, Vienna, Austria

F.L. Gaol et al. (Eds.): Proc. of the 2011 2nd International Congress CACS, AISC 145, pp. 53–61.
springerlink.com

locations is not far from that of biological organisms and socio-technical ecosystems. Traditional communication and integration approaches are ineffective in this context, since they fail to address several required features, e.g. flexibility for project based collaboration, heterogeneity in node capabilities, and management complexity of such systems. Therefore, it is reasonable to explore OC approaches for future PLMS development and architectures keeping in mind that PLMS are the major means for today's engineers to organize the exploding amount of product related information created with the different CAD/CAE applications, manage collaboration, and control virtual product development processes.

2 Challenges

2.1 Management of Structured and Unstructured Information

Today's engineers are to a great extend information workers. Engineers are heavily dependent on retrieving and using documents and existing models in order to fulfill various engineering design tasks. Exploring design concept alternatives during the early stage of the development is as important as learning from the original design process and understanding the rationale behind the decisions made (handling change requests) or searching for past designs when working on a similar product or problem (design reuse). Various studies show that engineers spend conservatively one third of their time retrieving and communicating information.

A huge amount of knowledge in is generated in order to describe a product and corresponding processes. Internal information is captured in form of documents and models, not only 2D/3D CAD models and drawings, FEA and simulation files but also office documents for specifications, reports, spreadsheets, technical documentation, and informal communication documents like notes, memos, and e-mails. This unstructured information has to be described by metadata attributes in order to be retrievable within business IT systems like PLMS. Moreover, there is also structured information, e.g. BOMs stored in database tables.

A second class of information which is getting increasingly important and relevant for product creation processes are external resources, e.g. legal issues, regulations, patent information, international standard documents and certifications, supplier and product catalogues. With the emergence of the usage of Web 2.0 technologies like wikis or blogs - internet and intranet based - even more sources of unstructured information are building blocks of the enterprise knowledge.

However, engineering documents and models are different to other domains' information resources due to syntax variations and semantic complexities of their contents. Abbreviations, e.g. SLA (stereo lithography, service level agreement) reflect company-specific or domain-specific naming conventions. Acronyms and synonyms are widely used and depend on enterprise or international standards, e.g. "Steel" = "St37-2" = "S235JRG2" = "S235JR+AR". There is also a wide range of domain-specific issues and the relationships among these issues:

- customer requirements and specifications
- functions and performances
- structure design and materials
- manufacturing process selections.

2.2 Complexity

Complexity is characterized by networked structures nonlinear behavior and means multi-causation, multi-variability, multi-dimensionality, interdependence with the environment and openness. In the area of product creation we are facing the three facets of complexity (in causal order):

- product and system complexity
- process and organization complexity
- IT landscape and tool complexity.

Complexity of products is caused by multiple instances and variants of a base product to meet requirements with respect to customization demands and differentiation of the target markets. Furthermore, nearly all products consist not only of mechanical components including pneumatic and hydraulic parts but increasingly use electronics, automatic control parts, and contain firmware/software. Because there are many different domains of expertise involved in product development tasks, process complexity also increases through dissemination over locations (countries, cultures) and distribution within the supply chain (organizations). The biggest challenge here is not only the integration of different technical disciplines all of them using specialized IT tools but the diversity and dynamics of the relationships between project partners, manufacturers, vendors and suppliers. Whereas a decade ago manufacturing distribution was dominant distributed engineering and collaborative design require even more extensive use and support of advanced IT systems which again leads to more complex IT landscapes with docents of data formats and interfaces relying on system platforms with a vast heterogeneity of hardware, operating systems and computer network topologies. Only biologically-inspired systems with "organic" properties seem to be capable of coping with complexity and interference problems.

2.2.1 The IT and Tool Oriented Perspective

As stated above, PLMS are the major means for today's engineers to organize the exploding amount of product related information. Within the last 15 years of the development of this category of enterprise information systems they have become quite mature. Nevertheless, PLMS in implemented practice - despite vendors' high gloss brochures - mainly encompasses the management of mechanical engineering design data including BOM management and exchange with Enterprise Resource Planning (ERP). PLMS deployment in the very majority of cases ends with Start of Production (SOP). PLMS supported multi-disciplinary project work as well as inter-enterprise project control within the different supply chain partners is truly an exception just like significant and measureable advancements within the processes.

Current PLMS do not cover the whole available information source relevant for specific tasks and deliver too much or irrelevant results on queries. Results are not sufficiently in a form that users can navigate through and explore. The strength of today's PLMS solutions are storing, computing, managing, indexing of textual or metadata based information. Weaknesses can be discovered in areas such as acquisition, discovery, aggregation, organization, correlation, analysis and interpretation.

PLMS and other enterprise business software systems have not delivered on its promise to fully integrate and intelligently control complex business and engineering processes while remaining flexible enough to adapt to changing business needs in co-operative product development tasks. Instead, most IT system environments are patchworks installed and interconnected by poorly documented interfaces and slovenly customized processes [2]. IT systems that were supposed to streamline and simplify business processes instead have brought a worrying level of complexity, containing tenth of databases and hundreds of separate software modules installed over years. Rather than agility, they have produced rigidity and unexpected barriers to flexibly adapt to organizational and process oriented changes.

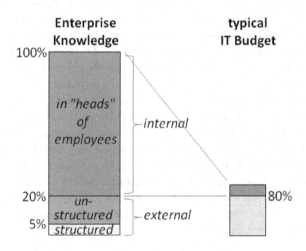

Fig. 1 Enterprise Knowledge and typically allocated IT budgets

To manage the growing complexity, IT departments of today's companies have grown substantially and spend 70% to 80% of their budgets just trying to keep existing business IT systems running. On the other hand, only 20 to 30% on relevant information for product development, production and other value adding core processes resides within a company's IT systems. In other words, 80% of a typical IT budget is spent for the management of 20% of the relevant data. Instead, significantly more money hast be spend to externalize internal knowledge (knowledge capturing), to create semantic enriched and context oriented data relationships leading to better information retrieval, to eliminate barriers of IT system usage (usability), and to establish means of stimulating information provision and externalization by individual engineers and experts. This is the only way to gain productivity in enterprise information and knowledge management.

2.2.2 The Engineer's Perspective (IT User)

Despite data explosion the individual knowledge decreases. Effects that can be observed are that the increasing dataset leads to reduced intake capacity of the human brain with consequences [3]:

- stronger information filters
- selective perception
- less-informed decisions
- only exciting information is noticed.

Current engineering practices are too week in terms of reuse of previous knowledge. A tremendous amount of time is wasted reinventing what is already known in the company or is available in outside resources. Redundant effort per employee is increasing and causing enormous cost as the complexity of enterprises and products increases. One reason behind this is that engineers in general do not make sufficient efforts to find engineering content beyond doing mere keyword/metadata searches within the PLMS environment. But this is also a matter of implemented procedures and regulations within the companies.

Engineers too much stick to tools like spreadsheets or "private" desktop databases. Certainly this is a problem of acceptance and benefits. If one has developed a spreadsheet which contains data for individual demands and covers needed functions, the effort for data maintenance directly pays off for the individual or the team using the spreadsheet.

Furthermore, developed functions directly correspond to the tasks and duties. Enterprise IT systems on the other hand are often intricately to use and literally seem absorb maintained data like a black hole without delivering a direct benefit to the user or only for other users down the road of the process chain. It is necessary to implement incentive system to motivate individuals to contribute to a cooperative knowledge by sharing information and eliminate islands of isolated information. But solving this issue is also a matter of further developing PLMS or enterprise IT systems in general to provide the demanded usability that common spreadsheet programs offer and that can be configured or customized to individual or role based needs. In the usability of enterprise PLMS increases user acceptance and grade of utilization also does. "Spreadsheet islands" can be eliminated.

3 Organic PLM Approach

The main challenge in terms of enterprise IT system architectures is to cope with complexity and dynamic changes. The growing complexity is notoriously difficult to handle in software systems. This has been known as the software crisis [4] for a long time. Traditional approaches of enterprise IT systems follow a hierarchical or centralized structure with some master or backbone system and others depending thereon. This proposition is also subject to change when project partners change or mergers & acquisitions lead to structural changes. Integration and interfacing projects for PLMS or other business critical IT environments are far away from being able to follow the pace of these changes.

Promising approaches to build applications from services - Service Oriented Architectures (SOA)- are not likely to do much better. The "Lego" idea behind this, the notion of reusable software works on a small scale but as software grows more complex, reusability in terms of services on a variety of granularity levels becomes difficult if not impossible. A unit of software code is not similar to other software code in terms of scale or functionality, as Legos are. Instead, software-code is widely various, semantic commonalities are rare and interfaces are heterogeneous.

Basic ideas of OC are self-organization and self-configuration. This can be established e.g. through use of agent based software systems but also – on a higher level – involving users as a part of the system configuring and organizing portions of applications as an integral part. The fundamental insight of the organic PLMS approach is that the complexity problem of products and processes cannot be solved through increasingly complex IT-Systems and connected development projects. An emphasis on simplicity, flexibility and efficiency is necessary.

The Organic PLM approach makes use of an infrastructure providing a technology for a flexible networking of different information fractals called Enterprise Mashups [5]. This is a strategic technology and will be the dominant model for the creation of composite enterprise applications according to various analysts. A mashup in the context of Web 2.0 is a hybrid WWW application that combines complementary elements from two or more sources to create one integrated experience. Content used in mashups is generally sourced from a third party via an API or from WWW feeds (e.g. RSS). Basically, the idea is to take multiple data

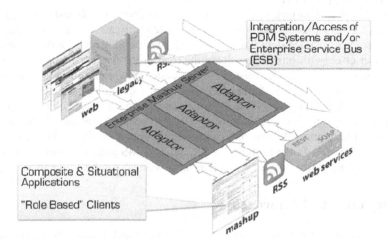

Fig. 2 Organic PLM approach through Enterprise Mashup Servers

sources or WWW services and compose them into something new or combined. In contrast to EAI (enterprise application integration) concepts or SOA, the unique point of this approach is that mashups allow business users and engineers to address their own information needs, to self-connect the data fragments in order to create information that answers their questions providing dynamic, user-specific views and customized filters. Mashups let users share their resulting services, making them a part of a services network in the sense of self-organization and self-optimization.

Advantages of enterprise mashup technology:

- Mashups are user driven, users are able consume public (enterprise spanning or WWW based) and local services and contents on demand.
- Users are able join in data from outside the enterprise to include external data in their work whereas SOA efforts are largely inwardly focused.
- The granularity of services and can be right-sized by the consumer without having the IT department to guess or make time consuming analyzes
- Composite and situational applications (role and/or project based) can be generated using a "Drag & Drop" configuration of the mashup server platform
- Interfaces/Adaptors to emerging data sources like Wikis, Blogs, and RSS.
  Disadvantages of enterprise mashup technology
- A consumer of a mashup service is not in control of the primary source of data. When sources are offline, a mashup is offline as well.
- Public accessible APIs for mashup services will limit the number of requests an application can make within certain period of time to avoid response problems.

4 Semantic Engineering Desktop

The second building block of a future IT working environment for engineers is based on semantic web technologies, also referred to as Web 3.0 technologies. Web 3.0 can be defined as a set of technologies that offer efficient new ways to help computers organize and draw conclusions from data whereas the term Web 2.0 is typically used to refer to a combination of

- improved communication between people via social-networking technologies
- improved communication between separate software applications (mashups) via open Web standards for describing and accessing data
- improved Web interfaces that mimic the real-time responsiveness of desktop applications within a browser window (Rich Internet Application – RIA technologies).

Web 2.0 technologies focus on social interaction and information acquisition. The intention of Web 3.0/Semantic Web is to improve the quality and transparency of data through provisioning of semantic relationships and ontologies. Web 3.0 - as seen by analysts like Gartner - is likely to decentralize enterprise wide information management. Self-optimization and context-awareness in the sense of OC can be accomplished by making use of ontologies and semantic networks within the

PLMS data sources as well as improving usability on PLMS client side. Given a three tier architecture as implemented in most of today's PLMS, additional layers between data layer and application layer comprise the means for semantic enrichment of managed information, i.e.

- consolidated and contextualized heterogeneous engineering documents and models,
- representation of knowledge in a more explicit, structured and navigatable manner,
- user-centric computer-aided tools and methods for a shift from text based/metadata based towards visual information retrieval.

to achieve high quality information retrieval of structured and unstructured knowledge assets. Several Technologies and research domains (Information Retrieval, Language Engineering and Natural Language Processing (NLP), Text Mining) have led to a multitude of commercially or public available software systems for this purpose. They have to be assembled to a coherent system. Ontologies can be used as a sophisticated indexing mechanism in order to structure an information repository mainly built from unstructured documents and to achieve better results in information retrieval systems.

5 Conclusion and Outlook

Sharing engineering information and providing access to relevant knowledge is a vital resource for enterprises. WWW and related technologies have had a tremendous momentum in the recent past. Web protocols, technologies, and middleware are well supported by various products evolving in the software industry and open source communities. Concentrating on the perspective of an engineer as a user of a sophisticated IT environment, PLM, Enterprise Content Management (ECM) and Personal Information Management (PIM) are converging or even merging towards a role based configurable "Semantic Engineering Desktop". Semantic Web Technologies offer new opportunities for enterprise IT aiming at establishing new approaches of handling information as resource and integrating different information sources. Organic Computing strives for a paradigm shift in looking at enterprise information systems: Hierarchical structures and integration approaches cannot be established in the same pace as changes of processes and organization occur. Instead networked relationships among different nodes of an IT environment and a demand (user) centric principle of looking at information resources are necessary.

References

1. Rochner, F., Müller-Schloer, C.: Emergence in Technical Systems. Special Issue on Organic Computing 47, 188–200 (2005)
2. Rettig, C.: The Trouble with Enterprise Software. MIT Sloan Management Review 49(1), 21–27 (2007)

3. Sternemann, K.-H.: Role based Clients in a Service Oriented Enterprise Architecture. In: Proceedings of the ProSTEP iViP Symposium 2008, Berlin, Germany, April 9-10 (2008)
4. http://en.wikipedia.org/wiki/Software_crisis (last viewed January 10, 2011)
5. Bitzer, S., Schumann, M.: Mashups: An Approach to Overcoming the Business/IT Gap in Service-Oriented Architectures. In: Nelson, M.L., Shaw, M.J., Strader, T.J. (eds.) AMCIS 2009. LNBIP, vol. 36, pp. 284–295. Springer, Heidelberg (2009)

Parameterized Path Based Randomized Oblivious Minimal Path Routing with Fault Tolerance in 2D Mesh Network on Chip

Mushtaq Ahmed, Vijay Laxmi, M.S. Gaur, and Yogesh Meena

Abstract. Fault tolerance has become one of the major concerns in routers or links as the transistor feature size is shrinking leading to gate scaling in Network on Chip (NoC). More number of processing elements (PE) are being incorporated for interconnection in system on chip, making it difficult to deliver packets to the destination. Overcome from permanent fault can be achieved using efficient routing algorithms whereas retransmission of faulty packet resolve transient faults in the network. Routing would be efficient, if it can handle multiple faults in the path while managing congestion in the network. In this paper a path based randomized oblivious minimal routing algorithm (FTPROM) with fault tolerance is proposed. FTPROM is derived from parameterized PROM is an oblivious routing. It uses probability functions to provide more diversity to route in the network and handles congestion in better way. Routing follows minimal path in presence of faults both as node and link failure. Simulation results show that the proposed algorithm is effective in terms of congestion handling, latency, and degree of adaptiveness.

1 Introduction

Network on Chip has become an accepted perspective of design of large scale systems. Limited resources are used for communication to minimize area and latency. Reliability of NoC increases by efficient fault tolerant routing [1]. Fault tolerance is unbounded as long as faulty nodes are connected outside in 2D mesh NoC. Fault-tolerant routing finds path from source (S)to destination (D) in presence of the faults with a certain degree of tolerance as long as faults does not partition the network [4].

Mushtaq Ahmed · Vijay Laxmi · M.S. Gaur · Yogesh Meena
Department of Computer Engineering,
Malaviya National Institute of Technology Jaipur, India
e-mail: {mushtaq,vlaxmi,gaurms,yogesh}@mnit.ac.in

F.L. Gaol et al. (Eds.): Proc. of the 2011 2nd International Congress CACS, AISC 145, pp. 63–70.
springerlink.com

J. Wu [2] presented deadlock free dimension order, Odd Even turn model based fault tolerant routing without virtual channels. Reconfiguration routing for DSPIN network using penalty provisions is proposed in [6]. Transient faults can be eliminated using duplicate packet transmission using broad-casting or flooding mechanism and N random walk [3], however this causes congestion in the network and increases overhead and hence causes increase in overall power consumption.

This paper is organized as follows. Section 2 presents the proposed fault tolerant (FTPROM) routing and describes path selection and fault tolerant mechanism. Section 3 describes simulation environment, experimental setup and result analysis. This is followed by concluding remarks and future work in section 4.

2 Fault Tolerant Path Based, Randomized, Oblivious, Minimal Routing

FTPROM is a oblivious, minimal path routing derived from parameterized PROM presented by *Cho* in [7]. PROM routes each packet separately via a path selected randomly from among all available minimal paths while considering the boundary and intermediate regions. Choices are given to constitute a random distribution over all possible minimal paths in the network. Turn model based fault tolerant routing algorithms like FTWF(Fault Tolerant West First), FTNL (Fault Tolerant North Last), FTNG (Fault Tolerant Negative First) etc. are deadlock free and works without need of any virtual channel(VC) [8], whereas *FTPROM* requires different allocation schemes for achieving deadlock free scenario and at least two virtual channels are required to make deadlock free routing. For *FTPROM* routing experiments we used four *VC*.

FTPROM provides more diversification of packet routing to enhances the performance in terms of congestion handling. We used parameterized PROM routing with fault tolerance for our experiments. Two different probabilities $P1$ and $P2$ for each path are used at every node for FTPROM. The f parameter as defined below is used in parameterized PROM.

Let n is the number of rows and cols in a $n \times n$ Mesh, D_x, D_y: x and y coordinate of destination node, C_x, C_y: x and y coordinate of the current node. Let $x_1 = |D_x - C_x|$, $y_1 = |D_y - C_y|$ then minimal rectangle size is $(x_1+1)*(y_1+1)$ and overall rectangle size will be n*n and therefore, f can be given as:

$$f = \left\lceil \frac{(x_1+1)*(y_1+1)}{(n*n)} \right\rceil \tag{1}$$

If x and y are the unit distance of source s and destination d. Packet is routed either in X direction or Y direction as per the ratio x + f : y + f Using this ratio the router determines the next hop. The ratio x + f : y + f is used if a packet arrives from north or south to a node in X-axis ingress, and it uses the ratio x : y + f for Y-axis ingress [7].

The path selection as per the growth of f is as shown in Figure 1-(a). As the values of f increases, less turns are allowed and traffic is pushed from the diagonal of the minimal-path rectangle towards the edges. Figure 1-(b) shows path selection by the fault tolerant Parametrized-FTPROM in presence of link failures.

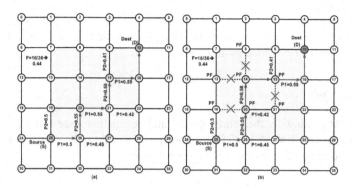

Fig. 1 (a) Path selection by the Parametrized PROM with respective to the available probability functions, and (b) Path selection by the fault tolerant Parametrized-FTPROM in presence of link failures.

Fault in *NoC* may occur anywhere and on the basis of fault location, following terms are defined **Potentially Faulty node(PF)**: node having local information about one or more faulty links. **Fault Regions**: those regions where faults lies in the intermediary links of a boundary and referred to as fault regions and **Fault Chains**: Multiple faults in the links of a boundary that constitutes the axis of the destination node.

2.1   Fault Tolerant FTPROM Routing Rules

FTPROM tries to push the packets towards boundary regions to avoid the congestion in the network. Either at the boundary or at intermediate tiles at every node two probability functions $P1$ and $P2$ are used for path selection, if it is available. At the initial source node the probability $P1$ and $P2$ are same and any one path which is available can be chosen. *FTPROM* tries to route packet in X-axis ingress if a packet arrives from North or South to a node i.e., Y-axis and routes packet in Y-axis ingress if a packet arrives from East or West to a node. i.e., X-axis. The routing decisions taken by the *FTPROM* are as per the following rules.

1. For each source and destination pair and boundary regions are defined and parameter f is calculated first.
2. If the current node is source node and if both neighbouring nodes are not *PF* nodes, the current router may choose any one for next-hop.
3. If the current node is source node and if one neighbouring nodes are *PF* nodes, the current router chooses other one for next-hop.

4. To avoid the faulty chains appearing at the axis of source node, packets are pushed toward the intermediary nodes using the priority functions.
5. If the current node is an intermediate node, these are two possibilities: 1. Flit arrives on the Y-axis ingress, 2. Flit arrives on the X-axis ingress, increasing probability of moving packets and checking of available link allows to avoid fault regions and faulty chains.
6. If one or both nodes are PF nodes, Next-hop is chosen with the available priority. However if one of the PF node lies on the axis of destination quadrant, FTPROM chooses the other PF node for next-hop.

2.2 Example

In the PROM routing, let the source node S and destination node D are 25 and 10 respectively in a 6×6 Mesh NoC. The parameter f function and probability function $P1$ and $P2$ for node 25 can be calculated as follows:

Parameter $f = 16/36$; $P1_{(25)} = \frac{y+f}{x+y+2f}$; $P2_{(25)} = \frac{x+f}{x+y+2f}$; $P1_{(25)} = \frac{2+0.44}{2+2+0.44}$; $P2_{(25)} = \frac{2+0.44}{2+2+0.44}$. This gives $P1_{(19)} = 0.5$ and $P2_{(19)} = 0.5$

As the initial node has same $P1$ and $P2$, arbitrarily any path can be chosen. Let us assume it is towards node 26. Again probability functions is calculated for node 26 as under: $P1_{(26)} = \frac{(y-1)+f}{x+(y-1)+f}$; $P2_{(26)} = \frac{x}{x+(y-1)+f}$; $P1_{(26)} = 0.45$; $P2_{(26)} = 0.55$. Probability function $P2_{(26)}$ is high therefore, path from node 26 to node 20 is selected. Now probability at node 20 is to be calculated $P1_{(20)} = \frac{(y-1)}{(x-1)+(y-1)+f}$; $P2_{(14)} = \frac{(x-1)+f}{(x-1)+(y-1)+f}$; $P1_{(20)} = 0.55$; $P2_{(20)} = 0.45$. Probability function $P1_{(20)}$ is higher that $P1_{(20)}$ therefore, path from node 20 to node 21 is selected which is having probability $P1_{(21)} = 0.42$ and $P2_{(21)} = 0.58$ and route packets towards node 15. Again at node 15 probability is calculated as $P1_{(15)} = 0.59$ and $P2_{(15)} = 0.41$ respectively. Packet is routed towards node 16 having higher priority and is lying at the axis of destination and finally packet reaches to destination. In case of presence of faulty nodes or links $FTPROM$ is applied. Let the path links 8-14, 13-14, 19-20 and 15-21 are made faulty. The $FTPROM$ takes routing decisions as follows:

1. At the source node 25 probability $P1$ and $P2$ calculated is same i.e. 0.5 therefore, next hop is selected randomly. let the selected node is 26.
2. At node $P_{(26)}$ faulty links are checked. If there is no faulty link probability $P1$ and $P2$ are checked and here in this case $P1_{(26)}$ is 0.45 and $P2_{(26)}$ is 0.55. Direction of packet move towards node 20.
3. At node 20, both the neighbouring node are faulty nodes and therefore path is selected as per probability $P1$ and $P2$ which is $P1_{(20)} = 0.42$ and $P2_{(26)} = 0.58$. Therefore, packet is routed towards node 14.
4. At the node 14 only one out going link is available and therefore is selected and packet is routed towards node 15.

5. At node 15 $P1$ and $P2$ is 0.59 and 0.41 respectively. Packet is routed towards node 16 which is at the axis of destination and link is available, therefore packet route upward and reaches to destination node 10 as shown in Figure 1-(b).

3   Experimental Setup and Result Analysis

We used extended version of NIRGAM simulator [9] for experimental analysis with fault tolerance. The first experiment is to explore the degree of tolerance of routing algorithm which dependents on region of interest [10]. In this experimental setup, a 7×7 two dimensional Mesh topology was selected as single source-destination pair. For our experiment, we chose node 9 as Source and node 33 as Destination. At source node 9 Bursty data generating pattern was applied for the burst length of three and simulated for 5000 cycles. Each time an additional link failure is introduced in the path to test the degree of tolerance within the boundary region of the source destination pair. Degree of fault tolerance observed is 10 in this experiment and varies from minimum of 2 links failure to a maximum of existence of one path.

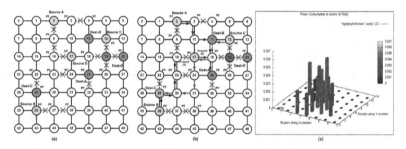

Fig. 2 (a) 13 faults, one node failure and four pairs of Source and destinations, (b) Path selected by *FTPROM* only two source destination pairs (7-25, 12-29) are successful in packet transmission, and (c) Power consumption in watts in the fix faults and fixed source and destination pairs.

For the second experimental setup four fixed source and destination pairs were chosen as shown in Figure 2-(a). Under 100 percent load condition, Bursty traffic patterns was applied with packet size of 30 and 10 bytes buffers at each port and 5 bytes flit size. Path chosen by the *FTPROM* is shown in Figure 2-(b). Only two source destination pairs $2 - 25$, $1 - 29$ as shown in Figure 2-(b) are successful to transmit packets with a throughput of 0.928976 Gbps, overall average latency 89.1198 (in clock cycles per flit), and an average latency of 13.2363(in clock cycles per packet per channel). Due to non availability of path, packets generated from two sources 7 and 36 could not reach the destinations. Figure 2-(c) shows total power consumption is 0.0469725 in watts for a fixed four source and destination pairs. We used Orion 2.0 library [11] for accurate power estimation in NIRGAM.

The third experiment was done to explore performance analysis of *FTPROM*. For this experiment a 7×7 Mesh was chosen. Total 13 numbers of faults $(3 - 4,$

Table 1 Table shows the failure link, path selected by FTPROM, packets generated and packets received and average latency

No-of-Links	Links Failed	Path Selected by FTPROM	Packets Generated	Packets Received	Average Latency
0	Nil	7-13-14-20-21-27-28	967	967	66.4367
1	14-20	7-13-14-15-21-27-28	967	967	66.4367
2	14-20, 15-21	7-13-14-15-16-22-28	966	966	66.4367
3	14-20, 15-21, 13-14	7-13-19-20-26-27-28	966	966	66.4367
4	14-20, 15-21, 13-14, 20-26	7-13-19-20-21-27-28	962	962	66.4365
5	14-20, 15-21, 13-14, 20-26, 19-20	7-13-19-25-26-27-28	962	962	66.4367
6	14-20, 15-21, 13-14, 20-26, 19-20, 7-13	7-8-14-15-16-22-28	967	967	66.4366
7	14-20, 15-21, 13-14, 20-26, 19-20, 7-13, 8-14	7-8-9-15-16-22-28	967	967	66.4366
8	14-20, 15-21, 13-14, 20-26, 19-20, 7-13, 8-14, 9-15	7-8-9-10-16-22-28	967	967	66.4366
9	14-20, 15-21, 13-14, 20-26, 19-20, 7-13, 8-14, 9-15, 9-10	7-8-9-Blocked	32	Nil	Infinite

$2-9, 9-16, 12-19, 16-17, 18-19, 19-20, 19-26, 22-29, 23-24, 25-32, 36-37, 37-38$) along with a fix node failure (node 19) were taken and initialized for simulation. Bursty traffic is applied at all available source nodes with 100 percent load traffic. Buffer size of 10 bytes for each input port and packet length of 30 bytes and 5 bytes flit size was introduced. We took average Burst length of 3 cycles and off time interval of 3 cycles between each Burst and simulation were carried out for 5000 cycles.

In Figure 3-(a) and (c) shows average latency under random and transpose traffic, 3(b) and (d) shows average throughput under random traffic and transpose traffic with Bursty data. Latency is calculated in clock cycles per flit. As *NIRGAM* is cycle accurate and every event handling consumes one cycle, number of simulation cycles for routing a flit from input port to output channel of the router may take about six to eight cycles. Latency in random traffic is almost less than 50 upto 80% of load injection increases 8% at 100% laod injection. in Transpose traffic where source and destinations are fixed more precise observation is observed. Latency remains almost same with the increase of load after 60% load. At 100% load injection a

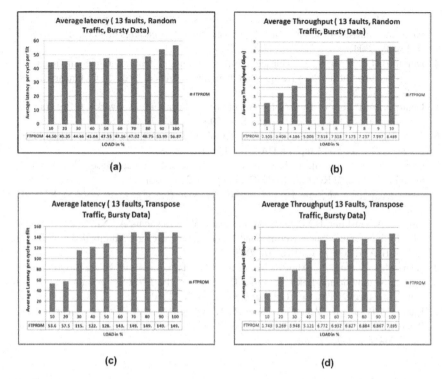

Fig. 3 Figure (a) and (c) shows average latency under random and transpose traffic, (b) and (d) shows average throughput under random traffic and transpose traffic with Bursty data pattern

good throughput of 8.489 Gbps and 7.395 Gbps is observed in random and transpose traffic respectivly.

4   Conclusions and Future Work

The proposed *FTPROM* is a minimal path, partially adaptive fault tolerant routing algorithm which can handle congestions and gives a good throughput. Future work requires exploration and comprehensive analysis of *FTPROM* routing with other minimal and non minimal routing techniques to compare efficiency in terms of area, latency, throughput and overall power consumption.

References

1. Lehtonen, T., Liljeberg, P., Plosila, J.: Fault Tolerant Distributed Routing Algorithms for Mesh Networks-on-Chip. In: ISSCS 2009, pp. 1–4 (2009)
2. Wu, J.: A fault tolerant and deadlock free routing protocol in 2D Meshes based on ODD-EVEN turn model. IEEE Transactions on Computers 52(9), 1154–1169 (2003)

3. Pirretti, M., Link, G.M., Brooks, R.R., Vijaykrishnan, N., Kandemir, M., Irwin, M.J.: Fault Tolerant Algorithms for NoC Interconnect. In: Proceedings of ISVLSI, pp. 46–51 (February 2004)

4. Dumitras, T., Kerner, S., Marculescu, R.: Towards on-chip fault tolerant communication. In: Proceedings of the 40th Design Automation Conference, DAC 2003, January 21-24, pp. 225–232 (2003)

5. Duato, J.: A Theory of Fault Tolerant Routing in Worm hole Networks. IEEE Transaction, Parallel and Distributed Systems 8(8), 790–801 (1997)

6. Zhen, Z., Alain, G., Sami, T.: A Reconfigurable Routing Algorithm for a Fault Tolerant 2D Mesh Network on Chip. In: DAC, pp. 441–446 (2008)

7. Cho, M.H., Lis, M., Shim, K.S., Kinsy, M., Devadas, S.: Path-Based, Randomized, Oblivious, Minimal Routing. In: NoCArc, pp. 23–28 (2009)

8. Glass, C.J., Ni, L.M.: The Turn Model for Adaptive Routing. J. ACM 41(5), 874–902 (1994)

9. Jain, L., Al-Hashimi, B.M., Gaur, M.S., Laxmi, V., Narayanan, A.: NIRGAM: A Simulator for NoC Interconnect Routing and Modelling. In: Design, Automation and Test in Europe (DATE 2007), April 16-20 (2007)

10. Gaur, M.S., Laxmi, V., Ahmed, M., et al.: Minimal path, Fault Tolerant, QoS aware Routing with node and link failure in 2-D Mesh NoC. In: IEEE Symposium Defect and Fault Tolerance DFTS 2010, October 6-8, pp. 60–66 (2010)

11. Kahng, A.B., Li, B., Peh, L.-S., Samadi, K.: ORION 2.0: A Fast and Accurate NoC Power and Area Model for Early-Stage Design Space Exploration. In: DATE 2009, pp. 423–428 (2009)

Radiation Study of SEE in ASIC Fabricated in 0.18μm Technology

Pan Dong, Long Fan, Suge Yue, Hongchao Zheng, and Shougang Du

Abstract. The research on the radiation effects of ASIC has been the hot point in the field of the international aerospace devices. This paper developed a test system for the single event effects (SEEs), which was applied to test the radiation effects of some domestic ASIC chips. Based on the result, SEEs under different work conditions: high temperature, normal temperature, low voltage, normal voltage were analyzed. Thus the voltage and temperature effects on SEE can be studied. The test showed that the increasing temperature and decreasing voltage could affect SEEs, meanwhile the sensibility of the ASIC circuit to SEE will be enhanced. The high temperature and low voltage is the worst work condition to the ASIC circuit, and there is no latch-up on the work condition of high temperature and high voltage.

1 Introduction

The semiconductor devices work in the radiation environments full of different energy particles in the space [1]. The high energy particles in the space may cause SEEs to the semiconductor components of the aerospace devices in the electronic system, and this influences the reliability and lifetime seriously of the semiconductor devices. Now more and more ASICs are applied in various kinds of spacecraft widely .Thus the SEE hardness assurance of the ASIC circuits may influence the accuracy and lifetime of aerospace devices directly. So it is very important to evaluate and test the SEE hardness assurance of the ASICs.

SEEs are one of the quotas to evaluate the radiation-harden ability of ASIC. SEE includes soft errors (the single event upset (SEU)) and hard errors (the single event latch-up (SEL)). SEEs are random, soft errors may cause the electronic system to operate wrongly, the hard errors may cause the electronic system to

Pan Dong · Long Fan · Suge Yue
NO.2 Siyingmen North Rd, Donggaodi, Fengtai District, Beijing, 100076, China
e-mail: dpgpyy @126.com, {meilfan,suge_yue}@yahoo.com.cn

F.L. Gaol et al. (Eds.): Proc. of the 2011 2nd International Congress CACS, AISC 145, pp. 71–78.
springerlink.com © Springer-Verlag Berlin Heidelberg 2012

lose functionality. Thus SEEs are one of the important facts that influence the in-orbit reliability of the aerospace devices [2].

It is well known that charge collected from an ion track by a sensitive node, the amount of the collected charges exceeds the critical amount, usually produce a SEU or lose logic function when the high energy particles get through the sensi-tive regions of the semiconductor devices.

SEL happens in the CMOS devices, for the CMOS devices are consist of spe-cial structure of PNPN as Fig. 1, which constitutes the parasitic controllable sili-con structure. In the normal state, the parasitic controllable silicon structure is high-resistance off. The high energy ions can make the parasitic controllable sili-con structure break over and there will be current flow. Due to the feedback func-tion of the controllable silicon structure, the amount of the passed current flow gets larger instantaneously, and then SEL occurs.

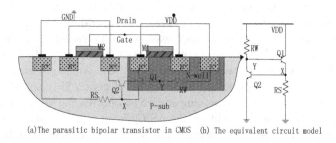

(a)The parasitic bipolar transistor in CMOS   (b) The equivalent circuit model

Fig. 1 Latch-up Model

In satellites and spacecraft, the temperature of the electronic system is con-trolled strictly in the range of 30°C~60°C. But in some special conditions, the temperature range may be expanded or narrowed. According to the semiconductor physical property, the semiconductor property is easy to be effected by the tem-perature. The physical property differs in different temperature conditions. If the environment temperature is changed, SEEs of the devices may differ. As the envi-ronment temperature increasing, SEE will be aggravated. Thus it is needed to study the temperature effect on SEE of the electronic devices. The work tempera-ture range of the aerospace devices is -55°C ~125°C according to the U.S. military standards. And some papers show that SEEs of the semiconductor circuits are not sensitive to the temperature. Thus the temperature of 125° is chosen to study the relationship between the temperature and SEEs [3].

The semiconductor characteristics will be influenced trough changing the vol-tage, so the change of the voltage may influence SEE. The voltage effects on SEE can be researched based on the study of the temperature.

The experiment in the paper was performed on the heavy ion cyclotron accele-rator in the Institute of Modern Physics, Chinese Academy of Sciences and the HI-13 tandem accelerator in the China Institute of Atomic Energy.

2 Experimental Details

2.1 ASIC

The ASIC in the paper is a corresponding chip. The main functions of the ASIC are data-processing and data-transmitting.

The functional block diagram is shown in Figure 2. The ASIC is fabricated in 0.18um 1p4m technology and the dose radiation-harden is finished using the ring gate structure. The single events radiation-harden is done through the circuit design and the ASIC is harden on the term of logic.

There are DFT test chains inside the ASIC, which are used to test the chip. This flip-flop radiation experiment was done based on these test chains. Figure 2 shows the characteristic parameters of the ASIC.

Design features		Process Features			Library Features	
Core Vdd	1.8V				Metal	
Core current	45mA	Starting Substrate	Epitaxial		levels	4
I/O Vdd	3.3V	(Baseline for class V)	3.5um		Flip	
I/O current	58mA				Chip	Yes
Core Lpoly	0.18um	Shallow				
Core tox	40Å	Trench	4000 Å		PLL	Yes
I/O Lpoly	0.35um	Isolation				
I/O tox	80 Å	(STI) Depth				

Fig. 2 ASIC Characteristic Parameters

2.2 Test System Structure

The test system of SEEs needs a radiation test board, the power control and current flow collection module, one power supply and two PCs. The ASIC device under test is put in the radiation test board, which is connected with the power control and current flow collection module outside the vacuum chamber through the 92 flange plate. Using RS232 to connect the test board with the upper computer, and the power controller and current flow collection model is controlled by the upper computer in application of the USB data lines. The upper computer inside test room is monitored remotely by the workers with PC through the 50m network line. The voltages of the I/O ports and the core are supplied by the outside power, and the current flow information monitored by the current flow monitor model is transmitted to the upper computer. Using one heater band assembled with the ASIC to heat the device one infrared temperature detector is added on the circuit. The detector and the heater band are connected with the temperature controller through the 92 flange plate. The block of the SEE test system is shown in Fig. 3.

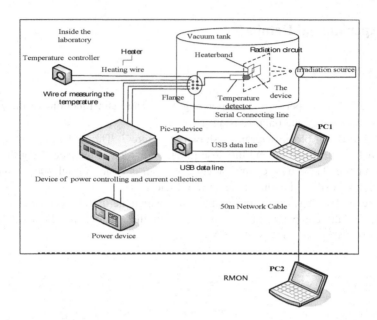

Fig. 3 Test System Block Diagram

Before the experiment, the ASIC was injected into hard errors and soft errors to verify the creditability of test system. The result shows that the test system could capture the fault data, which proved that the test system is very stable and reliable.

2.3 Test Methods

This paper mainly studies the SEEs of the ASIC from the following two types: SEU, SEL. The radiation-harden methods of the two types would be introduced in the following text.

1) Test method of the flip-flop upset

The flip-flop experiment was done by the scanning chain method. There are 9816 D flip-flops and 8 scanning chains in the ASIC circuit. The flip-flops with different clock domains and different edge-flip-flop types are distributed in different scanning chains. Input 1 or 0 into the 8 chains through FPGA, if the result was not accord to the expected, the result could be regarded as errors, and the error counter inside the FPGA adds 1. The upper monitor will read and show the error number recorded in the FPGA.

2) Latch-up test method

The performance of SEL is that the rapid increase of the current flow, so evaluate the current flow is one of ways to estimate whether there is SEL in the circuit. Using the millimeter and the Data acquisition card to monitor the current flow, if the current flow increases to more than three times and lasts increasing, the power should be cut off, and record one time of SEL. If the number of the particles gets to 10^7, the total latch-up rates is recorded.

3 Results

In the experiment, 1 and 0 were injected into the D flip-flop through the input port. The result shows that the SEU of the ASIC is not sensitive to the injected 0. To compare with the test result easily, all the D flip-flops were injected into 1 through the input ports. The particles and the effective LETs chosen in the experiment are shown in Table 1.

Table 1 Chosen charged particles

Particle	Energy(MeV)	Effective LET (MeV-cm2/mg)
Kr	504	39.6
I	240	58
Au	300	72.13
Bi	1985	101

The average number of injected particles was 8000/s in the experiment and the total ions is $1\times10^7/cm^2$ according to the rules of the USA army.

The ASIC circuit works in the following conditions: normal temperature and normal voltage, normal temperature and high voltage, normal temperature and low voltage, high temperature and normal voltage, high temperature and high voltage, normal temperature and low voltage. And the radiation experiment was finished in the vacuum environment [4]. The result was analyzed with the weibull fit, and Table 2 shows the analyze result. The result showed that there was no SEL on the six conditions.

Table 2 Test Result

SEE	Temperature	Core voltage	I/O voltage	Saturated cross section(cm^2)
SEU	25°C	1.8V	3.3V	~2.3x10^{-6}
SEU	25°C	1.62V	2.97V	~3.0x10^{-6}
SEU	25°C	1.98V	3.63V	~1.9x10^{-6}
SEU	125°C	1.8V	3.3V	~3.1x10^{-6}
SEU	125°C	1.62V	2.97V	~3.6x10^{-6}
SEU	125°C	1.98V	3.63V	~2.0x10^{-6}
SEL	125°C	1.98V	3.63V	0

4 Data Analysis

4.1 Voltage Effect

On the premise of steady temperature, if changing the work voltage, the ASIC SEU performs different outcomes, and the result was shown in the Fig. 4 and Fig. 5.The result showed that SEU hardness assurance on the condition of low voltage was obvious worse than on the condition of normal and high voltage, and SEU hardness assurance on the condition of high voltage was better than on the conditions of normal and low voltage. On the premise of normal temperature, the saturated cross section of the SEU on the condition of low voltage was 1.5 times of the saturated cross section on the condition of high voltage. Meanwhile on the premise of high temperature, the saturated cross section of the SEU on the condition of low voltage was 2 times of the saturated cross section on the condition of high voltage. Thus the decrease of the voltage could impel the saturated cross section of SEU to rise.

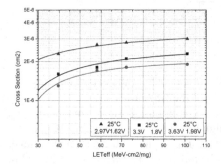
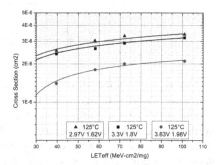

Fig. 4 Experiment result on the condition of 25°

Fig. 5 Experiment result on the condition of 125°

4.2 Temperature Effect

The temperature effect on the ASIC was compared though changing the work temperature on the premise of steady work voltage. The temperature effect on SEU was shown in Fig. 6 ~Fig.8. It can be seen from the three figures that the saturated cross section of SEU raises with the increase of the temperature. On the condition of normal voltage, the saturated cross section rises about 1.5 times, while the rise was not obvious on the conditions of low and high voltage. The reason may be that on the condition of low and high voltage, the temperature effect on SEU is not serious.

There were no SEE in the circuit on the above six conditions in the test,which proves that SEL hardness assurance of the circuit is very well.

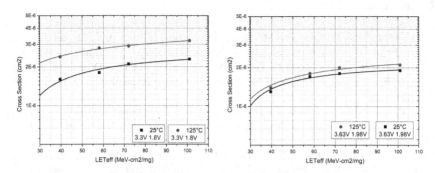

Fig. 6 Work With Normal Voltage   **Fig. 7** Work With High Voltage

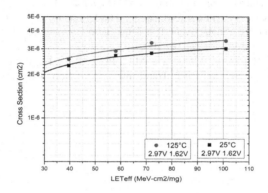

Fig. 8 Work With Low Voltage

5 Conclusions

The experiment result shows that the temperature can influence the SEU of the semiconductor devices [5]. The upset saturated cross section rises when the temperature is high, so the circuit sensibility to SEE is enhanced. And SEU hardness assurance is better on the condition of increasing voltage than on the condition of normal voltage, and the SEU hardness assurance is worse due to the decreasing voltage [6].

It can be seen from the experiment result that though the saturated cross section of SEU rises with the increase of the temperature, the rise amplitude is not obvious, which shows that this type of ASIC has good SEU hardness assurance because the temperature and the voltage has little effect on SEE. The result of the latch-up test was very good. There was no latch-up in the circuit on the worst work condition, which showed that SEL hardness assurance of this ASIC was very good [7].

References

1. Shoga, M., Jobe, K., Glasgow, M., Bustamant, M., et al.: Single Event Upset at Giga-hertz Frequencies. IEEE Transactions on Nuclear Science 41(6) (December 1994)
2. Schwank, J.R., Shaneyfelt, M.R., Felix, J.A., et al.: Effects of Total Dose Irradiation on Single-Event Upset Hardness. IEEE Transactions on Nuclear Science 53(4) (August 2006)
3. Laird, J.S., Hirao, T., Onoda, S., Mori, H., Itoh, H.: Temperature dependence of heavy ion induced current flow transients in Si -Epilayer devices. IEEE Transactions on Nuclear Science 49(3) (June 2002)
4. Gadlage, M.J., Ahlbin, J.R., Narasimham, B., et al.: Increased Single-Event Transient Pulse widths in a 90-nm Bulk CMOS Technology Operating at Elevated Temperatures. IEEE Transactions on Device and Materials Reliability 10(1) (March 2010)
5. Gadlage, M.J., Ahlbin, J.R., et al.: Increased Single-Event Transient Pulse widths in a 65-nm Bulk CMOS Technology Operating at Elevated Temperatures. IEEE Transactions on Device and Materials Reliability 10(1) (March 2010)
6. Chen, D., Buchner, S.P., Phan, A.M., et al.: The Effects of Elevated Temperature on Pulsed-Laser-Induced Single Event Transients in Analog Devices. IEEE Transactions on Nuclear Science 56(6) (December 2009)
7. Johnston, A.H., Hughlock, B.W., Baze, M.P., Plaag, R.E.: The Effect of Temperature on Single-Particle Latch up. IEEE Transactions on Nuclear Science 38(6) (December 1991)

Optimal Task Scheduling Algorithm for Parallel Processing

Hiroki Shioda, Katsumi Konishi, and Seiichi Shin

Abstract. This paper proposes an optimal task scheduling algorithm for parallel processing. The scheduling problem is formulated as a 0-1 integer problem, where a priority of processing is represented by constraints of the problem. A numerical example shows the effectiveness of the proposing scheduling.

1 Introduction

In parallel processing, a program is designed and realized as a set of computing tasks. Therefore, it is necessary to schedule these tasks to processing elements (PEs). This is called the scheduling problem and is known to be one of the most challenging problems in parallel processing. The goal is to assign tasks to PEs appropriately such that certain performance indices are optimized.

However, the problem has been described in a number of different ways in different fields, for example, a branch-and-bound algorithm [1,2], the task duplication-based scheduling (TDS) algorithm [3], task duplication based scheduling algorithm for the network of heterogeneous systems (TANH) [4], OIHSA (Optimal Insertion Hybrid Scheduling Algorithm) and BBSA (Bandwidth Based Scheduling Algorithm) [5], Optimal algorithms for scheduling divisible workloads on heterogeneous systems[6], and so on.

The CP/MISF [7] in OSCOR [8] is known as one of the most efficient heuristic scheduling algorithm. This algorithm provides the scheduling based on the priority

Hiroki Shioda · Seiichi Shin
Department of Mechanical Engineering and Intelligent Systems The University
of Electro-Communications 1-5-1 Choufugaoka Chofu-shi, Tokyo, 182-8585 Japan
e-mail: {shiota,shin}@se.uec.ac.jp

Katsumi Konishi
Department of Computer Science Kogakuin University 1-24-2 Nishi-shinjuku Shinjuku-ku, Tokyo, 163-8677 Japan
e-mail: konishi@kk-lab.jp

F.L. Gaol et al. (Eds.): Proc. of the 2011 2nd International Congress CACS, AISC 145, pp. 79–87.
springerlink.com
© Springer-Verlag Berlin Heidelberg 2012

of the tasks. The priority of the task is decided by the longest path length from the final task and by the number of post processing task. However, in the case of the existence of equal priority tasks, it is very difficult to obtain the optimal scheduling [9].

The authors presented the optimal scheduling algorithm of parallel processing based on the 0-1 integer problem framework [10]. Beaumont.et. al. [6] has provided several scheduling algorithms based on the 0-1 integer problem. However, its realization of the relationship among tasks is restricted to be simple and cannot express the complicated relationship since reference [6] is based on DLT (Disable Load Theory). On the other hand, we consider here a scheduling problem that has more complex priority relationship among tasks. In the references [3], [4], [5], the relation among tasks is formulated based on DAG (Directed Acyclic Graph) [11]. Therefore, we should make the DAG before scheduling tasks. The reference [10] only needs priority relation among tasks, because we formulate the problem as the 0-1 integer problem with the priority constraint. In our method, the priority is the workflow of processing tasks. In addition, our formulation can express the difference of the processing performance of each processing element and the difficulty of each task. Moreover, our formulation reduces delay time cased by idling of each PE.

This paper is intended to report the experimental result comparing CP/MISF [7] with our method [10]. The experiment have been conducted under conditions where [9] has been done.

2   Problem Setting

2.1   Notation

We consider the task scheduling with bounded parallelism. We assume that the number of PEs, the priority relation among tasks and the processing time of tasks are given. The processing time includes the communication time and the calculation time.

We utilize the following notations. n denotes the number of PEs. PE i means the ith PE. m denotes the number of tasks. Task j means the jth task. The processing time which takes for PE i to process task j is denoted by t_{ij}. A set of time steps is defined as $T = \{1,2,\ldots,m\}$. The indices i, j and k denote the subscriptions of PE, task, and time step, respectively.

2.2   Formulation

This paper deals with the following optimal task scheduling problem.

$$Minimize \qquad \varepsilon + s \tag{1}$$

Subject to

$$\sum_{k=1}^{m}\sum_{j=1}^{m} t_{ij} x_{ijk} \leq \varepsilon \qquad \forall i \tag{2}$$

$$\sum_{k=1}^{m} f_{ik} y_{ik} \leq s \quad \forall i \tag{3}$$

In this problem, $\mathcal{E}$ denotes the upper bound of the total processing time for each PE, and x_{ijk} is the binary variable defined by

$$x_{ijk} = \begin{cases} 1 & \text{if PE } i \quad \text{process task } j \text{ on time } k; \\ 0 & \text{otherwise.} \end{cases} \tag{4}$$

Equation (3) denotes the time required for processing all tasks on each PE. The variable s is the maximum time required for processing all tasks on each PE and f_{ik} is the time cost for PE i to process a task on the time step k. The time cost f_{ik} satisfies the following equation,

$$t_{ij} = f_{ik} \quad \forall i, \forall j, \forall k \tag{5}$$

y_{ik} is the binary variable as follows.

$$y_{ik} = \begin{cases} 1 & \text{if PE } i \quad \text{processes a task on} \\ 0 & \text{otherwise} \end{cases} \tag{6}$$

where

$$y_{ik} \geq x_{ijk} \quad \forall i, \forall j, \forall k \tag{7}$$

Equation (7) denotes that y_{ik} means making the reservation for processing.

In order to reduce the idling time, it is necessary to force processing on early time step. So we weight the time step k to formulate the constraints (3), (5), (6), and (7). However it takes more than these constraints for the reduction of idling time because discontinuous reservations create the idling time. Therefore, we formulate the constraint to make reservation continuously as follows.

$$\sum_{k=1}^{k-1} y_{ik} \geq (k-1) y_{ik} \quad \forall i, \forall k (k \geq 2) \tag{8}$$

The above inequality denotes that PE i can process task j by the time when y_{ik} becomes 0. Fig. 1 illustrates how (8) works in the case of $k = 5$ for instance.

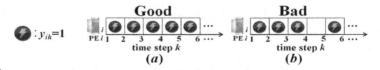

Fig. 1 Example of (16): (a) $4 \geq 4 \times 1$ (b) $3 \geq 4 \times 1$

The reservation for processing should be made continuously as shown in Fig.1, while tasks may be processed discretely. In addition, the tasks should be processed within the reservation interval. Therefore, by adding the constraints of (7) and (8), we can minimize the upper bound of the period of the all processing on each PE. To minimize s, our method reduces the delay time. The constraints of (3), (5), (6), (7) and (8) are illustrated by Figs. 2 and 3.

First, in order to stuff tasks up, we weight the time k. It is denoted by (5), (6), (7) and (8). Figure 2 illustrates the constraints of (5), (6), (7) and (8).

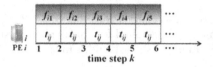

Fig. 2 Weighting the time step k

Second, we express the sum of f_{ik} by s. It is denoted by (3). Figure 3 illustrates the constraints (3), (5), (6), (7) and (8).

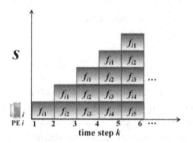

Fig. 3 Expressing the sum of f_{ik} by s

Each task has to be processed by the PEs within time step m. which is expressed as

$$\sum_{k=1}^{m}\sum_{i=1}^{n} x_{ijk} = 1 \quad \forall j .$$ (9)

Each PE can process only one task on each time step m (Fig.4), which is expressed as

$$\sum_{j=1}^{m} x_{ijk} \leq 1 \quad \forall i, \forall k .$$ (10)

Fig. 4 The constraint of processing

There is the priority relation in parallel processing. In this study, we formulate the relation as a constraint in the 0-1 integer problem. Figure 5 illustrates a priority relation to take the processing of task j' for instance. Task j' cannot be processed before the processing of p tasks (task 1, task 2,...,task p) complete.

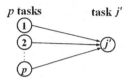

Fig. 5 The processing of task j'

Here we set the preconditions to formulate the priority relation such that the parallelism is preserved as shown below.

- The processing chance of p tasks is given to all PEs.
- p tasks do not have to be processed at the same time.
- After completing p tasks, the PEs can process task j' .

According to the above preconditions, the priority concerned with task j' can be formulated as

$$
\begin{cases}
\displaystyle\sum_{i=1}^{m}\sum_{j=1}^{p}\sum_{k=1}^{k-1} x_{ijk} \geq p\sum_{i=1}^{m} x_{ij'k} \, \forall k (k \geq 2) \\
\displaystyle 0 = \sum_{i=1}^{m} x_{ij'k} \quad (k=1)
\end{cases}
\tag{11}
$$

The left part of (11) denotes the processing of p tasks. The right part of (11) denotes the processing of task j' . The first equation in (11) formulates the situation $k \geq 2$. The second equation in (11) formulates the situation $k = 1$. Task j' cannot be processed at the time step $k = 1$. Therefore, the priority relation is denoted by the simultaneous equations in (11).

Finally, the scheduling problem in parallel processing is formulated. The objective function is (1) and the constraints are (2) - (11).

2.3   Processing Time Setting

The processing time of tasks depends on the processing performance of each PE and the difficulty of each task. This paper appreciates the processing time of tasks as the shortest processing time on each PE. In this case, the processing performance of each PE also can be evaluated by the shortest processing time. The shortest processing time for the ith PE is denoted by a_i as follows.

$$
t_{ij} = a_i \quad \forall i, \forall j
\tag{12}
$$

We express the difficulty of each task by the whole-number multiple b of a_i. This means that a task divide into b parts. Figure 6 illustrates the dividing process.

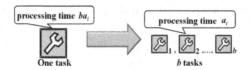

Fig. 6 Dividing task

The divided tasks are originally one task. Therefore, we unit the divided tasks as one task. Figure 7 illustrates the uniting process.

Fig. 7 Uniting the divided tasks

The process of Figs 6 and 7 can be formulated within the priority relation (11). Therefore, the difference of the processing time can be formulated as

$$\begin{cases} x_{i(j-1)(k-1)} = x_{ijk} & \forall i, \forall k (k \ge 2) \\ 0 = x_{ijk} & \forall i, (k=1) \end{cases} \quad \forall j = 2,...,b \qquad (13)$$

3 Numerical Experiment

This section deals with the problem of homogeneous multi-processor scheduling problem and compares the proposed method with CP/MISF scheduling in [7]. Figure 8 illustrates the relation among tasks on this experiment.

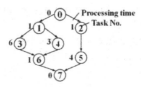

Fig. 8 The relation among tasks

3.1 CP/MISF

CP/MISF assigns task to idling PE in order of priority. The steps of decision of the priority are as shown below.

1. Define the longest path length from the final task on each task.
2. Determine the priority in order of the path length.

3. If the same priority tasks exist, determine the priority in order of the number of post processing tasks.

The priority of Fig. 8 is shown in table 1.

Table 1 The priority of Fig.8

Task No.	1	2	3	4	5	6
The longest path length	8	5	7	4	4	1
The number of post processing tasks	2	1	1	1	1	1
Priority	1	3	2	4	4	6

3.2 Proposed Method

In the numerical experiments, we use $i = 2$, $j = 16$, $k = 16$ and the priority relation among tasks in Fig. 5 for the validation of the formulation. Figure 9 illustrates the translation form of [9] to that of the proposed method.

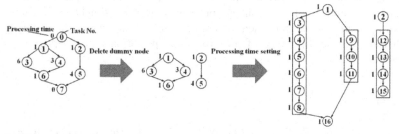

Fig. 9 The translation of the priority relation among task

3.3 Experimental Result

The linear problem solver GLPK for 64bit windows Version 4.39 is used in the experiment of our method.

Figures 10 and 11 illustrate the experimental results. Figure 10 and 11 show the result of CP/MISF and that of the proposed method, respectively.

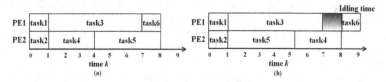

Fig. 10 The experimental result for CP/MISF: (a)assign task4 before task5 (b)assign task5 before task4

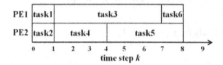

Fig. 11 The experimental result for our method

The results in Fig. 10 show that CP/MISF has possibilities to provide two kinds of result. In Fig. 10-(a), there exists idling time such that each PE does not do anything. The idling time is shown as a gray region. Therefore, the total processing time for each PE is not minimized. Figures 11 show that our method deletes the idling time and achieves the minimization of the total processing time for each PE.

4   Conclusion

This paper has compared our formulation based on 0-1 integer problem with CP/MISF in multi-processor. From numerical results, we can conclude that our proposed method based on 0-1 integer problem provides optimal task scheduling.

An advantage of the proposed method is that we can express the difficulty of each task and the difference of the processing performance of each processing element.

References

1. Viswanathkumar, G., Srinivasan, G.: A branch and bound algorithm to minimize completion time variance on a single processor. Computer&Operations Research 30, 1135–1150 (2003)
2. Rakrouki, M.A., Ladhari, T.: A branch-and-bound algorithm for minimizing the total completion time in two-machine flowshop problem subject to release dates. In: International Conference on Computer & Industrial Engineering (2009)
3. Park, C.-I., Choe, T.-Y.: An Optimal Scheduling Algorithm Based on Task Duplication. IEEE Trans. Computers 51(4), 44–448 (2002)
4. Bajaj, R., Agrawal, D.P.: Improving Scheduling of Tasks in a Heterogeneous Environment. IEEE Trans. on Parallel and Distributed Systems 15(2), 107–118 (2004)
5. Han, J.-J., Wang, D.-Q.: Edge Scheduling Algorithms in Parallel and Distributed Systems. In: Proc. International Conf. Parallel Processing, ICPP 2006 (2006)
6. Beaumont, O., Legrand, A., Robert, Y.: Optimal algorithms for scheduling divisible workloads on heterogeneous systems. In: IEEE Proc. International Paralell and Distributed Processing (2003)
7. Kasahara, H., Narita, S.: Practical Multiprocessor scheduling algorithms. IEEE Trans.Comput, C-33(11), 1023–1029 (1984)
8. Kasahara, H., Narita, S., Hashimoto, S.: OSCAR's Architecture. Trans. IEICE J71-D-I(8) (August 1988)

9. Sugeta, N., Rokusawa, K.: Decision of Task Priority Orders in CP/MISF Method. In: FIT 2006, p. 87 (2006); B_009 (in Japanese)
10. Shioda, H., Sawada, K., Shin, S.: Optimal Schedule of Parallel Processing. In: 2010 IRAST International Congress on Computer Applications and Computational Science (CACS 2010), pp. 360–363 (2010)
11. El-Rewini, H., Lewis, T.G.: Distiributed and Parallel Computing

Null Convention Logic Circuits Using Balanced Ternary on SOI

Sameh Andrawes and Paul Beckett

Abstract. We propose and analyze novel Ternary logic circuits targeting an asynchronous Null Convention Logic (NCL) pipeline where the Null value (i.e. Data not valid) is used to make the pipeline self-synchronizing and delay insensitive. A balanced Ternary logic system is used, in which the logic set {High Data, Null, Low Data} maps to voltage levels {$+V_{DD}$, $0V$, $-V_{DD}$}. Low power circuits such as Ternary to Binary converter, DATA/NULL Detector and Ternary Register are described based a 45nm SOI process technology that offers multiple simultaneous transistor thresholds.

1 Introduction

The International Technology Roadmap for Semiconductor (ITRS) [1] has identified a range of technical challenges that represent critical "road-blocks" to be solved to provide cheaper, faster and more capable embedded electronic systems into the future. Amongst the biggest challenges are the interrelated issues of power and performance in clocked synchronous systems. For example, the technique of switching off unused functionality has already become common in the synchronous domain [2].

Asynchronous techniques in general exhibit a number of key advantages over synchronous techniques, including an intrinsic lack of global clocks and their associated routing problems (power, delay) plus a high tolerance to variability in both manufacturing (dopant levels, line roughness etc.) and environment (voltage, temperature, low EMI). These characteristics will become increasingly useful at future nano-scale technology nodes.

Null Conventional Logic (NCL) [3] is a symbolically complete logic system that implicitly and completely expresses logic process and represents a convenient way

Sameh Andrawes · Paul Beckett
RMIT University, Melbourne, Australia
e-mail: s3300860@student.rmit.edu.au, pbeckett@rmit.edu.au

F.L. Gaol et al. (Eds.): Proc. of the 2011 2nd International Congress CACS, AISC 145, pp. 89–97.
springerlink.com © Springer-Verlag Berlin Heidelberg 2012

to describe asynchronous digital logic. Although NCL is inherently more complex than its conventional binary counterpart, it is delay-insensitive and correct-by-construction, so it will be straightforward to achieve a working system using assemblies of pre-built blocks. The synchronous timing closure problem is replaced by a need to carefully optimize the asynchronous pipeline for maximum throughput. NCL adds a third, NULL value to the Boolean set to create a three-value logic system. A NCL gate will only assert its output data when a complete set of valid data are present at the input, thereby enforcing a "completeness of input" criterion.

Most implementations of NCL to date have been based on a dual-rail binary signaling system, which can result in large and complex gate designs. In this paper, we present techniques to implement NCL systems using single rail ternary signaling. The objective is to minimize the number of interconnect wires along with the overall gate complexity, thereby reducing area and power consumption. We propose and analyze a number of circuits set up to support NCL that map the three NCL logic values {1, N, 0} onto balanced ternary levels {+VDD, 0V, -VDD}. The power and performance characteristics are based on a 45 nm Silicon on Insulator (SOI) technology that offers multiple simultaneous transistor threshold voltages.

The remainder of the paper proceeds as follows. In Section 2, we present some background to the NCL development. In section 3, we present simulation results for the proposed Ternary NCL circuits and in Section 4 we illustrate the concept with a simple NCL pipeline. Finally in section 5, we conclude the paper and identify future work.

2   Null Convention Logic

In this section, we briefly describe the main characteristic of NCL that differentiates it from a conventional synchronous binary approach and examine previous applications of single rail ternary.

As mention above, NCL is a symbolically complete logic system that implicitly and completely expresses logic processes without the need for time to be independently established (i.e., using a master clock). This is achieved by adding the control value NULL (i.e. DATA not valid) to the Boolean set to create a three-value logic system. A gate will only assert its output data when a complete set of valid (non-null) signals is present at its input.

Existing approaches to the design of NCL circuits have tended to be based on a pre-defined set of 27 fundamental majority logic gates [4] with hysteresis [5], [6], for which the number of inputs is limited to four or fewer. Both the basic structure of these majority logic gates [7] and the need to create duplicate signal paths to form a complete dual-rail logic system increases the area and design complexity in NCL systems. An underlying assumption is that the dual-rail signaling is mutually exclusive (i.e. "one-hot") [8]. However, the fact that there are two data paths introduces the possibility of a fourth (illegal) state. Within high speed, low power system on chip (SOC) implementations, it is possible for this illegal state to exist due to system noise.

In asynchronous NCL systems, the global clock is replaced by completion detection and local handshaking signals such as request for output (RFO) and request for input (RFI) as shown in Fig. 1. These signals, which are generated from the delay insensitive (DI) register, control the data flow through a NCL circuit [9]. An example of a dual rail NCL delay insensitive (DI) register implemented using TH22 gates is shown in Fig. 2. The register transfers its input DATA only when the RFI signal is Request for Data (RFD) and all inputs are valid. Further, it passes an input NULL value only when RFI signal is request for null (RFN) and all inputs are NULL; otherwise it holds its current output value, either DATA or NULL.

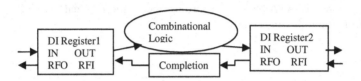

Fig. 1 Basic NCL system

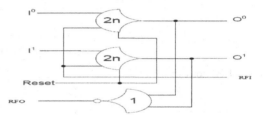

Fig. 2 Dual-rail NCL Register

A single N-bit register comprises N single-bit dual rail NCL registers and N completion detection circuits. The outputs of the N completion circuits are ANDed together to generate only one RFO signal that is connected to the previous register block. The completion detection circuit is implemented using a NCL NOR gate that generates a RFO signal for the current register which is a RFI signal for the previous register.

A NCL system can require large number of transistors which, in turn, may consume a lot of power even during the NULL cycle while the system is idle. The Multi-threshold CMOS (MTCMOS) technique of [10] avoids this by sending the gates into a sleep mode after each DATA cycle instead of propagating a NULL cycle. The RFO signal that indicates a register that it is either ready to process DATA or waiting for NULL is used to control its successor logic gates. When RFO is requesting null (RFN) the gates enter sleep mode and when the RFO is request for data (RFD) the gates are in active mode and ready to process data.

3 Ternary NCL

In this section, we describe the primary circuits forming the ternary NCL system. In this paper, the three NCL logic levels: DATA (Logic ONE), NULL and DATA

(Logic ZERO) are mapped to +VDD, 0V and –VDD respectively. The supply rail was set at 600mV (±300mV) which is the predicted "end-of-roadmap" supply value for low operating power technology [11]. As a result, transitions at the output are 300mV which minimizes the switching power at the cost of reducing the noise margin. As each data cycle is followed by a null cycle and vice versa, there are no direct transitions between logic values. Legal transitions occur only between DATA and NULL or vice versa.

The ternary NCL system comprises at least two ternary registers and one Multi-threshold logic gate between them. A ternary register is made up of a detector circuit that determines whether the input is DATA or NULL and the hold circuit that holds the input (DATA/NULL) according the request signal (RFD/RFN) that comes from the following stage.

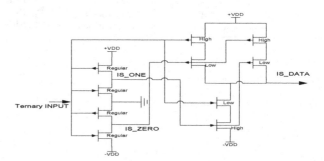

Fig. 3 Detector Circuit

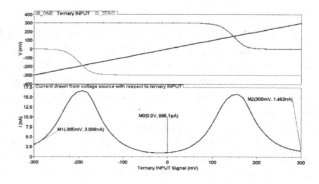

Fig. 4 Switching Thresholds for Ternary to Binary Interface and Supply Current vs. Ternary input signal.

3.1 Detector Circuit

The detector circuit (Fig. 3) is made up of a ternary to binary interface circuit that consumes low power during the null cycle [12] and generates IS_ONE and IS_ZERO signals. This is followed by a multi-threshold circuit that detects whether the input ternary logic is data or null and produces the appropriate output (i.e., IS_DATA). The switching threshold (V_{SW}) of the interface circuit can be calculated in the conventional way as [13]:

Where k_P and k_N are the gains of the P and N transistors (related to the transistor width on length ratios and process issues such as mobility), while $V_{THP/N}$ is the threshold of the individual P/N transistors. The switching threshold has to be

$$V_{SW} = \frac{V_{THN} + \sqrt{k_P / k_N}\,(V_{DD} + V_{THP})}{1 + \sqrt{k_P / k_N}} \tag{1}$$

adjusted carefully to keep both of the two supply current peaks away of the NULL value (i.e. 0V), as shown in Fig. 4, to maintain low power during the null cycle.

Table 1 Output of Detector Circuit

Ternary INPUT Signal	IS_ZERO	IS_ONE	IS_DATA
-1	0	1	1
NULL	-1	1	-1
1	-1	0	1

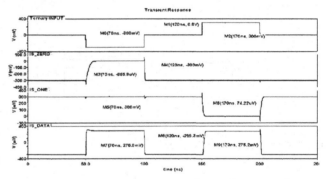

Fig. 5 Detector Circuit output waveform

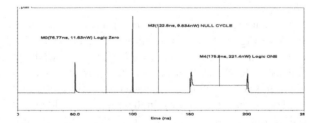

Fig. 6 Detector Circuit Power Consumption.

The Multi-threshold circuit employs transistors with two different threshold values (low and high) at -100mV and -400mV for PMOS, 100mV and 400mV for NMOS. The circuit (Fig. 3) uses three signals, Ternary input, IS_ZERO and IS_ONE, to detect whether the input is DATA or NULL and comprises one pull-down network (PDN) and two pull-ups (PUN). In the PDN, the low-threshold

NMOS is connected to the Ternary signal while the high-threshold device is connected to the IS_ONE signal. This network is switched on only when the ternary signal is 0V (i.e. NULL value) and the signal IS_ONE = 300mV (i.e. logic one), resulting in an output -300mV (i.e. logic low when the ternary input is NULL).

Similarly, a low-threshold PMOS is connected to IS_ZERO signal and a high-threshold one is connected to the ternary signal. This PUN circuit is on only when the IS_ZERO signal is 0V and the ternary signal is -0.3V and produces an output of around 275mV (i.e. logic one when the ternary input is DATA). Finally, a low-threshold PMOS is connected to the IS_ONE signal while a high-threshold device connected to IS_ZERO so that this branch yields an output 275mV (i.e. logic one when the ternary input is DATA) when the IS_ZERO signal and IS_ONE signals are -300mV and 0V, respectively. The outputs of the detector with respect to the ternary input signal are shown in Table I while Fig. 5 shows its output waveform. Fig. 6 shows the power consumption of the detector circuit, demonstrating that the NULL cycle has the lowest power consumption (i.e. 9.8nw), compared to the other two data levels.

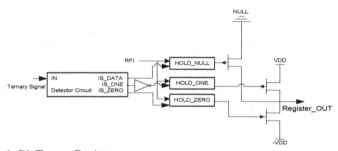

Fig. 7 Single Bit Ternary Register

3.2 Ternary Register

The Ternary register circuit is formed from one detector and three hold circuits, as shown in Fig. 7. The hold circuit is a conventional C-element gate followed by a latch circuit to hold the output signal (i.e. DATA/NULL) according to the request signal (i.e. RFD/RFN). It can be seen that the Hold One circuit differs from the others in that its output is connected directly to the output of the C-element gate (i.e. before the latch circuit). In the other two hold circuits, the output is connected to the output of the latch circuit. The HOLD NULL circuit is connected to both the RFI signal and the IS_DATA signal, so they become NULL and RFN respectively. When the output of the HOLD NULL circuit is at -300mV, the low-threshold PMOS connecting the ternary output signal to 0V (i.e. the NULL value) is activated and holds this value until the register receives both DATA and RFD (Fig. 8).

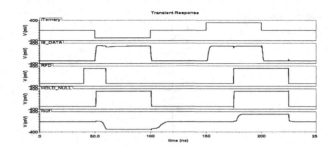

Fig. 8 Hold Null Waveform

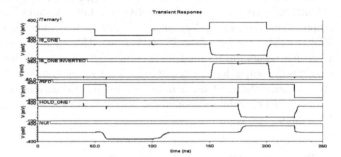

Fig. 9 Hold One Waveform

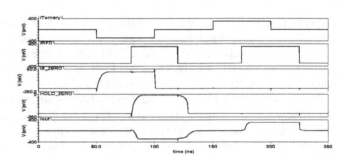

Fig. 10 Hold Zero Waveform

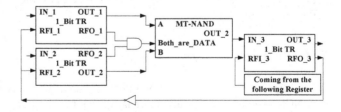

Fig. 11 Ternary NCL Pipeline

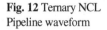

Fig. 12 Ternary NCL
Pipeline waveform

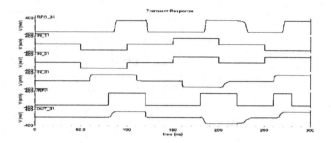

The HOLD ONE signal is 0V only when the ternary signal and RFI are logic one and RFD respectively, which activates a low threshold PMOS connecting the output ternary signal to V_{DD} (i.e. 300mV) as illustrated in Fig. 9. Finally, the HOLD ZERO combines the IS_ZERO and RFI signals so that its output is 0V when the inputs are 0V and RFD, which activates a low threshold NMOS that connects the output ternary signal to logic ZERO as shown in Fig. 10. The particular implementations of these hold circuits guarantees that the ternary register is hazard free and exhibits monotonic output behavior, which is crucial to correct NCL circuit operation.

4 Ternary NCL Asynchronous Pipeline

In this section we illustrate the idea of the Ternary NCL Asynchronous system by implementing a simple pipeline comprising two ternary registers separating a single logic gate, in this case a multi-threshold NAND gate (Fig. 11). The NAND gate generates an output only when the Both_are_DATA signal is high, while when it is low the NAND gate is in sleep mode and the output is NULL (i.e. 0V).

Fig. 12 shows the waveform of the Ternary NCL asynchronous system. It illustrates that when the output register (i.e. OUT_3) is NULL, its RFO_3 signal is RFD which is connected to the two input ternary register (i.e. RFI_1, RFI_2) requesting them to pass DATA to their output only if they exhibit valid DATA at that time. If they are NULL, RFO_3 stays high (i.e. RFD). When the outputs of both of the input registers are valid DATA, the multi-threshold gate is activated and yields a valid DATA that is connected to IN_3. When RFI_3 is RFD, the output register pass the DATA to OUT_3 and hold it until it receives both NULL and RFN.

5 Conclusion

In this paper, we have proposed and analyzed some new circuits supporting Ternary NCL systems, specifically a ternary detector and a register block. The ternary circuits are simple enough to replace their multi-rail binary counterparts.

At this stage, only small pipeline has been implemented to illustrate the concept. For this work to proceed, we need to develop a complete NCL pipeline. The ultimate intent is for these circuits to form one part of a larger reconfigurable system supporting ternary NCL that can be applied to high-performance DSP systems.

References

1. International Technology Roadmap for Semiconductors (2009), `http://www.itrs.net/Links/2009ITRS/Home2009.htm`
2. Calhoun, B.H., et al.: Design Methodology for Fine-Grained Leakage Control in MTCMOS. In: International Symposium on Low Power Electronics and Design, Seoul, Korea, pp. 104–109 (2003)
3. Fant, K.M., Brandt, S.A.: Null Convention Logic System (1994)
4. Dugganapally, I.P., et al.: Design and Implementation of FPGA Configuration Logic Block Using Asynchronous Static NCL. In: IEEE Region 5 Conference, pp. 1–6 (2008)
5. Kakarla, S., Al-Assadi, W.K.: Testing of Asynchronous NULL Conventional Logic (NCL) Circuits. In: IEEE Region 5 Conference, pp. 1–6 (2008)
6. Smith, S.C.: Design of an FPGA Logic Element for Implementing Asynchronous NULL Convention Logic Circuits. IEEE Transactions on Very Large Scale Integration (VLSI) Systems 15, 672–683 (2007)
7. Zahrani, A.A., et al.: Glitch-free Design for Multi-threshold CMOS NCL Circuits. Presented at the Proceedings of the 19th ACM Great Lakes symposium on VLSI, Boston, MA (2009)
8. Joshi, M.V., et al.: NCL Implementation of Dual-Rail 2^S Complement 8x8 Booth Multiplier using Static and Semi-Static Primitives. In: IEEE Region 5 Technical Conference, pp. 59–64 (2007)
9. Smith, S.C.: Speedup of self-timed digital systems using Early Completion. In: Proceedings of IEEE Computer Society Annual Symposium on VLSI, pp. 98–104 (2002)
10. Bailey, A., et al.: Multi-Threshold Asynchronous Circuit Design for Ultra-Low Power. Low Power Electronics 4/3, 337–348 (2008)
11. International Technology Roadmap for Semiconductors: Focus C Tables, Process Integration, Devices, & Structures, PIDS (2009), `http://www.itrs.net/links/2009ITRS/Home2009.htm`
12. Beckett, P.: Towards a balanced ternary FPGA. In: International Conference on Field-Programmable Technology, FPT 2009, pp. 46–53 (2009)
13. Kang, S.-M., Leblebici, Y.: CMOS Digital Integrated Circuits: Analysis and Design. McGraw-Hill (1996)

Application of Computer Capacity to Evaluation of Intel x86 Processors

Andrey Fionov, Yury Polyakov, and Boris Ryabko

Abstract. Computer[1] capacity was recently suggested as a new theoretical metric for computer performance evaluation. To show the benefits of this approach, we determine the capacities of well-known processors of Intel x86 family and compare the results to the metrics obtained from some available benchmarks. We show that our theoretical evaluation of computer performance conforms well to that given by the benchmarks.

1 Introduction

The problem of computer performance evaluation attracts much research because various aspects of performance are the key goals of any new computer design, see, e.g., [5, 6]. Simple performance metrics, such as the number of integer or floating point operations executed per second, are not adequate for complex computer architectures we face today. A more appropriate and widely used approach is to measure performance by execution time of specially designed programs called benchmarks. The main issues of benchmarking are well known. First, it is very difficult, if ever possible, to find an adequate set of tasks (in fact, any two different researchers suggest quite different benchmarks). Then, when a benchmark is used at the design stage, it must be run under a simulated environment which slows down the execution in many orders of magnitude, making it difficult to test various design decisions in the time-limited production process. Quite often, benchmarking is applied to ready devices for the purposes of evaluation and comparison. Here,

Andrey Fionov · Yury Polyakov · Boris Ryabko
Siberian State University of Telecommunications and Information Sciences,
Kirov St. 86, Novosibirsk 630102 Russia
e-mail: a.fionov@ieee.org, polyakov1987@yandex.ru, boris@ryabko.net

[1] The work is supported by the Russian Foundation for Basic Research; grant no. 09-07-00005.

the benchmarks produced by a hardware manufacturer may be suspected of being specially tuned just to facilitate sales. The benchmarks suggested by independent companies are prone to be outdated when applied to technologically novel devices. All these appeal to objectivity of evaluation results.

Computer capacity, as a new theoretical metric for computer performance evaluation, was first suggested in [3, 4], where it was applied to Knuth's MMIX machine. This metric is based on the number of different tasks that can be executed in a given time. It was shown that the upper bound of computer capacity

$$\hat{C}(I) = \log Z_0,$$

where Z_0 is the greatest positive root of the characteristic equation

$$Z^{-\tau(u_1)} + Z^{-\tau(u_2)} + \cdots + Z^{-\tau(u_s)} = 1. \tag{1}$$

To construct the characteristic equation we must know the execution times $\tau(u_1), \ldots, \tau(u_s)$ of instructions from the instruction set $I = \{u_1, u_2, \ldots, u_s\}$.

In this paper, to show a practical application of the concept, we provide calculations of computer capacities for some processors of x86 family. In order to make the results more comparable through different generations we calculate the capacities assuming that the processors perform their "most generic" tasks, e.g., 16-bit tasks in 16-bit computers and 32-bit tasks in 32-bit computers.

We extend the concept of computer capacity [3, 4] to modern computer architectures that incorporate such features as cache memory, pipelines and parallel processing units, as well as multicore and hyperthreading technologies. We also compare the results to the metrics obtained from some available benchmarks. We show that our theoretical evaluation of computer performance through capacities conforms well to that given by the benchmarks.

2   Computer Capacity of Intel 80286 Processor

To show how we estimate the capacity of 80286 processor, we first write the characteristic equation (1) and then give necessary comments.

$$\frac{9}{Z^2} + \frac{5}{Z^3} + \frac{1}{Z^5} + \frac{1}{Z^{14}} + \frac{1}{Z^{16}} + \frac{2}{Z^{19}} + \tag{2}$$

$$8\left(\frac{1}{Z^3} + \frac{1}{Z^5}\right) + 4\left(\frac{12}{Z^2} + \frac{2}{Z^{21}} + \frac{1}{Z^{22}} + \frac{1}{Z^{25}}\right) + \tag{3}$$

$$2^{15}\left(\frac{2}{Z^5} + \frac{12}{Z^7} + \frac{2}{Z^{24}} + \frac{1}{Z^{25}} + \frac{1}{Z^{30}}\right) + \tag{4}$$

$$2^{19}\left(\frac{2}{Z^7} + \frac{12}{Z^9} + \frac{2}{Z^{26}} + \frac{1}{Z^{27}} + \frac{1}{Z^{32}}\right) + \tag{5}$$

$$8 \cdot 8 \left(\frac{1}{Z^2} + \frac{1}{Z^3} \right) + 4 \cdot 4 \left(\frac{7}{Z^2} + \frac{2}{Z^2} \right) + \tag{6}$$

$$8 \cdot 2^{15} \left(\frac{2}{Z^5} + \frac{3}{Z^7} \right) + 4 \cdot 2^{15} \left(\frac{7}{Z^7} + \frac{2}{Z^6} \right) + \tag{7}$$

$$8 \cdot 2^{19} \left(\frac{2}{Z^7} + \frac{3}{Z^9} \right) + 4 \cdot 2^{19} \left(\frac{7}{Z^9} + \frac{2}{Z^8} \right) + \tag{8}$$

$$8 \cdot 2^{15} \cdot \frac{1}{Z^3} + 4 \cdot 2^{15} \cdot \frac{7}{Z^7} + 8 \cdot 2^{19} \cdot \frac{1}{Z^5} + 4 \cdot 2^{19} \cdot \frac{7}{Z^9} + \tag{9}$$

$$2^{15} \left(\frac{1}{Z^5} + \frac{1}{Z^7} + \frac{1}{Z^9} \right) + 2^{15} 2^{15} \left(\frac{1}{Z^7} + \frac{1}{Z^{10}} \right) + \tag{10}$$

$$\sum_{i=1}^{4} \frac{32}{Z^{7+i}} + \frac{32}{Z^3} + \sum_{i=1}^{4} \frac{5}{Z^{8+i}} + \frac{5}{Z^4} = 1. \tag{11}$$

The equation seems huge, but, in fact, it is simple and can be easily solved, e.g., by the bisection method.

Let us give some explanations. Denote, as earlier, by τ the number of processor cycles required to execute some instruction. A good resource for information on timing is [1]. The first line (2) arises from operations without operands. There are 9 ops with $\tau = 2$, 5 ops with $\tau = 3$, 3 single ops with $\tau = 5$, 14, and 16, and two with $\tau = 19$. The next three lines (3)–(5) are devoted to single operand instructions. There are 2 stack operations (PUSH and POP) that can be applied to any of 8 registers (AX, ..., DI) with $\tau = 3$ and $\tau = 5$, respectively (the first term in (3)). There are 12 operations DEC, ..., SHR ($\tau = 2$), 2 ops MUL, IMUL ($\tau = 21$), and DIV ($\tau = 22$), IDIV ($\tau = 25$), all of whom can be used with only 4 main registers (the second term in (3)). The same operations can be applied to memory locations. The total size of addressable memory for Intel 80286 is 2^{20} bytes or 2^{19} 16-bit words. However, only one segment (2^{15} words) can be accessed directly, which is reflected by (4). The execution time of instructions grows. To make access to the whole memory, a change of segment is required, which adds 2 extra cycles for each instruction (Line (5)). Similarly, Lines (6)–(9) are due to two-operand instructions, Line (10) arises from string operations, and the last line (11) reflects jump and loop instructions.

The solution of the equation (2)–(11) $Z_0 \approx 65.49$, so $\hat{C}(\text{i80286}) = \log Z_0 \approx 6$ bit/cycle. By taking into account that the typical clock frequency for Intel 80286 was 12.5 MHz, we conclude that $\hat{C}(\text{i80286}) = 75$ Mbit/s.

3 Computer Capacities of Advanced Intel Processors

The characteristic equations for modern processors are usually 5–10 times longer than that for Intel 80286, and we do not provide them here for the lack of space. We shall only discuss the main differences.

Let us consider one of the first versions of Pentium 4 [2] with the following features: clock frequency 1.5 GHz; 8K bytes L1 cache with access time 1 cycle; 256K bytes L2 cache with access time 7 cycles; and 512M bytes of RAM with access time 35 cycles. We assume that the processor operates in 32-bit mode and performs 32-bit tasks. In contrast to 80286, there are caches, parallel units and an instruction pipeline, so we must assess the effect of these architectural features.

There are 8 registers and 2^{27} 32-bit words in memory available, 2^{16} words being cached in L2 and 2^{11} of them cached in L1 cache (L1 and L2 are not exclusive). If the address hits L1 cache, the execution time is 1 cycle. Otherwise, if the address hits L2 cache, the execution time is $1 + 7$ cycles. If the address is not cached, the execution time is $1 + 7 + 35$ cycles. So the corresponding part of characteristic equation for one instruction looks like this:

$$8 \cdot 2^{11} \cdot \frac{1}{Z^1} + 8(2^{16} - 2^{11})\frac{1}{Z^8} + 8(2^{27} - 2^{16} - 2^{11})\frac{1}{Z^{43}}.$$

All instructions are considered similarly.

The other major difference from 80286 processor are FPU/MMX and XMM blocks that can operate concurrently with the "main" part that performs basic operations. First, we compute capacities separately for all these blocks. Then we can estimate the total capacity taking into account that FPU and MMX operations are mutually exclusive and giving preference to MMX as having greater capacity (since we are interested in an upper bound).

The next difference from 80286 is the pipeline processing combined with branch prediction. We take into account the pipline by separately considering predicted and mispredicted jumps with corresponding timings.

Finally, we address the problem of parallelism which is essential in hyper-threading and multicore technologies. Here we have a number of execution flows (threads) competing for shared resources such as execution units or memory. We associate a thread with a separate processor. Again, we observe what happens at every time instant. We assume that only one processor has access to all shared resources and the other processors can only execute instructions operated upon their private resources. The overall capacity is obtained as the sum of capacity of the first processor and reduced capacities of the others.

4   Comparison of Some x86 Intel Processors

By using the techniques explained in the previous sections, computer capacities of a number of x86 based computers were estimated. The results are shown in Table 1.

Table 1 Computer Capacities of Some x86 Family Processors

CPU	Clock Freq., MHz	RAM Size, M byte	L1 Cache, K byte	L2 Cache, K byte	Computer Capacity, Gbit/s
Pentium E5300	2600	2048	32	2048	280
Pentium E6500	2930	2048	32	2048	315
Core 2 Duo E7500	2930	4096	32	3072	315
Core 2 Duo E8400	3000	4096	32	6144	323
Core 2 Quad Q8300	2500	4096	32	4096	538
Core 2 Quad Q9450	2660	4096	32	12288	572
Core 2 Quad Q9550	2830	4096	32	12288	609

It is interesting to compare the estimates of computer capacities shown in Table 1 to the results of known benchmarks that are exemplified by SiSoftware Sandra 2010 Pro and PassMark. We make simple normalization of the results output by the benchmarks and plot the results as the diagrams shown in Figures 1–2. The normalization was done by considering the mark for Pentium E5300 as the unity within each benchmark. Similar normalization was done for computer capacity.

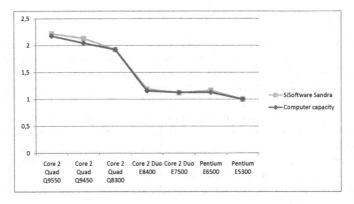

Fig. 1 Computer capacity against SiSoftware Sandra 2010 Pro

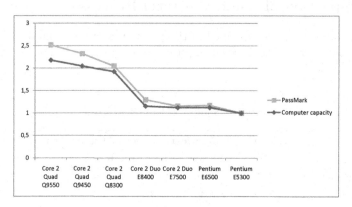

Fig. 2 Computer capacity against PassMark

We can see that the graphs of benchmarks generally show the similar tendency in performance evaluation as the computer capacity when proceeding from one processor to another.

5   Conclusion

We presented the application of a new approach to evaluation of computer performance based on the notion of computer capacity. It is purely analytical and does not require any simulators or hardware for testing in order to produce an evaluation of performance. Yet the results are quite comparable to those of benchmarks.

References

1. Fog, A.: Lists of instruction latencies, throughputs and microoperation breakdowns for Intel, AMD and VIA CPUs. Copenhagen University, College of Engineering (2010), http://www.agner.org/optimize/
2. Intel 64 and IA-32 Architectures Software Developers Manual, vol. 1, 2. Intel Corp. (2011)
3. Ryabko, B.: Applications of Information Theory to analysis of efficiency and capacity of computers and similar devices. In: Proc. IEEE Region 8 SIBIRCON-2010, Irkutsk Listvyanka, Russia, July 11-15, pp. 11–14 (2010)
4. Ryabko, B.: Using Information Theory to study efficiency and capacity of computers and similar devices. Information 1, 3–12 (2010)
5. Stallings, W.: Computer Organization and Architecture: Designing for Performance. Prentice-Hall (2009)
6. Tanenbaum, A.S.: Structured Computer Organization. Prentice-Hall (2005)

Channel Height Estimation for VLSI Design

Yuannian Liu

Abstract. Given an instance of channel routing problem (CRP) in VLSI design, the number of tracks used in a solution for the instance is called the channel height of the solution. The objective of CRP is to minimize the channel heights, that is, to find a solution with minimum channel height. In an instance of CRP, HCG and VCG denote the horizontal and vertical constraint graphs, respectively. Let GM be the graph obtained from HCG by adding edges whose ends are connected by a directed path in VCG. Pal et al. first gave lower bounds on the channel heights in terms of the clique number of GM, and presented algorithms to find such lower bounds. In this paper, we find some interesting theoretic properties, about the structure of the cliques in GM, which can be used to improve Pal's algorithms. So far, little is known about upper bounds on the channel heights. We find that CRP can be translated into an orientation problem on HCG with arcs in VCG oriented and keeping directed acyclic, and it is also proved that the channel height is determined by the longest directed path in the orientation. Moreover, we show that a lemma on the lower bound in [2] is incorrect and thus another lemma is given to modify it.

1 Introduction

In the design of VLSI chip, a design is said to exhibit routing *congestion* when the demand for the routing resources in some region exceeds their supply [12], that is, some region can not supply enough space to route and thus makes the design failed. Therefore, designer must estimate and reserve enough before detailed routing. This is known as the congestion estimation. When the routing region is a channel, the congestion estimation comes to the channel height estimation. *Channel routing problem* (CRP) is a routing problem where interconnections are made within

Yuannian Liu

Center for Discrete Mathematics and Theoretical Computer Science, Fuzhou University, Fuzhou, P.R.China

e-mail: yuannian82@163.com

F.L. Gaol et al. (Eds.): Proc. of the 2011 2nd International Congress CACS, AISC 145, pp. 105–111.
springerlink.com   © Springer-Verlag Berlin Heidelberg 2012

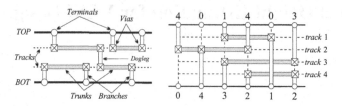

Fig. 1 Illustrations and examples for CRP

a rectangular region (channel) without obstructions. Each channel supplies a limited number of tracks for wires to route. *Channel height of a solution* is the number of tracks used in the solution, *channel height of an instance* is the minimal number of tracked used for all solution. The main objective of the channel routing is to obtain a solution with the minimum channel height. CRP is a key problem in VLSI chip design [1, 3] and has been known to be NP-hard [6, 5]. In 1990, the computation of a non-trivial lower bound for channel height estimation was proposed as an open problem by Kaufmann and Mehlhorn [4].

In 1998, Pal et al. [3] gave a non-trivial lower bound for no-dogleg reserved 2-layer Manhattan model of CRP and an $O(n^4)$ time algorithms to compute a non-trivial lower bound in 2- or 3-layer channel routing. In 2007, Pal et al. [2] presented a mimetic algorithm to compute a nontrivial lower bound in 2-layer channel routing. In this paper, we explored some new properties of the horizontal and vertical constraint graphs, which are used to improve the algorithm given by Pal et al. in [3]. Also, we correct an error in [2].

2 Definitions

As illustrated in Fig. 1, a *channel* is defined by a rectangular region and two rows of terminals along its top and bottom sides. Each terminal is assigned a number between 0 and N. All of these numbers represent the *netlist*. Terminals having the same label i ($1 \leq i \leq N$) must be connected by net i, while zeros indicate that no connection to the terminal. The netlist is usually represented by two vectors TOP and BOT. $TOP(k)$ and $BOT(k)$ represent the grid points on the top and bottom sides of the channel in column k, respectively (In Fig. 1, $TOP = (401403)$ and $BOT = (043212)$). The task of channel routing is to specify for each net a set of horizontal wire segments (*trunks*) and vertical wire segments (*branches*) that interconnect the net. Since trunks and branches of a net belong to different layer in the model, they may be connected by *vias* when desired.

A solution of a CRP is said to belong to the *no-dogleg reserved 2-layer Manhattan* (NR2M) model if: (i) exactly two layers for wires to lay on, one for branches and the other for trunks, and (ii) each net has at most one horizontal wire segment (no-dogleg).

Fig. 2 Illustrations for constraint graphs

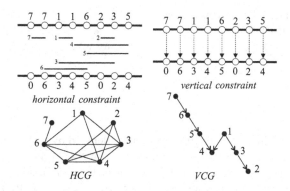

A solution of CRP in NR2M model should satisfy two constraints (cf. Fig. 2):

1. horizontal constraint: trunks of different nets must not overlap;

2. vertical constraint: branches of different nets must not overlap, i.e., the trunk of a net connected to the upper terminal at a given column must be placed above the trunk of another net connected to the lower terminal at that column.

These two constraints can be characterized by two graphs [8, 9, 1]: the *horizontal constraint graph* (*HCG*) and the *vertical constraint graph* (*VCG*) respectively.

HCG is an undirected graph where vertex $v_i \in V(HCG)$ represents net i and edge $v_i v_j \in E(HCG)$ if and only if the trunks (horizontal segments) of net i and net j overlap as closed intervals (i.e., the trunks of the two nets can not be laid out on the same track). It is easy to see *HCG* is an interval graph. *VCG* is a directed graph where vertex $v_i \in V(VCG)$ represents net i and (v_i, v_j) is an arc of *VCG* if and only if net i has a terminal on the top and net j on the bottom in the same column (i.e., the trunk of v_i must be above the trunk of v_j).

3 Lower Bounds by Clique Numbers

Given an instance of CRP with horizontal constraint graph HCG and the vertical constraint graph VCG. Let G_V be the digraph obtained from *VCG* by adding all the arcs (u, v), whenever there is a directed path from u to v in *VCG*. Denote by UG_V the underlying graph of G_V, that is, the undirect graph obtained from G_V by ignoring orientations on all arcs (see Fig. 3). Set $G_M = HCG + UG_V$ and $G_P = G_M \setminus HCG$. We note that G_V must be free of directed cycles, otherwise there would be no solution in NR2M model. Denote by v_{max} the number of vertices in the longest directed path of *VCG*. For each column of the instance, the number of nets crossing the column is called the *density* of the column. The maximum density over all the columns is the *channel density*, denoted by d_{max}, of the instance. The total number of tracks required is called the *channel height* of the instance. For example, in the instance showed in Fig. 2, the densities of columns are $1, 2, 3, 4, 4, 4, 4, 2$, so $d_{max} = 4$; while $v_{max} = 4$ (for directed path $7 \to 6 \to 5 \to 4$).

Fig. 3 Illustrations for constraint graphs

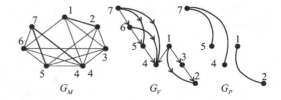

The *clique number* of graph G, denoted by $\omega(G)$, is the number of vertices in a maximum complete subgraph of G. In NR2M model, the channel height is at least $\max\{d_{max}, v_{max}\}$, which is known as the *trivial lower bound* of channel height [10]. In [3], Pal et al. proved that $\omega(G_M)$ gives a non-trivial lower bound of channel height, which can be found in an $O(n^4)$ time algorithm for a channel of n nets. Basically, the algorithm in [3] is to find a large chordal subgraph in G_M, by adding edges, one by one, from G_P to HCG, without violating chordality. Then, the chordal subgraph gives a lower bound of the channel height.

We give some interesting properties on the cliques in G_M which can be used to improve Pal's algorithms in [3] and [2]. Let K be a clique of G_M. We show that $K \backslash HCG$ can have at most one nonempty component, each pair edges (if any) of which are adjacent or connected by another edge in the component. Let $G_P = G_M \backslash HCG$. Instead of take all the edges of G_P into account, the algorithms can do it component by component of G_P to find the maximum clique of G_M. Moreover, for a component B of G_P we also find that if B is a tree, then $\omega(HCG+B) \le \omega(HCG)+2$ and etc. More results will be deduced in the follows.

A *chord* of a cycle C in a graph is an edge not in C, but joining two vertices of C. A *chordal graph* is a simple graph in which every cycle of length greater than three has a chord. An *interval graph* is one whose vertices are intervals and two vertices are adjacent if and only if the two intervals intersect. It is known that interval graphs form a subclass of chordal graphs. We say that a graph has a *transitive orientation* if each edge can be assigned a one-way direction such that in the resulting oriented graph, if (a,b) and (b,c) are arcs, then (a,c) is an arc. In the following, we list some known results on these graphs.

Lemma 1 (Ghouilà and Alain [7]). *The complement of an interval graph has a transitive orientation.*

Let G be a grpah. A *stable set* in G is a set of vertices no two of which are adjacent. The number of vertices in a maximum stable set of G is called the *stability number* of G, denoted by $\alpha(G)$. A *vertex cover* of G is a set of vertices which together meet all edges of G. The number of vertices in a minimum vertex cover of G is called the *covering number* of G, denoted by $\beta(G)$.

Lemma 2 (Gallai [11]). *For any graph G, $\alpha(G) + \beta(G) = |V(G)|$.*

Theorem 1. *For a clique K of G_M, let K_H be the spanning subgraph of K with $E(K_H) = E(K) \cap HCG$ and let K_P be the complement of K_H (so, $E(K_P) = E(K) \cap E(G_P)$). Then,*

(i) Any two edges of K_P are either adjacent or joined by another edge in K_P;
(ii) Any induced cycle in K_P has length at most 4;
(iii) $|V(K)| = \omega(K_H) + \beta(K_P)$.

Proof. (i) Suppose that $u_1 u_2, v_1 v_2$ are two non-adjacent edges that are not joined by another edge in K_P. Then, there is an induced cycle on $\{u_1, u_2, v_1, v_2\}$ in the complement of K_P, namely K_H. But, K_H is an induced subgraph of HCG, which is an interval graph, and thus a chordal graph. This contradiction proves (i).

(ii) Let Q be an induced cycle of K_P with length q. By (i), $q \leq 5$, and therefore it suffices to show that $q \neq 5$. Suppose, to the contrary, that $q = 5$. Since K_P is the complement of K_H, and by Lemma 1, K_P has a transitive orientation. Under such an orientation, any two adjacent edges on Q, say adjacent at vertex v, must have orientations either both into v or both out of v. This holds for every vertex v of Q, which is impossible since q is odd.

(iii) Note that $|V(K)| = |V(K_P)|$. By Lemma 2, $|V(K_P)| = \alpha(K_P) + \beta(K_P) = \omega(K_H) + \beta(K_P)$. □

By Theorem 1, any clique of G_M results from the addition of one component of G_P, instead of the addition of the whole G_P. Thus, in an algorithm to find the clique number of G_M, which gives a lower bound for the channel height of an instance of CRP, we only need to consider the addition of components of G_P to HCG, one by one.

Theorem 2. *For a clique K of G_M, let K_H be the spanning subgraph of K with $E(K_H) = E(K) \cap HCG$ and let K_P be the complement of K_H. Then, for any nonempty component B of K_P,*

(i) If B is a star, then $|V(K)| = \omega(K_H) + 1$;
(ii) If B is a double star, then $|V(K)| = \omega(K_H) + 2$;
(iii) If B has exactly one cycle, then $|V(K)| \leq \omega(K_H) + 3$;

Proof. By Theorem 1 (i), B is the only nonempty component in K_P, and hence $\beta(K_P) = \beta(B)$. Therefore, by Theorem 1 (iii), it suffices to show $\beta(B) = 1$ in (i), $\beta(B) = 2$ in (ii), and $\beta(B) \leq 3$ in (iii).

(i) Since B is a star, the center vertex of B is a vertex cover, and thus $\beta(B) = 1$.

(ii) For a double star, the two centers of the double star is a vertex cover, and thus $\beta(B) \leq 2$. It is easy to check that $\beta(B) \geq 2$, and so $\beta(B) = 2$.

(iii) Let Q be the unique cycle of length q in B. Clearly, Q must be an induced cycle. By Theorem 1 (ii), $q = 3$ or 4. If $q = 3$, by Theorem 1 (i), every edge of B has at least one end in Q, and therefore, the three vertices of Q is a vertex cover of B. If $q = 4$, similarly by Theorem 1 (i), every edge of B has at least one end in Q, and moreover, at most one of any two nonadjacent vertices in Q is joined to vertices in $V(B) \setminus V(Q)$, and thus, there is a vertex in Q adjacent to no vertex in $V(B) \setminus V(Q)$. It follows that the other three vertices of Q form a vertex cover of B. In either case, we have that $\beta(B) \leq 3$. □

4   Upper Bounds by Longest Paths

Given an instance of CRP with the graphs HCG, VCG and G_M, as defined in Section
3. Let $n_1, n_2, \ldots, n_k$ be the nets in the instance, a solution for the instance is a track
assignment $T : \{n_1, n_2, \ldots, n_k\} \to \{1, 2, \ldots, m\}$ such that: for any edge $uv \in E(G_M)$,
$T(u)$ and $T(v)$, the tracks assigned to u and v by T, are distinct, and $T(u) < T(v)$
whenever there is a directed path from u to v in VCG. The channel height of the
solution is $|T(G_M)| = |\{T(v) : v \in V(G_M)\}|$.

The results in Section 3 are about the lower bounds of channel heights. The following theorem addresses upper bounds of channel heights.

Theorem 3. *Let D be an acyclic digraph obtained from G_M, by extending the orientation from VCG to G_M. Denote by $l(D)$ the number of vertices in the longest
directed path of D. Then there is a solution with channel height $l(D)$.*

Proof. We use induction on $l(D)$ to show that there exists a track assignment T with
$|T(D)| = l(D)$. For $l(D) = 1$, both HCG and VCG are empty, and clearly $|T(D)| =
l(D)$.

Assume that it holds for $l(D) = k$ ($k \geq 1$). Consider that $l(D) = k + 1$. Let $S =
\{v_1, v_2, \ldots, v_t\}$ be the set of all terminals of the longest directed paths in D. So, since
D is directed acyclic, each vertex of S has no out-neighbor. Let $D' = D - S$. Then,
$l(D') = k$ and by the induction, $|T(D')| = l(D') = k$. Using an additional track on the
bottom for all members in S, we have that $|T(D)| = |T(D')| + 1 = k + 1 = l(D)$.   □

By Theorem 3, any acyclic orientation of G_M, extended from VCG, yields an upper
bound for the channel height. The next theorem shows that the exact value of the
channel height can be obtained by choosing such an orientation with the longest
directed paths as short as possible.

Theorem 4. *The channel height is $\min_{D \in \mathscr{D}} l(D)$, where $l(D)$ is the number of vertices of the longest directed path in D and $\mathscr{D} = \{D$ is an acyclic oriented graph on
G_M with $VCG \subseteq D\}$.*

Proof. Let $n_1, n_2, \ldots, n_k$ be the nets in an instance of CRP with channel height h.
We number the tracks in the channel from top to bottom as $1, 2, \cdots, h$. Suppose that
$T : \{n_1, n_2, \ldots, n_k\} \to \{1, 2, \cdots, h\}$ is a solution of the CRP. Thus, $T(n_i) = t$ means
that the trunk of net n_i is assigned to track t. For any acyclic oriented graph D on
G_M, extended from VCG, by Theorem 3, we have that $h \leq l(D)$.

Let D^* be the acyclic oriented graph obtained from G_M as follows: for each edge
$uv \in E(G_M)$, if $T(u) < T(v)$, orient the edge from u to v; otherwise, from v to u. Since
$T(u) \neq T(v)$ for each $uv \in E(G_M)$, we see that D^* is indeed an oriented graph on G_M.
Since T is a solution, which must satisfy the vertical constraint condition, we have
that $VCG \subseteq D^*$. By the definition of D^*, for any $x, y \in V(D^*)$, if there is a directed
path from x to y, then $T(x) < T(y)$, which implies that D^* is acyclic. Therefore,
$D^* \in \mathscr{D}$. Moreover, if $x_1 \to x_2 \to \cdots \to x_t$ is a directed path in D^*, then $1 \leq T(x_1) <
T(x_2) < \cdots < T(x_t) \leq h$, which means that $t \leq h$. In particularly, $l(D^*) \leq h$. It follows
that $h = \min_{D \in \mathscr{D}} l(D)$, as claimed by the theorem.   □

5 Correction on a Known Lower Bound

Lemma (Lemma 1 in [2]) *For a pair of directed paths (chain) from source to sink vertices, with length difference less than or equal to one and at least one with length v_{max}, if source vertices, or sink vertices, or both pairs are connected in G_M, then at least one extra track is essentially required to route the channel.*

The above lemma was given by Pal et al. [2] in 2007 without a proof and it is incorrect. One can check that instance with $TOP = (0230340)$ and $BOT = (1124055)$ is a counter-example for theirs lemma. Correcting the lemma, we have the following one.

Lemma 3. *Let $v_1 \to v_2 \to \cdots \to v_k$ and $u_1 \to u_2 \to \cdots \to u_l$ $(k \geq l)$ be two directed path of VCG. The channel height is at least $k + 1$ if: $k = l$ and $v_i u_i \in G_M$ for some $1 \leq i \leq k$; or $k - 1 = l$ and $v_j u_j, v_{j+1} u_j \in G_M$ for some $1 \leq j \leq k - 1$.*

References

1. Sait, S.M., Youssef, H.: VLSI Physical Design Automation: Theory and Practice. McGraw-Hill, New York (2006)
2. Pal, R.K., Saha, D., Sarma, S.S.: A Mimetic Algorithm for Computing a Nontrivial Lower Bound on Number of Tracks in Two-Layer Channel Routing. Journal of Physical Sciences 11, 199–210 (2007)
3. Pal, R.K., Pal, S.P., Pal, A.: An Algorithm for Finding a non-Trivial Lower Bound for Channel Routing. Intergration, the VLSI Journal 25, 71–84 (1998)
4. Kaufmann, M., Mehlhorn, K.: Routing Problems in Grid Graphs. In: Korte, B., Lovasz, L., Promel, H.J., Schrijver, A. (eds.) Paths, Flows, and VLSI-Layout, Algorithms and Combinatorics, vol. 9, pp. 165–184. Springer, Heidelberg (1990)
5. Korte, B., Vygen, J.: Combinatorial Optimization: Theory and Algorithms. Springer, Heidelberg (2006)
6. Gerez, S.H.: Algorithms for VLSI Design Automation. John Wiley & Sons, Inc. (1999)
7. Alain Ghouilá, H.: Caractérisation Des Graphes non Orientés Dont on Peut Orienter les Arrêtes de Manièreà Obtenir le Graphe d'une Relation d'ordre. C.R.Acad. Sci. Aaris 254, 1370–1371 (1962)
8. Burstein, M.: Channel Routing. Layout Design and VeriTcation, 133–167 (1986)
9. Yoshimura, T., Kuh, E.S.: Efficient Algorithms for Channel Routing. IEEE Trans. CAD Integrated Circuits and Systems 1, 25–35 (1992)
10. Shenvani, N.: Algorithms for VLSI Physical Design Automation, 3rd edn. Kluwer Academic Publishers (1999)
11. Gallai, T.: Über extreme Punkt-und Kantenmengen. Ann. Univ. Scibudapes. Eötvös Sect. Math. 2, 2133–2138 (1959)
12. Saxena, P., Shelar, R.S., Sapatnekar, S.S.: Routing Congestion in VLSI Circuits. Springer, Heidelberg (2007)

High Speed BCD Adder

C. Sundaresan, C.V.S. Chaitanya, P.R. Venkateswaran, Somashekara Bhat,
and J. Mohan Kumar

Abstract. In recent trends managing, using and processing a high volume of data
has become part of real time computation. As the importance of decimal arithmetic is growing in all financial, internet-based and commercial applications, more attention has to be paid for fast decimal data processing. Despite the widespread use
of binary arithmetic, decimal computation remains essential for many applications, Decimal numbers are not only required whenever numbers are presented for
human inspection, but also often used when fractions are involved. In this paper
new design of high speed Binary Coded Decimal (BCD) addition is discussed.

1 Introduction

In electronics system, binary coded decimal (BCD) is an encoding method for the
decimal numbers in which each digit can be represented in its own binary format.
Decimal fractions are pervasive in human endeavors, yet most cannot be
represented by binary fractions.

Extensive work has been done on building adders for BCD arithmetic and different adders have been proposed [9, 10, 11, 12]. Still, the major consideration
while implementing BCD arithmetic will be to enhance its speed as much as possible which is being addressed in this paper.

This paper introduces and analyses various techniques for high speed addition
of higher order BCD numbers which form the core of other arithmetic operations
such as multi-operand addition [3, 4], multiplication [5] and division [6]. A new
architecture for the fast decimal addition is proposed, based on which architectures
for higher order adders such as ripple carry and carry look-ahead adder is derived.

The rest of paper is organized as follows: Section 2 gives overview of the BCD
arithmetic. In section 3 details on the proposed algorithm for fast BCD addition.
Section 4 discusses about the High speed BCD adder which has been developed

C. Sundaresan · C.V.S. Chaitanya
Manipal Centre for Information Science
e-mail: {sundaresan.c,chaitanya.cvs}@manipal.edu

F.L. Gaol et al. (Eds.): Proc. of the 2011 2nd International Congress CACS, AISC 145, pp. 113–118.
springerlink.com   © Springer-Verlag Berlin Heidelberg 2012

using proposed algorithm for fast BCD addition. In section 5 area, timing and power results for reduced delay BCD adder and suggested adder has been presented. Last section presents our conclusion.

2 Overview of BCD Arithmetic

BCD is a decimal representation of a number directly coded in binary, digit by digit [9]. For example the number $(9527)_{10} = (1001\ 0101\ 0010\ 0111)_{BCD}$. It can be seen that each digit of the decimal number is coded in binary and then concatenated, to form the BCD representation of the decimal number.

To use this representation all the arithmetic and logical operations need to be defined. As the decimal number system contains 10 digits, at least 4 bits are needed to represent a BCD digit. Considering a decimal digit A, the BCD representation is given by $A_4A_3A_2A_1$. The only point of note is that the maximum value that can be represented by a BCD digit is 9. The representation of $(10)_{10}$ in BCD is $(0001\ 0000)_2$.

Addition in BCD can be explained by considering two decimal digits A and B with BCD representations as $A_4A_3A_2A_1$ and $B_4B_3B_2B_1$ respectively. In the conventional algorithm, these two numbers are added using a 4-bit binary adder. It is possible that the resultant sum can exceed 9 which results in an overflow. If the sum is greater than 9, the binary equivalent of 6 is added to the resultant sum to obtain the exact BCD representation.

3 Proposed Algorithm for Fast BCD Addition

The existing algorithm [13] for addition of two BCD digits uses the carry network for the generation of carry for each digit. As the number of digits increase in both BCD digits, the number of levels in carry network will increase which in turn increases the delay in calculating the final carry out. In the proposed adder, three signals named C_G, C_P, and P will be computed using input operands. C_G is the carry generate signal, C_P is the carry propagate signal and P shows that the sum of two corresponding digits is greater than or equal to 9. The above three signals are used to generate the Carry output without using the carry network. Signal P to next decimal digit and signal P from previous digit in a decimal number can be used to determine whether carry will be suppressed in corresponding digit location or it will be propagated to the next digit. The output of the carry network and the output of the carry suppression or propagation logic is ORed to generate the value which will be used in the correction step. The equation for value (V) used in the correction step is as follow

$$V = (C_n) + (P_n\ \&\ P_p\ \&\ C_p)$$

C_n: Output carries to next digit, P_n: Signal P to next digit, P_p: Signal P from previous digit, C_p: Input carry from previous digit.

The digital logic which implements the above algorithm is discussed in the following section.

4 Higher Valence BCD Addition

The existing conventional BCD adder [9] is simple in operation, but very slow due to the ripple carry effect. In BCD additions following cases are considered:

Case 1: The sum of two BCD digits is smaller than 9.
Case 2: The sum of two BCD digits is greater than 9.
Case 3: The sum of two BCD digitals is exactly 9.

For the first two cases, the incoming carry has no effect on determining the output carry. Therefore, the carry output will be independent of carry input. On the other hand for case 3, the output carry will depend on the input carry. Case 2 and Case 3 can be represented by a Carry Generate (C_G), Carry propagates (C_P) and P signals, respectively.

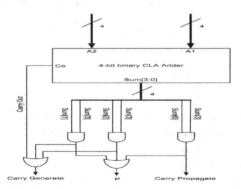

Fig. 1 Carry generate and carry propagate

The computation of the CG, CP and P signals of a digit are shown in Fig 1. The signals CG, CP and P are calculated using the Sum and Carry outputs of the CLA adder in first stage using following equations.

$$CP = SUM[3] . SUM[0] \tag{1}$$

$$CG = Cout + (SUM[3].(SUM[2]+SUM[1])) \tag{2}$$

$$P = SUM[3] . (SUM[2]+SUM[1]+SUM[0]) \tag{3}$$

In High speed BCD adder, the carry input is no more connected to the adder block as in reduced delay BCD adder [13], instead it is directly connected to the carry look ahead block which will generate carry to next stage. Because of the above modification the first stage adders will not depend on the carry input. Also the output carry will be independent of the PG logic. Output carry from CLA block is calculated from the following equation.

$$C_{i+4}=CG_{4+i}+P_{4+i}(CG_{3+i}+P_{3+i}(CG_{2+i}+P_{2+i}(CG_{1+i}+P_{1+i}Ci))) \qquad (4)$$

Where, i = 0,4,8,........

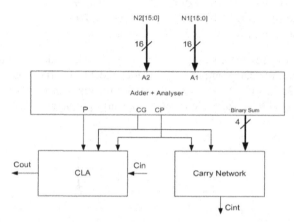

Fig. 2 Adder and analyzer block.

Carry network can be a parallel prefix network which performs their operations in a constant time irrespective of the length of inputs. In this paper Kogge-Stone [8] prefix network is used as carry network. The output of the carry network is used for correction of the output.

Carry suppressor will take in P value from the present digit, P value from the previous digit and the carry input and determine whether to suppress the carry or to propagate to the next digit. Carry suppressor also generates the correction value which is to be added to the sum value to convert it back to the BCD equivalent.

The integrated block diagram of Adder, Analyzer, CLA and Carry network is shown in Figure 2. The correction to the sum output of first stage adder is done by adding 0, 1, 6 or 7 to it. As in [13], the generation of correction value will depend on the present digit carry output and previous digit carry output. Figure 3 shows the block diagram for High Speed BCD adder which includes CLA adder in the first stage, combination of PG logic, CLA to generate next stage carry and Carry suppressor in the second stage and Correction logic in third stage.

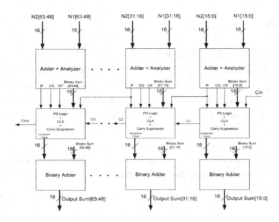

Fig. 3 Higher valence BCD adder

5 Simulation Results

The reduced delay BCD adder [13] and High speed BCD adder are designed to support 64-bit and 128-bit decimal addition with BCD operands. They are synthesized using the TSMC 65nm library. Both the adders are implemented in Verilog HDL. They are synthesized using Synopsys Design Compiler synthesis tool and TSMC 65nm library. The synthesis results are presented in Table 1 and Table 2.

Table 1 Synthesis result of high speed BCD adder and Reduced delay BCD adder.

Parameters	High Speed(64)	Reduced Delay(64)	High Speed(128)	Reduced Delay(128)
Gate Count	556	452	1112	950
Area	1343.52	1216	2687.04	2517.11
Dynamic Power	389.14μW	365.83μW	780.99μW	748.65μW
Cell leakage Power	37.2μW	35.9μW	74.48μW	73.3 μW
Delay	0.37ns	0.88ns	1ns	1.14ns

According to the Table 1, the proposed High speed BCD adder will take less delay when compared to reduced Delay BCD adder. When the area is under consideration, then High speed BCD adder will occupy more area when compared to reduced Delay BCD adder. However Delay of MRDBCD is higher than the suggested adders. The high speed addition can be compromised with little area and power overheads.

6 Conclusion

This paper describes the design high speed BCD adder to perform decimal addition. The new proposed adder has the shortest delay among the decimal adders examined in this paper. The new decimal adder improves the delay of BCD addition by increasing parallelism. The adder has been designed using the Verilog HDL and verified for different corner case inputs. Also the adder has been synthesized with 65nm CMOS library.

Acknowledgments. We would like to thank Vidyanand, system administrator, Tanmay M Rao and Arjun Ashok Post Graduate students for their assistance in VLSI Lab.

References

1. Buchholz, W.: Fingers or Fists? (The Choice of Decimal or Binary Representation). Communications of the ACM 2(12), 3–11 (1959)
2. Cowlishaw, M.F.: Decimal Floating-Point: Algorism for Computers. In: Proceedings of 16th IEEE Symposium onComputer Arithmetic, pp. 104–111 (June 2003)
3. Kenney, R.D., Schulte, M.J.: Multioperand Decimal Addition. In: Proc. IEEE CS Ann. Symp. VLSI, pp. 251–253 (February 2004)
4. Kenney, R.D., Schulte, M.J.: High-speed multioperand decimal adders. IEEE Transactions on Computers 54(8), 953–963 (2005)
5. Erle, M.A., Schulte, M.J.: Decimal Multiplication via Carry-Save Addition. In: Proc. IEEE 14th Int'l Conf. Application-Specific Systems, Architectures, and Processors, pp. 348–358 (June 2003)
6. Parhami, P.: Computer Arithmetic: Algorithms and Hardware Designs. Oxford Univ. Press, New York (2000)
7. Hwang, I.S.: High Speed Binary and Decimal Arithmetic Unit. Unit. United States Patent (4,866,656) (1989)
8. Kogge, P.M., Stone, H.S.: A Parallel Algorithm for The Efficient Solution of a General Class of Recurrence Equations. IEEE Trans. on Computers C-22(8) (August 1973)
9. Mano, M.M.: Digital Design, 3rd edn., pp. 129–131. Prentice Hall (2002)
10. Schmookler, M.S., Weinberger, A.W.: High Speed Decimal Addition. IEEE Transactions on Computers C-20, 862–867 (1971)
11. Shirazi, B., Young, D.Y.Y., Zhang, C.N.: RBCD: Redundant Binary Coded Decimal Adder. IEEE Proceedings, Part E 136(2), 156–160 (1989)
12. Thompson, J.D., Karra, N., Schulteb, M.J.: A 64-Bit Decimal Floating-Point Adder. In: Proceedings of the IEEE Computer Society Annual Symposium on VLSI, pp. 297–298 (February 2004)
13. Bayrakci, A.A., Akkas, A.: Reduced Delay BCD Adder. IEEE (2007)

Research on Handover Algorithm for LEO Satellite Network

Ye XiaoGuo, Wang RuChuan, and Sun LiJuan

Abstract. Low Earth Orbit (LEO) satellite networks are playing an important role in current global mobile communications. An optimized handover scheme can greatly improve performance in respect of end-to-end communication delay and link bandwidth utilization of LEO satellite networks. Various handover schemes in LEO satellite networks are researched in this paper. Ground-satellite link handover and re-computing routes problem is analyzed in detail. Then, a link handover algorithm for LEO satellite mobile communication network is proposed. Simulation results show that the satellite link handover algorithm proposed has less end-to-end delay, and is more stable and customizable.

1 Introduction

Low Earth Orbit (LEO) satellite networks can provide a seamless global coverage and consistent services by connecting with inter-satellite links (ISLs). With low end-to-end delay and high bandwidth, LEO satellite networks become an important part of next generation mobile communication system[1, 2, 3]. Handover scheme research of LEO satellite is an important research topic. However, due to high-speed moving of satellite, the satellite of ground-satellites access point experiences frequently handover so that the former ground-satellite link invalidate. Then, the communication interrupted continues using new established ground-satellite links and

Ye XiaoGuo · Wang RuChuan · Sun LiJuan
College of Computer, Nanjing University of Posts and Telecommunications, Nanjing, Jiangsu 210003, China

Jiangsu High Technology Research Key Laboratory for Wireless Sensor Networks, Nanjing, Jiangsu 210003, China

Key Lab of Broadband Wireless Communication and Sensor Network Technology (Nanjing University of Posts and Telecommunications), Ministry of Education Jiangsu Province, Nanjing, Jiangsu 210003, China
e-mail: xgye@njupt.edu.cn

F.L. Gaol et al. (Eds.): Proc. of the 2011 2nd International Congress CACS, AISC 145, pp. 119–124.
springerlink.com   © Springer-Verlag Berlin Heidelberg 2012

new inter-satellites routes. As for inter-plane cross-seam inter-satellite links, more frequent handover processing is required[4, 5].

At present, there have already much research on switching and handover technology in LEO satellite communication networks. Spot-beam handover and dynamic channel allocation in LEO satellite network is discussed in detail in[6, 7]. Ref.[8] provided a handover management scheme in Low Earth Orbit satellite networksa satellite constellation design is proposed in[9]. Ref.[10] provided an optimized polar constellations scheme for redundant earth coverage.

We proposed a novel method to optimize the process of satellite link switching, by combining routes recomputing scheme to improve the performance of satellite mobile communication networks.

2   Analysis of LEO Satellite Constellations

In order to provide global seamless coverage of LEO satellite constellations, a sophisticated constellation should be designed well based on multi-coverage. There exist two typical LEO satellite constellations with global coveragewhich are the polar orbit constellation and the rosette orbit constellation.

Especially, the polar orbit satellite constellations can usually cover the global area including the polar and high latitudinal areas. Iridium system uses a polar orbit constellation with 66 satellites at altitude of 780 km, with 11 satellites being placed in each orbital plane and inclined at 86.4°.

Ref.[11] provides a good idea for studying satellite handover scheme especially in the LEO satellite networks. Each satellite has four inter-satellite links in Iridium constellation. However, it's too hard to build inter-satellite links between the satellites in the first and sixth orbit because of their high-speed relative movement.

3   Satellite Link Handover Scheme

Relative movement of satellite and user terminal will inevitably course ground-satellite links handover. As a result, we propose a satellite links handover scheme based on analyzing the movement regular of satellites, which consists of ground-satellites links handover scheme and route re-computing scheme.

3.1   Ground-Satellites Link Handover Scheme

The basic idea of ground-satellites link handover scheme is that, ground station or satellites communication terminal operate handover checking regularly, checking whether the elevation between ground and satellite is smaller than the specified minimum elevation or not, usually per 10 seconds. When the elevation is smaller than the specified minimum elevation (usually setting to 8.2°), or the received signal is lower than a certain minimum value, ground-satellites link handover is initiated. Algorithm 1 describes the ground-satellites link handover algorithm.

Algorithm 1. Ground Satellite link handover algorithm

$flag_{found}$ =false; //used to represent whether find handover access satellite
for *every inter* − *satellite link* s_1 *connected to node*$_{groundStation}$ **do**
$\quad$ $node_{peer}$=satellite on the other side of s_1;
$\quad$ $elev = elevation_{nodepeer}$;
$\quad$ **if** $elev \leq minElevation$ **then**
$\quad\quad$ destroy *link* between $node_{groundSation}$ and $node_{peer}$;
$\quad\quad$ //initial value of $flag_{changed} =$ false
$\quad\quad$ $flag_{changed} =$ true;
$\quad$ **end**
$\quad$ **if** $flag_{linkDestroyed} = true$ **then**
$\quad\quad$ **if** $flag_{handoverOptimizationNext} = true$ *and* $elev > minElevation$ **then**
$\quad\quad\quad$ $flag_{found}$ =true;
$\quad\quad\quad$ $node_{peer} = next$;
$\quad\quad$ **end**
$\quad\quad$ **if** $flag_{found}$ =false **then**
$\quad\quad\quad$ $node_{peer} = node_{maxElevation}$;
$\quad\quad\quad$ $flag_{found}$ =false;
$\quad\quad$ **end**
$\quad\quad$ **if** $flag_{found}$=true **then**
$\quad\quad\quad$ build inter-satellite link between node and peer;
$\quad\quad\quad$ config protocol stack of interface;
$\quad\quad\quad$ $flag_{linkChanged}$=true;
$\quad\quad$ **end**
$\quad$ **end**
end
if $flag_{recomputeChoice}$ =true **then**
$\quad$ recomputeRoute;
else if $flag_{linkChanged}$ =true **then**
$\quad$ recomputeRoute;
end
set timer for next handover check;

3.2 *Recompute Routes Scheme*

The basic idea of recomputing routes scheme is to recompute the costs of the links in satellites constellation network, and to update topology status information and route table at last. Algorithm 2 describes the recomputing routes scheme.

4 Simulation Experiment

4.1 *Simulation Environment*

We extend the satellite simulation module of NS-2[12], and simulate the ground-satellites link handover scheme and recompute routes scheme proposed in this paper. The simulation environment includes 66 Iridium LEO satellite constellation

---

Algorithm 2. Recomputing routes algorithm

---

for *every satellite node* **do**
    for *every link to node s_1* **do**
        if $status_{s1} = UP$ **then**
            //compute link bandwidth
            $bandwidth = getBandwidth(nodetx, noderx)$;
            if $flag_{Delay} = true$ **then**
                // compute link propagation delay
                $delay = prop_{delay}(nodetx, noderx)$;
            else
                $delay = 1$;
            end
            if $algorithmChoice = 0$ **then**
                $cost = delay$;
            else if $algorithmChoice = 1$ **then**
                $cost = 1/bandwith$;
            end
            update cost of link s1;
        end
    end
    computeRoutes;
    updateRtTables;
end

---

and a number of satellite mobile communication terminal ground station. Assuming that two end-users are respectively located at Beijing(E116.39°,N39.91°)and NewYork (W73.94°,N40.67°). The performance and characteristics such as end-to-end average propagation, delay as well as jitter and handover frequency are analyzed in detail as follow.

4.2 Simulation Results and Analysis

From the simulation results, we can get the delay value of CBR traffic from ground node T0 to node T1. When using hops as a metrix of recalculating routes, the propagation performance of handover-optimized strategy and no-optimization handover strategy are shown in Figure 1 and Figure 2.

The simulation results shows, when adopting hops as a scale of recalculating routes, the total average propagation of handover optimized strategy is 0.06235 seconds, a little larger than that of non-handover optimized scheme, and so do jitter. The switching frequency of ground-satellite links handover and ISL handover in case of handover optimized strategy is lower than that of non-handover optimized strategy.

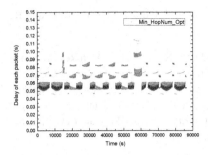

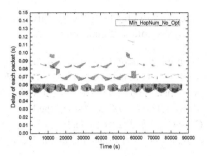

Fig. 1 Delay of minimum hops and handover optimized scheme

Fig. 2 Delay of minimum hops and non-
handover optimized scheme

The propagation performance adopting delay as a scale of recomputing routes of
handover optimized strategy and non -handover strategy are shown in Figure 3 and
Figure 4.

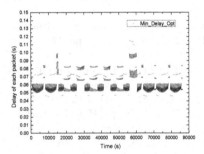

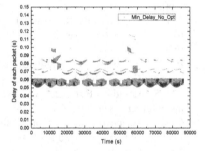

Fig. 3 Delay of minimum delay and handover optimized scheme

Fig. 4 Delay of minimum delay and non-
handover optimized scheme

The simulation results shows, when adopting delay as a scale of recalculating
routes, delay performance of optimization handover strategy is larger than strategy
with no-optimization handover. This is because optimized scheme aims to reducing
the switching frequency ,but non-handover optimized scheme always selects the
GSL satellite which has the maximum elevation to be next access satellite when
handover.

5 Conclusion

In this paper, we proposed a LEO mobile satellite network handover scheme including ground-satellite links handover scheme and recompute routes scheme. The
simulation indicated that the proposed handover scheme can improve performance

in respect of end-to-end communication delay and link bandwidth utilization and it is more stable and customizable.

Acknowledgment. This work was supported in part by the National High-Tech Research and Development Plan of China(No. 2010AA7010201), the National Natural Science Foundation of China(No.60903181), Jiangsu Province High-school Natural Science Research Plan(No. 09KJB520009, 09KJB520010), the Postdoctoral Science Foundation of Jiangsu Province of China(No. 1002022B),and a Project Funded by the Priority Academic Program Development of Jiangsu Higher Education Institutions(yx002001).

References

1. Del Re, A., Pierucci, L.: Next-generation mobile satellite networks. IEEE Communications Magazine 40(9), 150–159 (2002)
2. Zahariadis, T., Vaxevanakis, K.G., Tsantilas, C.P., et al.: Global romancing in next-generation networks. IEEE Communications Magazine (2), 145–151 (2002)
3. Karapantazis, S., Papapetrou, E., Pavlidou, F.N.: Multiservice On-Demand Routing in LEO Satellite Networks. IEEE Transactions on Wireless Communications 8(1), 107–112 (2009)
4. Gkizeli, M., Tafazolli, R., Evans, B.: Performance analysis of handover mechanisms for non-geo satellite diversity based Systems. In: IEEE Globecom 2001, pp. 2744–2748 (2001)
5. Bottcher, A., Werner, B.: Strategies for handover control in low earth orbit satellite systems. In: IEEE 44th Vehicular Technology Conference 1994, pp. 1616–1620 (1994)
6. Cho, S., Akyildiz, I.F., Bender, M.D., et al.: A new spotbeam handover management technique for LEO satellite networks. In: IEEE GLOBECOM 2000, pp. 1156–1160 (2000)
7. Cho, S.: Adaptive dynamic channel allocation scheme for spotbeam handover in LEO satellite networks. In: 52nd IEEE VTS-Fall VTC 2000, pp. 1925–1929 (2000)
8. Nguyen, H.N., Lepaja, S.: Handover management in low earth orbit satellite IP networks. In: IEEE GLOBECOM 2001, pp. 2730–2734 (2001)
9. Walker, J.G.: Satellite constellations. J. British Interplanetary Soc. 37, 559–571 (1984)
10. Rider, L.: Optimized polar constellation for redundant earth coverage. Journal of the Astronautical Sciences 33(2), 147–161 (1985)
11. Liu, G., Gou, D., Wu, S.: The study on handover in the LEO satellite constellations. Journal of China Institute of Communications 25(4), 151–159 (2004)
12. McCanne, S., Floyd, S.: The LBNL network simulator, ns-2 (October 1,1997/ May 18, 2007), http://www.isi.edu/nsnam/ns

Mobility Support in IPv4/v6 Network

Zheng Xiang and Zhengming Ma

Abstract. IETF has specified Mobile IPv4 and Mobile IPv6 in RFC3344 and RFC3775 respectively, but not yet discussed Mobile IPv4/v6 in any published RFC. This paper proposes a solution to Mobile IPv4/v6 problems. In the solution, a gateway called Mobile IPv4/v6 translation gateway (MIPv4/v6-TG) is introduced to bridge between IPv4 network and IPv6 network, which is made up of a traditional NAT-PT gateway and a Mobile IP application level gateway (MIP-ALG) built upon the NAT-PT gateway. MIP-ALG maintains MIP table, a data structure, which is formed by entries. We use the MIP table to realize the communication between the IPv4 entities and the IPv6 entities. The creation, usage and update processes of MIP table are described in this paper. Through the experiment we find that the MIPv4/v6-TG can realize the basic communication in IPv4/v6 mixed network. The measurement of performance test proves the solution's efficiency. And it can work compatibly with RFC3344 and RFC3775.

Keywords: Mobile IP, NAT-PT, Mobile IP application level gateway (MIP-ALG), Mobile IPv4/v6 translation gateway (MIPv4/v6-TG), MIP Table.

1 Introduction

The growing demand of mobile communication through IP has raised great enthusiasm to develop a new protocol that allows mobile nodes to remain reachable while moving from one place to another. Such kind of protocol is called Mobile IP. IETF specified the latest version of Mobile IPv4 [1] and Mobile IPv6[2] in RFC3344 and RFC3775 respectively. These two protocols work very well in pure

Zheng Xiang
College of Medical Information Engineering, Guangdong Pharmaceutical College
Guangzhou, 510006, China

Zhengming Ma
College of Information Science and Technology Sun Yat-sen UniversityGuangzhou,
510006, China
e-mail: {rousseau2000,wartime2000}@163.com

F.L. Gaol et al. (Eds.): Proc. of the 2011 2nd International Congress CACS, AISC 145, pp. 125–131.
springerlink.com   © Springer-Verlag Berlin Heidelberg 2012

IPv4 network and pure IPv6 network. However, both of them can not be directly applied to IPv4/v6 mixed networks. As the current network evolves gradually from IPv4 to IPv6, more and more mobile nodes need to roam in IPv4/v6 mixed networks. So we need to develop a transition scheme to support Mobile IP in IPv4/v6 mixed networks.

As a transition scheme, the solution should be developed based on the transition mechanisms for IP. Currently, three transition ways are recommended by IETF, namely, dual stacks (RFC4213) [3], tunneling (RFC3053) [4], and NAT-PT (RFC2766) [5][6]. There have been a lot of studies on Mobile IP in IPv4/v6 mixed networks, many of which are based on NAT-PT. An Internet Draft published by IETF [7] presents a solution to a situation of Mobile IPv4/v6, in which Home Agent (HA) and Mobile Node (MN) are in IPv6 network, Correspondent Node (CN) is in IPv4 network and NAT-PT gateway is located between IPv4 network and IPv6 network.

[8] studies the situation where both HA and CN are in IPv6 network and MN is in IPv4 network. A border router is placed between IPv4 network and IPv6 network to connect the two networks.

This paper proposes a scheme to solve Mobile IPv4/v6 problems. The solution is based on Mobile IPv4/v6 Translation Gateway (MIPv4/v6-TG) which is made up of NAT-PT and a Mobile IP-Application Level Gateway (MIP-ALG). And we will introduce the data structure, MIP table, which is maintained by MIP-ALG, and describe in details how to create, use and update MIP table entries. We use the MIP table to realize the communication between the Mobile IPv4 entities and the Mobile IPv6 entities in mixed IPv4/v6 networks.

2 Mobile IPv4/v6 Operations

Mobile IPv4/v6 uses a traditional NAT-PT and a MIP-ALG added on the NAT-PT to form a new device named MIPv4/v6-TG. On IPv6 network side, MIPv4/v6-TG acts as one of the Mobile IPv6 entities (HAv6, MNv6 or CNv6) to combine with other MIPv6 entities inside the IPv6 network to form a complete MIPv6 model described in RFC3775. Inside the IPv6 network, the registration process and communication process can be performed as specified in RFC3775. Similarly, on IPv4 network side, MIPv4/v6-TG acts as one of the Mobile IPv4 entities (HAv4, MNv4 or CNv4) to combine with other MIPv4 entities inside the IPv4 network to form a complete MIPv4 model described in RFC3344. Inside the IPv4 network, the registration process and communication process can be performed as specified in RFC3344.

As mentioned above, MIPv4/v6-TG will act as different MIP entities according to different combinations of the IP versions of HA, MN and CN. When MIPv4/v6-TG intercepts a MIP-related message or datagram, however, it can not decide which role it should act, if it does not know the IP versions of the three entities. Therefore, MIPv4/v6-TG should keep the information of the IP versions of the three entities and the bindings that are used when it acts as a particular MIP entity. In our solution, we introduce a new data structure called MIP table to solve this problem. We use 3-bit binary numbers to indicate the location where Mobile node, Home agent and Correspondent node sit. If the communication entity location in the IPv6 network we record it as 1, otherwise we record it as 0.

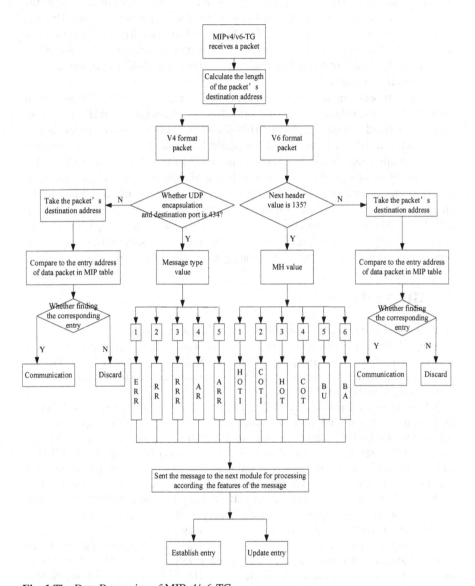

Fig. 1 The Data Processing of MIPv4/v6-TG

The three MIP entities have eight kinds of IP version combinations, from 000 to 111.Two of which correspond to two scenarios where all of the three entities are located in IPv4 network or IPv6 network and MIPv4/v6-TG needn't do anything. So we need six kinds of MIP table entries. At any time, there may be many MIP sessions, which correspond to the same scenario and thus use the same kind of MIP table entry. MIPv4/v6-TG should be able to distinguish them so that the messages and datagrams can be sent to the right place. Therefore, each MIP table entry has four kinds of entrances: MIPv6 Message Entrance, MIPv6 Datagram Entrance, MIPv4 Message Entrance and MIPv4 Datagram Entrance. And we can use the MIP table to realize the communication between the Mobile IPv4 entities and the Mobile IPv6 entities.

The data processing of MIPv4/v6-TG is listed in Fig.1. MIPv4/v6-TG is located between IPv4 networks and IPv6 networks. It receives packets from IPV4 network and IPv6 network constantly. When MIPV4/v6-TG receives a packet, it determines the packet is an IPV4 packet or an IPv6 packet according to the packet destination address length firstly. Then uses intercepting mechanisms to determine the packet is Mobile IP messages whether or not. If it is a Mobile IP message, and then sent the packet to the update module or to the communication module according to the characteristics of the message. If it is not a Mobile IP message, then take the packet's destination address and compare to the entry address of data packet in MIP table. If finding the corresponding entry, the packet will be sent to the communication module. If not finding the corresponding entry, the packet will be discarded.

3   MIP Table

An MIP table entry should, typically, have the following fields.

Type: A three-bit field indicating the IP versions of HA, MN and CN respectively. For each bit, a value of 1 indicates the MIP entity is located in IPv6 network, while a value of 0 indicates the MIP entity is located in IPv4 network. Type value varies from 001 to 110.

MIPv6 Message Entrance: A 128-bit field through which a particular MIP table entry can be found and accessed when a MIPv6 message is intercepted by MIPv4/v6-TG. Usually, this field is set to the IPv6 home address. When a MIPv6 message is intercepted by MIPv4/v6-TG, MIPv4/v6-TG takes out the IPv6 home address from the intercepted message and uses it as an index to search the MIP table. In this process, MIPv4/v6-TG compares the IPv6 message entrance field of each entry with the IPv6 home address. If an entry is found, MIPv4/v6-TG will use the information recorded in the entry to process the intercepted message. If no entry is found, MIPv4/v6-TG will create a new entry.

MIPv6 Datagram Entrance: A 128-bit field through which a particular entry can be found and accessed when a MIPv6 datagram is intercepted by MIPv4/v6-TG. If an intercepted packet is not a MIPv6 message, MIPv4/v6-TG will take out its destination address and use the address as an index to search the MIP table. In this process, MIPv4/v6-TG compares the IPv6 datagram entrance filed of each entry with the address. If an entry is found, the intercepted packet is a MIPv6 datagram, and will be processed based on the information recorded in the entry. If no entry is found, the intercepted packet is not a MIPv6 datagram.

MIPv4 Message Entrance: A 32-bit field through which a particular entry can be found and accessed when a MIPv4 message is intercepted by MIPv4/v6-TG. On intercepting a MIPv4 message, MIPv4/v6-TG takes out its destination address and uses the address as an index to search for the corresponding MIP table entry. It is based on the MIP table entry that MIPv4/v6-TG processes the intercepted MIPv4 message.

MIPv4 Datagram Entrance: A 32-bit field through which a particular entry can be found and accessed when a MIPv4 datagram is intercepted by MIPv4/v6-TG. If an intercepted packet is not a MIPv4 message, MIPv4/v6-TG will take out its destination address and use the address as an index to search the MIP table. In this process, MIPv4/v6-TG compares the IPv4 datagram entrance filed of each entry with the address. If an entry is found, the intercepted packet is a MIPv4 datagram, and will be processed based on the information recorded in the entry. If no entry is found, the intercepted packet is not a MIPv4 datagram.

Cached Bindings: Bindings of home address and care-of address that are used by MIPv4/v6-TG when it acts as a particular MIP entity. Entries may have no binding, one binding, or two bindings. This depends on the types of the entries.

Source Port: A 16-bit field that records the source port value of Registration Request message, extended Registration Request message, or Agent Request message. When MIPv4/v6-TG intercepts such kind of message, it fills this field with the source port value. This field is used as the destination port when MIPv4/v6-TG sends back a reply later.

Destination Port: A 16-bit field that records the destination port value of Registration Request message, extended Registration Request message, or Agent Request message. It will be uses as the source port when MIPv4/v6-TG sends back a reply later.

State: A 1-bit field indicating whether the entry is completed or not. There are two possible states for each entry: 1 (finished) and 0 (unfinished). An unfinished state indicates that the entry is now being created or updated.

Lifetime: A 16-bit field indicating entry lifetime. The meaning of this field depends on the value of the State field. When the State is 0 (unfinished), Lifetime means the entry should be completed before the time expires.

4   Performance Test Based on IXP2400 Network Processor

This paper gives the result of the channel throughput in data layer and packet delay chiefly in order to evaluate the ability to process data packets of the gateway. Throughput test results is shown in Fig.2.Abscissa represents test group of different lengths, vertical axis represents the throughput. One curve represents the packet from IPv4 network to IPv6 network. The other curve represents the packet from IPv6 network to IPv4 network. Fig.2 shows that the channel throughput reaches the minimum value which is 750Mbps from IPv4 network to IPv6 network when the packet's length is 64B. The increase of the packet length affect the throughput reduced gradually along with the test packet's length become longer. From IPv6 network to IPv4 network the channel throughput reach the maximum value which is about 1000Mbps when the packet's length is 64B.But the throughput changes little along with the test packet's length become longer.

Packet delay test results is shown in Fig.3. Abscissa represents test group of different lengths, vertical axis represents the delay time. One curve represents the packet from IPv4 network to IPv6 network. The other curve represents the packet from IPv6 network to IPv4 network. Fig7 shows that the delay time will become longer along with the increase of the test packet's length whether from IPv6 network to IPv4 network or from IPv4 network to IPv6 network. And they will have the same delay time when the packet's length reaches one value.

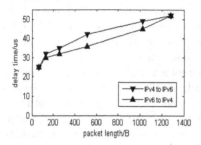

Fig. 2 Throughput Test Results

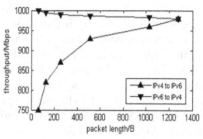

Fig. 3 Packet Delay Test Results

5   Conclusion

The key to our solution is a gateway called Mobile IPv4/v6 translation gateway, which is bridged between IPv4 network and IPv6 network and consists of a traditional NAT-PT gateway and a Mobile IP application level gateway built upon the NAT-PT gateway. MIP-ALG is responsible for maintaining MIP table, a newly introduced data structure. With the help of this gateway and the MIP table, RFC3344 and RFC3775 can be reused in IPv4 network and IPv6 network respectively. In this way, the Mobile IP entities in IPv4 network and the Mobile IP entities in IPv6 network can be transparent to each other.

Compared with other solutions, our solution has three main advantages. Firstly, it can work compatibly with RFC3344 and RFC3775. This is very important in that it makes Mobile IP in IPv4/v6 mixed networks possible without any update to the existing networks. Secondly, our solution introduces MIP table. We can use the MIP table to realize the communication in IPv4/v6 mixed network easily. Thirdly, the creation, usage and update of the MIP table entries is easily too. Through the experiment we find that the MIPv4/v6-TG can realize the basic communication in IPv4/v6 mixed network. The measurement of performance test proves the solution's efficiency.

Acknowledgment. This work is supported in part by The Medium and Small Enterprise Technology Innovation Fund of The Ministry of Science and Technology (No. 09C26214402150), China.

References

1. Perkins, C.E. (ed.): IP Mobility Support for IPv4. RFC3344, IETF (August 2002)
2. Perkins, C.E.(ed.): Mobility Support in IPv6. RFC3775, IETF (June 2004)
3. Nordmark, E.: Basic Transition Mechanisms for IPv6 Hosts and Routers. RFC4219, IETF (October 2005)
4. Durand, A.: IPv6 Tunnel Broker. RFC3053, IETF (January 2001)
5. Nordmark, E.: Stateless IP/ICMP Translation Algorithm (SIIT). RFC2765, IETF (February 2000)
6. Tsirtsis, G., Srisuresh, P.: Network Address Translation-Protocol Translation (NAT-PT). RFC2776, IETF (February 2000)
7. Lee, J.C.: Considerations for Mobility Support in NAT-PT. Internet Draft, IETF (June 2005)
8. Liu, C.: Support mobile IPv6 in IPv4 domains. In: IEEE 59th Vehicular Technology Conference, VTC 2004-Spring (2004)

Finding Security Vulnerabilities in Java Web Applications with Test Generation and Dynamic Taint Analysis

Yu-Yu Huang, Kung Chen, and Shang-Lung Chiang

Abstract. This paper investigates how to combine techniques of static and dynamic analysis for finding security vulnerabilities in Java web applications. We present a hybrid analyzer that employs test case generation and dynamic taint analysis to achieve the goal of no false negatives and reduced false positives.

1   Introduction

Web applications are prone to various kinds of security vulnerabilities. Ac cording to OWASP [1], injection, such as SQL injection (SQLI) [2], and cross-site scripting (XSS) [3] are two major kinds of vulnerabilities identified by Both of them have the same root cause: the lack of proper validation of user's input. An attacker can exploit unchecked input by injecting malicious data into a web application and using the data to manipulate the application.

In the case of SQL injection, the hacker may embed SQL commands into the data he sends to the application, leading to unintended actions performed on the back-end database. Similarly, in a cross-site scripting attack, the hacker may embed malicious JavaScript code into dynamically generated pages of trusted web sites. When executed on the machine of a user who views the page, these scripts can steal user cookies, deface web sites, or redirect the user to malicious sites.

Fig. 1 displays a simple Java web program in servlet form that can easily be exploited by the cross-site scripting technique. This code snippet obtains a user name (`name`) and a user id (`id`) from user input by the method invocations `req.getParameter("name")`(Line 2) and `req.getParameter("id")` (Line 3), respectively. These two pieces of user input are not checked before they are used in other places in the program. In particular, they are sent back to the browser verbatim via the invocations of the `writer.println` method in Line

Yu-Yu Huang · Kung Chen · Shang-Lung Chiang
Department of Computer Science, National Chengchi University, Taiwan
e-mail: {97753022,97753023}@nccu.edu.tw, chenk@cs.nccu.edu.tw

F.L. Gaol et al. (Eds.): Proc. of the 2011 2nd International Congress CACS, AISC 145, pp. 133–138.
springerlink.com   © Springer-Verlag Berlin Heidelberg 2012

10 and 15, respectively. To attack the program, the hacker can enter malicious JavaScript code into the name and id fields, whose values will be set to variables
s1 and s2, and then sent to the user browser for execution.

```
0  public class XSS_Ex extends HttpServlet { void
1  doGet(HttpServletRequest req, HttpServletResponse resp) {
2    String name = req.getParameter("name");
3    String id = req.getParameter("id");
4    PrintWriter writer = resp.getWriter();
5    String s1 = name;
6    String s2 = id.substring(1);
7    int sum = 1;
8    if (s1.matches(".[A-Z][^0-9]+")){
9       if (s2.matches("[A-Z][0-9]+")) {
10         writer.println(s1+s2); //sink
11      } else {
12      for (int i=0; i<s2.length(); i++)
13        sum++;
14      if (sum>10)
15          writer.println(s2); //sink
16   } else {
17      LinkedList c1 = new LinkedList();
18      c1.addLast(name);
19      c1.addFirst("123");
20      Object[] array = c1.toArray();
21      List c2 = java.util.Arrays.asList(array);
22      writer.println(c2.get(0)); //sink
23   }
24 }}
```

Fig. 1 Motivating Program

Such unchecked input vulnerabilities are also called taint-style vulnerabilities
as they can be modeled by the *tainted object propagation* problem in which
tainted information from an untrusted "source" (e.g., req.getParameter)
propagates, through data and/or control flow, to a high-integrity "sink" (e.g.,
writer.println) without being properly checked or validated [4].

To address the tainted object propagation problem, the research community has
devised many methods using the technique of program analysis, including static
and dynamic analyses. Both of them have their own pros and cons. Specifically,
static analysis supports high coverage detection of vulnerabilities, but usually
causes many false positives due to inevitable analysis imprecision. As for dynamic
analysis, although it produces highly confident results, yet it may cause false negatives without complete test cases.

To avoid false negatives and to reduce false positives, we present in this paper a
hybrid approach to finding taint-style vulnerabilities in a Java web application.
Section 2 describes our hybrid approach in detail. Section 3 presents the set of microbenchmarks we devised to evaluate our approach. Section 4 discusses related
work and Section 5 briefly concludes this paper.

2   The Hybrid Vulnerability Analyzer

In the tainted object propagation problem, the program path that connects a source point and a sink point is referred to as a *vulnerable path*. Our analyzer finds such taint-style vulnerabilities in two stages: test case generation and dynamic taint analysis. Fig. 2 displays the structure and main execution steps of the two stages.

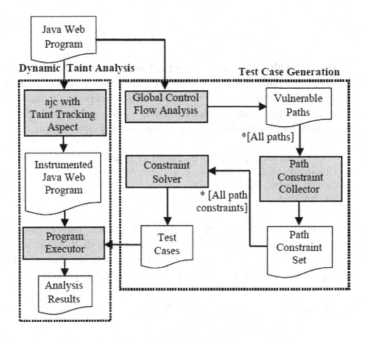

Fig. 2 Structure and Execution Steps of the Hybrid Analyzer

2.1   Test Case Generation

In this stage, we generate the test cases for the dynamic taint analyzer to run. The goal of this stage is to obtain test cases that cover all vulnerable paths. We di-vide the task into three main steps, namely path finding, path constraint collection and constraint solving. We describe them as follows.

Firstly, we conduct a global control flow analysis on the program to create a complete call graph of the program. Then we traverse the call graph in a topologi-cal order to find all program paths, including those inter-procedural ones. Vulner-able paths are identified during the traversal. For example, there are in total six paths in the program of Fig. 1, but only four of them are vulnerable. Table 1 shows the specification of the six paths.

Table 1 All Paths Found in the Motivating Program

Path	Path in line numbers	Vulnerable
1	1-8, 9, 10, 11	Yes
2	1-8, 9, 10, 13, 14, 15, 16	Yes
3	1-8, 9, 10, 13, 14, 15	No
4	1-8, 9, 10, 13, 15, 16	Yes
5	1-8, 9, 10, 13, 15	No
6	1-8, 9, 18-23, 24	Yes

Secondly, for each vulnerable path, we apply the techniques of symbolic execution [5] to generate the path constraints over program input variables such that input data satisfying these constraints will result in the execution of that path. For example, after performing symbolic execution over path 1 in a backward traversal, we obtain the following constraints: four equations in conjunctive form.

```
s2 \in RegularExpr(/[A-Z][0-9]+/) &&
s1 \in RegularExpr(/.[A-Z][^0-9]+/) &&
s2 == id.substring(1) &&
s1 == name_0;
```

Finally, we employ the Kaluza string solver [5] to solve path constraints and obtain the test cases. For example, given the constraints of the four vulnerable paths in table 1, Kaluza will find solutions for three of them, as shown in Table 2, and report that path 4 is infeasible ('@' represents a wildcard character).

Table 2 Test Cases Generated for the Motivating Program

Test Case	Path	Input values (name, id)
1	1	(name = @H@, id=@A4)
2	2	(name=@H@, id=@@@@@@@@@@@@@)
3	6	(name=@, id=dontcare)

2.2 Dynamic Taint Analysis

In this stage, we run the program on each test case generated in the previous stage and track the dynamic dataflow of tainted objects throughout the execution. If a tainted object reaches any sensitive sinks, the underlying vulnerable path is considered a verified (true) vulnerability. To accomplish this task, we need to instrument the program with code that tracks the propagation of tainted objects. Here we utilize the instrumentation mechanism of the AspectJ language [6]. In particular, we developed a taint tracking aspect and instrument the program with the aspect using the AspectJ compiler, *ajc*.

Our taint tracking aspect maintains a repository of tainted objects, called *taint-set*. Initially, only objects from user input are deposited into the taintset. As the instrumented program executes, the aspect will trace the propagation of tainted objects, resulting in addition or deletion of objects from the taintset. The detailed design of such aspects can be found in the third author's thesis [7].

Applying the taint tracking aspect to the three test cases listed in Table 2, we get the report in which the paths associated with test case 1 and test case 2 are recorded as true vulnerabilities, whereas path 6 associated with test case 3 is not. This is valid since, in Figure 1, the argument passed to the sink method (`writer.println`) in Line 22 is the constant string "123", which is obviously innocuous. By contrast, a static analyzer for Java will most likely report this (path 6) as a vulnerability, which is indeed a false positive.

3   Evaluation

We have done a preliminary evaluation of our hybrid approach by applying it to a set of 22 micro-benchmark programs. Branches are the common examples of false negative results for dynamic analyzers; and data structures often mislead static analyzers to give false positive results. Therefore, besides method calls and loops, the foci of our microbenchmarks are branches and data structures such as arrays and collections. Detailed descriptions of these microbenchmarks can be found in the first author's master thesis [8].

As we expected, our hybrid analyzer finds all embedded vulnerabilities in the microbenchmarks, i.e., no false negative (F.N.) results. But, in three benchmarks, the use of data structures to propagate tainted objects has made our static control flow analyzer produce some false positive (F.P.) results.

4   Related Work

Monga et al [9] presented a hybrid analysis framework in which static analysis techniques are employed to reduce the runtime overhead of the instrumentation-based online taint monitor module. Our hybrid approach is partly inspired by the work of Kiezun et al [10]. They also combine test generation and taint analysis to detect security vulnerabilities, but their novel test generator is based on dynamic analysis that combines concrete and symbolic execution. By contrast, our test generator is based static analysis.

5   Conclusion

In this paper, we presented an approach for detecting taint-style vulnerabilities in Java web applications through hybrid analysis techniques. Our proposal applies control flow analysis and symbolic execution to generate test data that exhaust all potentially vulnerable paths in the target program. Then, we employ an aspect-based dynamic taint analysis to run the program with the test data and track the flow of tainted data for verifying the vulnerabilities. The preliminary results indicate that our approach can attain the goal of reducing false positive results and eliminating false negative results.

References

[1] OWASP. OWASP Top 10 for (2010),
    https://www.owasp.org/index.php/Top_10_2010-Main
[2] Halfond, W.G., Viegas, J., Orso, A.: A Classification of SQL-Injection Attacks and Countermeasures. In: Proc. IEEE Int'l Sym. on Secure Software Engineering (March 2006)
[3] CERT. Advisory CA-2002: Malicious HTML Tags Embedded in Client Web Requests (2002)
[4] Livshits, V.B., Lam, M.S.: Finding security vulnerabilities in Java applications with static analysis. In: Proc. 14th Usenix Security Symposium, pp. 271–286 (August 2005)
[5] Saxena, P., Akhawe, D., Hanna, S., Mao, F., McCamant, S., Song, D.: A Symbolic Execution Framework for JavaScrip. In: IEEE Sym. on Security and Privacy (May 2010)
[6] Laddad, R.: AspectJ in Action. Manning Publications Co. (2003)
[7] Chiang, S.L.: A Hybrid Security Analyzer for Java Web Applications, Master Thesis, National Chengchi University, Taiwan (July 2010)
[8] Huang, Y.Y.: Test Case Generation for Verifying Security Vulnerabilities in Java Web Applications, Master Thesis, National Chengchi University, Taiwan (July 2011)
[9] Monga, M., Paleari, R., Passerini, E.: A Hybrid Analysis Framework for Detecting Web Application Vulnerabilities. In: Proc. Workshop on Software Engineering for Secure Systems (IWSESS 2009), pp. 25–32 (2009)
[10] Kieżun, A., Guo, P.J., Jayaraman, K., Ernst, M.D.: Automatic creation of SQL injection and cross-site scripting attacks. In: Proc. the 31st International Conference on Software Engineering (May 2009)

A Scalable Distributed Multimedia Service Management Architecture Using XMPP[*]

Xianli Jin

Abstract. In this paper, we propose a scalable distributed multimedia service management architecture using XMPP. We first study the XMPP and investigate the limitations of related multimedia service management models. Then we describe the scalable distributed multimedia service management architecture in detail along with a video conferencing system case using this model. The performance simulation results show that the proposed distributed multimedia service management architecture achieves great scalability.

1 Introduction

Ever since its emergence in the early 1990s, the Internet has radically changed the way of accessing and exchanging information among desktops around the globe. Today, almost every information-hungry business banks heavily upon the Internet. The success of business in the 21st century depends not only on procuring up-to-date

Xianli Jin

College of Computer, Nanjing University of Posts and Telecommunications, Nanjing, Jiangsu 210003, China

Jiangsu High Technology Research Key Laboratory for Wireless Sensor Networks, Nanjing, Jiangsu 210003, China

Institute of Computer Technology,Nanjing University of Posts and Telecommunications, Nanjing, Jiangsu 210003, China
e-mail: jxl@njupt.edu.cn

[*] The work reported in this paper is supported by a project Funded by the Priority Academic Program Development of Jiangsu Higher Education Institutions(yx002001), the Natural Science Research Project of Higher Education of Jiangsu under Grant No. 09KJD520007, and the Startup Fund for Distinguished Scholars introduced into Nanjing University of Posts and Telecommunications, China.

F.L. Gaol et al. (Eds.): Proc. of the 2011 2nd International Congress CACS, AISC 145, pp. 139–145.
springerlink.com
© Springer-Verlag Berlin Heidelberg 2012

information but also on procuring it fast. Instant messaging (IM) [1] is such an Internet-based protocol application that allows one-to-one communication between users employing a variety of devices. Over the years, instant messaging has proven itself to be a feasible technology not only to fun-loving people but to the world of commerce and trade, where quick responses to messages are crucial. With the wide acceptance it commands, instant messaging is looked upon by the business world as an ideal tool for promoting its interests more than a convenient way of chatting over the Internet.

It is in consideration of such a scenario that in this paper we propose a scalable distributed multimedia service architecture using XMPP (Extensible Messaging and Presence Protocol, an instant messaging protocol) [2, 3, 4]. Applications in certain domain and context could be developed based on this architecture. Furthermore, it is fine designed to adapt various media types and heterogeneous network conditions. Because of its extensibility, it is able to meet the exacting demands of today's multimedia e-commerce environment.

2   XMPP Overview

XMPP is a protocol for streaming Extensible Markup Language (XML) elements in order to exchange structured information in close to real time between any two network endpoints. It provides a generalized, extensible framework for exchanging XML data, used mainly for the purpose of building instant messaging and presence applications.

XMPP is not wedded to any specific network architecture, but to date it usually has been implemented via a client-server architecture wherein a client utilizing XMPP accesses a server over a TCP connection, and servers also communicate with each other over TCP connections. This architecture mainly has four entities: Server (managing connections and routes from or sessions for other entities, in the form of XML streams to and from authorized clients, servers, and other entities), Client (connecting directly to a server and using XMPP to take full advantage of the functionality provided by a server and any associated services), Gateway (a special-purpose server-side service whose primary function is to translate XMPP into the protocol used by a foreign (non-XMPP) messaging system, as well as to translate the return data back into XMPP) and Network (a network of servers that inter-communicate).

The sessions between any two entities using XMPP is within an XML stream. The start of an XML stream is denoted unambiguously by an opening XML <stream> tag, while the end of the XML stream is denoted unambiguously by a closing XML </stream> tag. During the life of the stream, the entity that initiated it can send an unbounded number of XML elements over the stream. XMPP contains three top-level XML elements: <message/>, <presence/>, and <iq/> elements. The <message/> element is used to contain messages that are sent between two entities. The <presence/> element provides availability information about an entity. Info/Query (<iq>) structures a rudimentary conversation between any two entities and allows them to pass XML-formatted queries and responses back and forth.

3 Scalable Distributed Multimedia Service Management Architecture Based on XMPP

3.1 Service Architecture

Figure 1 shows the scalable distributed multimedia service management architecture using XMPP. The architecture mainly has three entities: *service participant system (SPS), service manager system (SMS), delivery network (DN).*

Service participant system (SPS) is responsible for connecting directly to a *service manager system (SMS)* and taking full advantage of the functionality provided by the *service manager system (SMS)* to provide or consume any associated multimedia services.

A service participant is either an *End-Service Provider (ESP)*, or an *Intermediate-Service Provider (ISP)*, or an *End-Service Consumer (ESC). ESP* is an entity with jurisdiction over a domain that contains a system that predominantly provides multimedia services to service consumers. *ISP* which can be regarded as a service agent or broker, provides adjunct services and conveys information among *ESPs* and *ESCs. ESC* consumes the multimedia services provided by *ESPs* and *ISPs.*

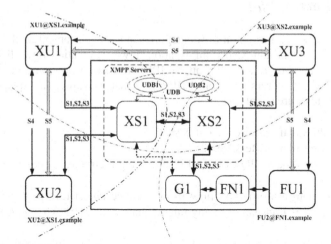

The symbols are as follows:
o XU1, XU2, XU3 = XMPP client users which may be a service provider or a service consumer
o XS1, XS2 = XMPP servers   o UDB1, UDB2 = Database about users' information
o G1 = A gateway that translates between XMPP and the protocol(s) used on a foreign (non-XMPP) messaging network
o FN1 = A foreign messaging network   o FU1 = A client user on a foreign messaging network
o S_i = An information flow between two entities (i = 1,2,3,4,5)

Fig. 1 The scalable distributed multimedia service management architecture using XMPP

Service manager system (SMS) is responsible for managing connections and routes from or sessions for other entities, in the form of information flows (e.g. XML streams) to and from authorized service participants and other SMSs. Also, it assumes responsibility for the storage of data that is used by service participants (e.g., service register information of *ESPs*).

Delivery network (DN) is a network of entities that inter-communicate, enabling the transfer of information between any two entities.

Figure 2 provides a high-level overview of three models included in this architecture (where "-" represents communications that use XMPP and "=" represents communications that use any other protocol).

XU1---XS1---XU2   **XU1--- XS1 --- XS2 ---XU3**
(a) The basic client-server model   (b) The multi-server model
XU3--- XS2 --- G1===FN1===FU1

(c) The agent model

Fig. 2 Three models included in the Architecture

3.2 *Information Flows*

There are two basic information flow categories in the management model: XMPP information flow and non-XMPP information flow. Information flows between a XMPP server and any other entity are XMPP-information flow, and information flows between two users are non-XMPP information flow. The specific five information flows are as follows:

S1: S1 is the stream <Presence> of XMPP.
S2: S2 is the stream <Message> of XMPP.
S3: S3 is the stream <IQ> of XMPP, providing a structured request-response mechanism.
S4: S4 is control-information flow from a source object to a peer destination object. The flow is transparent to any intermediate object through which the flow passes.
S5: S5 is content-information flow from a source multimedia service provider user to a destination multimedia service consumer user. The flow is transparent to any intermediate object through which the flow passes. The flow does not appear to alter the behavior of the information source and destination objects. E.g., audio,video, or data transferred between two users.

(a) Pull   (b) Push

Fig. 3 Signaling procedures of multimedia service

3.3 Procedures of Multimedia Service

Figure 3 shows the procedures of multimedia service which can be divided into four stages: service initiation, join service, specific service content and quit the service.

Stage 1: The main purpose of service initiation is to register the service provided by ESP/ISP. ESP or ISP first sends service register request including the full information of the service to the nearest SMS. The SMS notices it is a service register request and then sends back an *OK* response indicating permission. When the ESP/ISP receives *OK*, it will start the service and send an *ACK* to notify the SMS.

Stage 2: There are two ways to join a service: pull way (showed in figure 3a) and push way (showed in figure 3b). In the former case, the ISP/ESC first sends the service request to SMS2 where it registered. Then SMS2 notices that SMS1 has the service registered and forwards the request to SMS1. SMS1 receives the request and sends back the result with the service provider's address to the ISP/ESC through SMS2. Receiving the service provider's address, the ISP/ESC can send the service request directly to the service provider. When the handshake *OK* and *ACK* are completed, the specific service starts. In the latter case, the service is first triggered by the service provider by sending its address to the service consumers invited through SMS1 and SMS2. The handshake process is similar to the pull way.

Stage 3: The procedure of this stage depends on the specific multimedia service application. Information flows S4 and S5 are used in the stage to meet the specific requirements.

Stage 4: Quitting the service is quite simple, which is completed only by a *BYE* request from anyone of the participants. The whole process in the lifecycle of a media session is shown in Figure 3.

4 Case Study and Performance Evaluation

According to our architecture, we designed and implemented a scalable distributed video conference system. The development of the system used the multimedia middleware proposed in [5]. Figure 4 shows the network structure of the video conference system, where node server is responsible for forwarding the local user's video streams to other node servers and forwarding streams from other node servers to local users. And conference manager server is responsible for the management of conferences and corresponding with other conference manager servers.

Figure 5 shows the simulation results of the comparison of the processing load of the centralized conferencing model and that of the distributed conferencing model. As shown in Figure 5, the processing load of the centralized model drastically increases as the number of participants increase, while in the scalable distributed model, the processing load is almost constant. This illustrates that the distributed model achieves greater scalability than that of the centralized model.

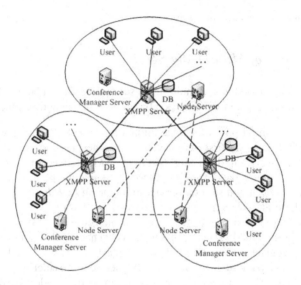

Fig. 4 Networks structure of the scalable distributed video conference system

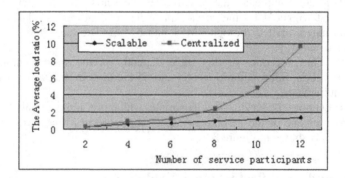

Fig. 5 The average load ratio

5 Conclusion

In this paper, we have proposed a scalable distributed multimedia service architecture using XMPP. The performance simulation results show that the proposed distributed multimedia service management architecture performs better in scalability than that of the centralized conferencing model. Furthermore, our architecture can also be applied to other distributed networks such as wireless sensor networks.

References

1. Extensible Messaging and Presence Protocol(XMPP): Core/Introduction. RFC, 3920–3921
2. Saint-Andre, P.: Extensible Messaging and Presence Protocol (XMPP): Core. RFC 3920, IET (October 2004)
3. Saint-Andre, P.: Extensible Messaging and Presence Protocol (XMPP): Instant Messaging and Presence. RFC 3921, IETF (October 2004)
4. JABBER INC. Core XMPP Protocol Stack[EB/OL] (August 15, 2009), http://xmpp.org/protocols/
5. Zhang, J., Ma, H.: An Approach to Developing Multimedia Services Based On SIP. In: Proceedings of ICCC 2004, vol. 1, pp. 95–101 (2004)

Sensor-Level Real-Time Support for XBee-Based Wireless Communication

Mihai V. Micea, Valentin Stangaciu, Cristina Stangaciu, and Constantin Filote

Abstract. The ZigBee standard is focused on low-cost, low-power, wireless mesh networking, having a wide applicability mainly in the field of wireless sensor networks. A growing number of such applications require real-time behavior, both at the wireless communication and at the sensor levels. This paper proposes a solution to the problem of providing sensor-level real-time support for wireless platforms using ZigBee-based devices such as the XBee module. The discussion of the experimental results proves the predictable behavior of the XBee sensor platform used as a case study.

1 Introduction

Intelligent wired and wireless sensor networks play a key role in a large number of systems and applications currently developed. Many applications require timely response of the sensors and their network, as outdated information could induce control or decision errors with a high potential to harm people or the environment. Real-time sensor networks are used in space, avionics, automotive, security and fire monitoring applications, among many other examples.

A significant amount of research has been recently focused on providing real-time behavior to wireless sensor networks (WSNs) operation. In [1], a scheduling mechanism based on guaranteed time slots (GTSs) is proposed, to provide timing guarantees for message exchange within ZigBee and IEEE 802.15.4 wireless

Mihai V. Micea · Valentin Stangaciu · Cristina Stangaciu
Computer & Software Engineering Dept., Politehnica University of Timisoara, Romania
e-mail: mihai.micea@cs.upt.ro, {stangaciu,certejan}@gmail.com

Constantin Filote
Computer & Automation Dept., Stefan cel Mare University of Suceava, Romania
e-mail: filote@eed.usv.ro

F.L. Gaol et al. (Eds.): Proc. of the 2011 2nd International Congress CACS, AISC 145, pp. 147–154.
springerlink.com   © Springer-Verlag Berlin Heidelberg 2012

networks [2]. Priority toning strategy is described in [3], as a soft real-time solution to wireless communication over the same type of networks. These works, among many others, focus on solving the real-time problem at medium access control (MAC) level.

The real-time routing aspect has also been investigated. In [4], the authors introduce the real-time power-aware routing protocol (RPAR), which allows the application to specify packet deadlines, leading to dynamic adjustment of the transmission power and the routing decisions. A heuristic solution to finding energy-efficient routes for messages with soft timing specifications is proposed in [5].

Despite the large number of operating systems (OS) proposed for WSNs, providing hard real-time support for the sensor platform still remains an open problem. For example, preemptive multitasking mechanisms are provided by the Nano-RK OS [6]. On the other hand, task preemption could issue predictability problems. Many other WSN operating systems, [7], [8], do not provide native real-time scheduling mechanisms.

A solution to this problem, of providing hard real-time support for wireless sensor platforms, is presented in this paper. The two topical aspects are discussed, based on a case study sensor platform with the XBee wireless module: the operating system level support and the communication application (interface driver).

2 XBee-Based Wireless Sensor Platform

An example sensor platform, able to exchange information over ZigBee wireless networks [2], is the Wireless Intelligent Terminal (WIT) for the CORE-TX platform [9]. For the wireless access, its communication daughterboard uses an XBee module [10] from Digi International, Inc., as shown in Fig. 1. This module has been chosen for its features such as low cost, reduced dimensions, good range (up to 100 m indoor and 1500 m outdoor) and ease of use.

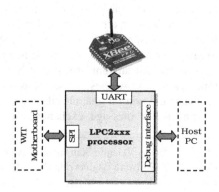

Fig. 1 Main blocks of the WIT communication daughterboard

An ARM7-based LPC2xxx [11] is used as the main processing unit of the daughterboard. It communicates with the XBee module through the UART interface, with a host PC to exchange debug information, and with the WIT motherboard, through the Serial Peripheral Interface (SPI). As a result, the onboard processor has to execute at least the following main tasks: (1) control of the XBee module and wireless data exchange, (2) communication with the motherboard, which generates and processes the wireless information at the WIT level, (3) exchange and process of debug information with a host PC, as a temporarily development stage, and (4) various local processing operations.

3 Sensor-Level Real-Time XBee Support

At sensor level, real-time support must be equally provided by the operating environment and by the communication application (or interface drivers). This section presents the sensor-level operating support and the design principles of the real-time XBee wireless communication driver implemented on the CORE-TX WIT.

3.1 Sensor-Level Operating Support

The operating environment plays a key role in providing the required level of timeliness for the sensor platform behavior. Such a real-time operating system is the HARETICK kernel [12], developed in the DSPLabs, Timisoara.

HARETICK (Hard REal-TIme Compact Kernel) is designed to provide execution support and maximum predictability for critical or hard real-time (HRT) applications on sensing, data acquisition, signal processing and embedded control systems. The kernel provides two distinct execution environments: a) *HRT context*, as a non-preemptive framework for the execution, with the highest priority, of the HRT tasks, and, b) *Soft real-time (SRT) context*, for the execution of SRT (or non real-time) tasks, in a classical, preemptive manner.

The HRT context is based on three essential elements: the real-time clock (RTC), which is assigned to the highest interrupt level, the real-time executive (HDIS), and the scheduler task (HSCD). HDIS has the role of framing the execution of any HRT task, thus it is composed by a prefix (HDIS_Pre) and a suffix (HDIS_Suf). HDIS_Pre is called by the RTC interrupt (it *is* actually the RTC interrupt handler) each time a HRT task is scheduled for execution. HDIS_Suf programs the RTC interrupt to activate at the instant the next HRT task is scheduled. The HRT scheduler currently developed and extensively tested implements the so called Fixed-Execution Non-Preemptive (FENP) algorithm [12], providing strong guarantees that all the HRT tasks meet their temporal specifications, even under the worst–case operating conditions.

A particular model has been introduced to specify all the HRT tasks on the HARETICK kernel [12]. The ModX (eXecutable MODule) models the periodic, modular HRT task, with complete and firm temporal specifications, scheduled and executed in the non-preemptive HRT context of the kernel. Among the key temporal parameters of the ModX M_i, are its period, T_{Mi}, and worst case execution time (WCET), C_{Mi}. All these parameters have to be known or derived at the off-line application analysis stage, when the feasibility tests are performed, prior to executing the system on the target platform.

3.2 Real-Time XBee Driver

Due to the hard requirements regarding execution predictability, timing and code analysis, conventional architectures and programming strategies have to be adapted when writing a real time application. One of the most important aspects to be considered when writing a real time task is to provide the means to analyze and calculate its worst case execution time (WCET) [13], [14].

The real-time XBee driver has been designed based on a layered approach, following the ISO OSI Model principles and considering the particular communication protocol stack which interfaces the target processor to the XBee module. Furthermore, the driver is also divided between the two execution contexts provided by the HARETICK kernel, the HRT and the SRT contexts. The general architecture of the XBee driver is depicted in Fig. 2 (left).

Each of the three layers managing the communication protocol has a transmission (Tx) and a reception (Rx) component, with similar functionalities. The first two layers are directly related to the timing requirements of the communication interface (the UART) and, therefore, have been designed to be scheduled and executed as the XBee ModX, within the HRT context of the HARETICK kernel.

Layer 1 handles the physical interface control and the raw data at word level (where a word has a 10 bit structure: START bit + DATA byte + STOP bit).

Layer 2 manages the XBee frames structure, composed of a start byte (0x7E), a Frame Length field (2 bytes), payload bytes and a checksum tail byte. This layer implements a distinct but similar finite-state automaton, for the Tx and the Rx components. The basic diagram of the frame receive automaton at the layer 2 is depicted in Fig. 2 (right). After successfully extracting the data from a frame, the receive automaton releases the data buffer to layer 3 for further interpretation and processing. Due to the fact that layer 2 operates (along with layer 1) as a ModX, in the HRT context of the kernel, while layer 3 is executed within the SRT context, their data communication is based on guarded (or flagged) buffers [15].

Layer 3 has several important roles, including (a) handling the user/application data, (b) constructing and decoding the XBee status and control data, and (c) application programming interface, providing the required functions and configuration parameters to the upper (application) layer. For example, the XBee modem status data is structured over 2 bytes: <API Id> (0x8A) + <Status byte>. <Status byte> can have the following values: 0 for "Hardware reset"; 1 for "Watchdog timer reset"; 2 meaning "Associated" to a Personal Area Network (PAN); 3 for "Disassociated"; 4 for "Sync lost"; 5 meaning "Coordinator realignment"; 6 for "Coordinator started". Layer 3 is also capable of handling the various AT commands featured by the XBee protocol.

Since layer 3 handles data frame exchanged through buffers, both with the application layer and with layer 2, it is not directly affected by the timing specifications of the physical interface. Therefore, layer 3 task has been designed as a SRT task, and, as a result, its execution does not need to be added to the HRT processor utilization, which should be kept as low as possible for a feasible schedule of this context.

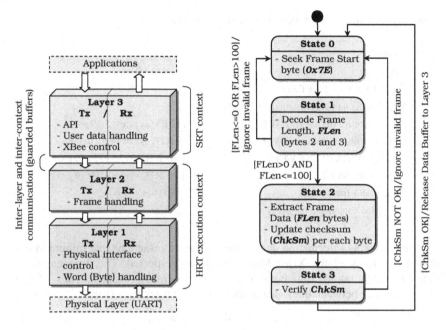

Fig. 2 General architecture of the real-time XBee driver (left) and the Layer 2 frame receive automaton (right)

4 Discussion of the Experimental Results

The WIT communication daughterboard presented in Section II and in Fig. 1 has been used to measure and validate the real-time behavior of the sensor-level communication. A set of HRT and SRT tasks have been programmed, analyzed and implemented to provide the basic data exchange capabilities. The tasks directly interacting with the hardware components which require timely behavior have been implemented as ModXs: the HSCD (HRT Scheduler), the XBee ModX (containing the first two layers of the proposed driver), the SPI and the Debug ModXs. The tasks in the SRT execution context include the XBee_Layer3 task, the SPI_Layer4 task and a set of data processing and general debug tasks. They are scheduled with a simple loop algorithm.

Table 1 presents the experimental setup, along with some of the most important parameters taken into consideration or calculated at the offline system/application analysis step, prior to launching it on the target platform. The worst execution scenario has been considered the case when the maximum possible data is transferred through the XBee module, thus involving the use of all the corresponding buffers at their maximum capacities.

Table 1 System and XBee ModX Configuration Parameters

Parameter	Value
CPU Clock	58.9824 MHz
HARETICK Real-Time Clock (RTC)	14.7456 MHz
Total ModXs on target system (HSCD included)	4
Prefix executive (HDIS_Pre) WCET	112 RTC cycles (7.59 µs)
Suffix executive (HDIS_Suf) WCET	68 RTC cycles (4.61 µs)
Scheduler ModX (HSCD) WCET	4052 RTC cycles (274.79 µs)
XBee ModX WCET	4568 RTC cycles (309.79 µs)
XBee UART transfer rate	57600 bps
XBee ModX minimum frequency	360 Hz

To avoid losing information during the wireless transactions, the minimum execution frequency of the XBee ModX has to be respected. This frequency has been calculated taking into consideration the operating characteristics of the UART interface between the XBee module and the target processor:

$$F_{\min \text{XBee}} = TR_{\text{UART}} / (SizeW_{\text{FIFO}} \cdot WLen_{\text{UART}}) \tag{1}$$

where TR_{UART} is the UART transfer rate (57600 bps), $SizeW_{\text{FIFO}}$ is the size in words of the hardware UART FIFO buffer (16 words), and $WLen_{\text{UART}}$ is the number of bits per UART frame (10 bits/word). From the minimum frequency of the XBee ModX, the maximum value of its period as a task can be derived:

$$T_{\max \text{XBee}} = 1 / F_{\max \text{XBee}} = 2777.77 \ \mu s \tag{2}$$

and the minimum processor utilization of the XBee ModX can be calculated:

$$U_{\min \text{XBee}} = \frac{C_{\text{XBee}}}{T_{\max \text{XBee}}} = \frac{WCET_{\text{XBee}}}{T_{\max \text{XBee}}} = 11.15\% \tag{3}$$

Table 2 shows the worst case and the typical execution times measured during the experiments, for the HRT context of the system. As seen in the table, the typical execution of the HSCD ModX is equivalent to the worst case, as the scheduler applies each time the same algorithm to a constant set of tasks (ModXs) and temporal parameters.

It is worth mentioning that, although the HDIS executive is distinctly shown from the execution times of the ModXs in Table 1 and Table 2, the actual (total) WCET of any ModX includes also the HDIS execution time. This total WCET is considered when the ModX is scheduled (or at the offline analysis step).

Table 2 Measurement Results for the Worst Case and Typical Execution Times

Parameter	Approx. Value
XBee ModX WCET	305 µs
Xbee ModX typical execution time	75 µs
Scheduler ModX (HSCD) WCET	271 µs
HSCD typical execution time	271 µs
Debug ModX WCET	41 µs
SPI ModX WCET	21 µs
HDIS executive (Prefix + Suffix) WCET	11.76 µs

Fig. 3 shows one of the system operation timeframes captured with a logic analyzer. It depicts the execution of the HRT context, with its 4 ModXs (HSCD, SPI, XBee and Debug) and the two components of the executive (HDIS_Pre and HDIS_Suf), which frame the executions of each ModX. The SRT context operation has also been captured.

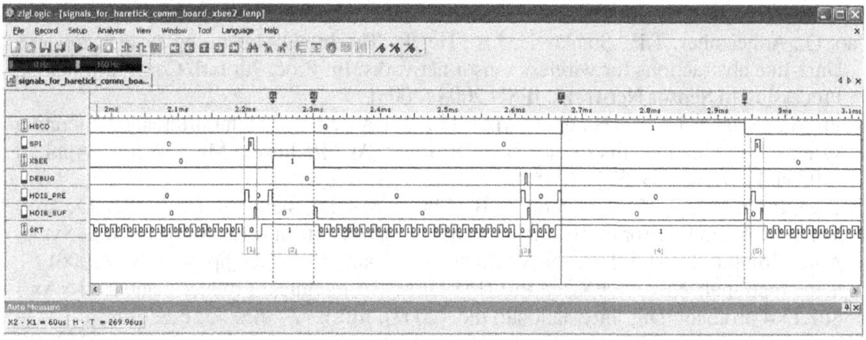

Fig. 3 Logic analyzer capture of the system operation

Several interesting intervals have been marked in the figure. The execution of the SPI ModX, prefixed by the HDIS_Pre component of the executive, and ended by the HDIS_Suf, is highlighted with (1) and (5). Interval (2) marks the execution of the XBee ModX (elapsing approx. 60 µs, here). It also is framed by HDIS_Pre and HDIS_Suf. The Debug ModX is outlined at (3), while (4) corresponds to the execution of the HRT scheduler (HSCD), measured here at around 270 µs.

The SRT context captured in Fig. 3 shows a cyclic behavior, due to the implemented scheduling policy. The diagram also outlines the intervals in which the SRT tasks are interrupted by the higher priority HRT context: (1) – (5).

Acknowledgments. This work is supported in parts by the Romanian Ministry of Education and Research, through the grant PNCDI II PDP 22-137/2008-2011.

References

Francomme, J., Mercier, G., Val, T.: A simple method for guaranteed deadline of periodic messages in 802.15.4 cluster cells for control automation applications. In: Proc. IEEE Conf. Emerging Technologies and Factory Automation (ETFA 2006), pp. 270–277 (2006)

IEEE Standard: Specific requirements part 15.4: Wireless medium access (MAC) and physical layer (PHY) specifications for low-rate wireless personal area networks (LR-WPANs). IEEE-SA (2006),
http://standards.ieee.org/about/get/802/802.15.html

Kim, T.H., Choi, S.: Priority-based delay mitigation for event-monitoring IEEE 802.15.4 LR-WPANs. IEEE Communication Letters 10(3), 213–215 (2006)

Chipara, O., He, Z., Xing, G., Chen, Q., et al.: Real-time power-aware routing in sensor networks. In: Proc. 14th Intl. Wshop. Qual. Service (IWQoS 2006), New Haven, CT, pp. 83–92 (2006)

Pothuri, P.K., Sarangan, V., Thomas, J.P.: Delay-constrained, energy-efficient routing in wireless sensor networks through topology control. In: Proc. IEEE Intl. Conf. Networking, Sensing and Control (ICNSC 2006), Ft. Lauderdale, FL, pp. 35–41 (2006)

Eswaran, A., Rowe, A., Rajkumar, R.: Nano-RK: An energy-aware resource-centric RTOS for sensor networks. In: Proc. 26th IEEE Intl. Real-Time Systems Symp. (RTSS 2005), Miami, FL, pp. 256–265 (2005)

Cao, Q., Adbelzaher, T.F., Stankovic, J.A., He, T.: The LiteOS operating system: towards Unix-like abstractions for wireless sensor networks. In: Proc. 7th Intl. Conf. Information Processing in Sensor Networks, IPSN 2008 (2008)

Bhatti, S., Carlson, J., Dai, H., et al.: MANTIS OS: An embedded multithreaded operating system for wireless micro sensor platforms. ACM/Kluwer Mob. Netw. Appl. J (MONET) 10(4), 563–579 (2005)

Cioarga, R.D., Micea, M.V., Ciubotaru, B., Chiciudean, D., Stanescu, D.: CORE-TX: Collective Robotic Environment - the Timisoara Experiment. In: Proc.3rd IEEE Intl. Symp. Appl. Comput. Intell. Inform (SACI 2006), Timisoara, Romania, pp. 495–506 (2006)

Digi International: XBeeTM/XBee-PROTM OEM RF modules: Product manual v1.xAx - 802.15.4 protocol. Digi International, Inc. (2007), http://www.digi.com

NXP Semiconductors: UM10114: LPC21xx and LPC22xx user manual. Rev. 03. NXP Semiconductors N. V (2008),
http://www.nxp.com/documents/user_manual/UM10114.pdf

Micea, M.V., Cretu, V.I., Groza, V.: Maximum predictability in signal interactions with HARETICK kernel. IEEE Trans. Instrum. Meas. 55(4), 1317–1330 (2006)

Puschner, P.: Algorithms for dependable hard real-time systems. In: Proc. 8th Intl. Wshop. Object-Oriented Real-Time Dependable Systems (WORDS 2003), pp. 26–31. IEEE Press (2003)

Wilhelm, R., Engblom, J., et al.: The worst-case execution time problem (overview of methods and survey of tools). ACM Trans. Embed. Comput. Syst. 7(3), 36–88 (2008)

Micea, M.V., Certejan, C., Stangaciu, V., et al.: Inter-task communication and synchronization in the hard real-time compact kernel HARETICK. In: Proc. IEEE Intl. Wshop. Robotic Sensors Environments (ROSE 2008), Ottawa, Canada, pp. 19–24 (2008)

An Ontology Based Personalizing Search Measure with the Protection of Original Engine Integrity

Xiao-dong Wang and Qiang (Patrick) Qiang

Abstract. IR techniques are widely used to locate surfers' interesting information resource on the Web. Every IR engine has its immanent advantage, while user's retrieval hobby is also diverse. Thus, to satisfy individual, personalizing search becomes more and more popular. But, regardless of familiar environment for users, exist personalizing search measures just present a mixed result list from different engines. The combination destroys the integrity of original IR engine and pilot user to a strange interface. Aim to overcome the outstanding shortcoming, this paper presents an Ontology based personalizing search measure, which builds user profile and offers not a rigid combination result but the most appropriate engine with respect to query, so that the original engine integrity is protected. Some primary experiments illustrate that the measure is feasible and valuable.

1 Introduction

In order to find out some useful Web information resource, Information Retrieval (IR) techniques are developed exponentially since last century 90's. It is not difficult to observe that different IR engines have their own predominance. In different search missions, an individual usually has diverse IR engine using preference. For example, a user maybe prefer to Baidu to find Chinese fictions than Google, while

Xiao-dong Wang
The Telecommunication Engineering Institute, AFEU, 1 East Fenghao Road,
Xi'an 710077, China
e-mail: hellowxd@mail.nwpu.edu.cn

Qiang Patrick Qiang
The Great Valley School of Graduate Professional Studies, The Pennsylvania State University, 30 East Swedesford Road Malvern, PA 19355, USA
e-mail: qzq10@psu.edu

F.L. Gaol et al. (Eds.): Proc. of the 2011 2nd International Congress CACS, AISC 145, pp. 155–160.
springerlink.com   © Springer-Verlag Berlin Heidelberg 2012

he likes Google more to search some English technical literature. It is mainly due to the differences design of engines as well as users' hobby distinction. In order to better satisfy the personalized requirements and provide more relevant results, user has to frequently switch his search among these engines. It is an undesirable experience to most of us. Obviously, how to resolve the problem is on agenda.

2 Design of the Measure

There already exist some personalizing search methods [1], which just return a combination of several IR engines results for users. Different form them, we design an Ontology based novel measure which recommends a most appropriate and integral engine to a query rather than a rigid mixed list. With the help of user's profiles, it provides personalizing service and protects original engine integrity.

It mainly includes three modules as following

- Learning process module: build user profile by learning the IR engine matching score with respect to a topic form user's search histories;
- Recommendation process module: calculate the matching score of IR engine with respect to user query;
- Loading process module: load the most appropriate IR engine and carry out the corresponding search.

3 Implement of the Evaluation Measure

The whole process of our measure is illustrated in Fig 1. Once a query comes, it uses the favor profile to select the best IR engine for user. On the other hand, the learning process keeps running on the background to analyze user's personalizing search hobby, which is accumulated in the database.

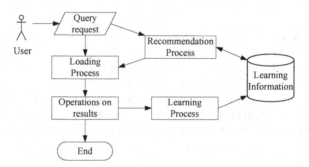

Fig. 1 The Process of Search Engine Based Evaluation Measure.

1. Learning the IR engine matching score with respect to a topic form user's search histories

According to reference [2], in personalized search systems, the user modeling component, which builds user profile, can affect WebPages search in three phases, i.e. retrieval process, query modification and re-ranking. In this paper, we will use and reform the last one. Here, the main distinct is that it will be used not to re-rank but recommend an appropriate engine. The recommend process takes use of Ontology. Though there exist many Ontologies on the Web, in view of universality we take WordNet[3] as the fundamental Ontology in this paper. Moreover, from a user's action to the engine feedback list, his or her interest (or favor) can be profiled. In addition to the query t has been mapped to a topic, and then it is possible to build a relationship between the topic and the favor. We take IR engine matching score as metric to quantize the relationship volume. The subsequent description explains how to calculate IR engine matching score in two steps.

- Step 1: Calculating user's interest (or favor) on a Webpage

Our measure profiles user interest from search histories. After searched result list is provided, user's subsequent browsing operations expose his interest. We consider three operations are important, namely mouse action, browse time and total search time. From them, we can know how much a Webpage p_j user is interested in, under the topic (mapped to a node of Ontology and depicted in N) with respect to t. The relationship is formalized to a triples: $<$ User, Topic, Webpage $>$. With time pass, system will collect many pieces of triples and then get the profile.

A score will be assigned to every triple to quantize the users' interests. For p_j, the more mouse action, longer the browsing time, and shorter total search time means the more interested in it. In the other words, the score of the triples must be higher. The total score is calculated by formula (1):

$$s_{p_j} = \alpha_1 \cdot A_{mouse} + \alpha_2 \cdot T_{browse} + \frac{\alpha_3}{T_{ter\,min}} \tag{1}$$

Where s_{p_j} is the score of <user, N_t, p_j>. A_{mouse} is the value of mouse action user performs on p_j. For each mouse action, A_{mouse} is accumulated 1. T_{browse} is the time user browses p_j. $T_{ter\,min}$ is the total time from browsing of p_j to the termination of the search mission. $\alpha_1, \alpha_2, \alpha_3$ is the weight of $A_{mouse}, T_{browse}, T_{ter\,min}$ respectively, $\alpha_1, \alpha_2, \alpha_3 \geq 0$, $\alpha_1 + \alpha_2 + \alpha_3 = 1$. For the convenient sake, only if $s_{p_j} > \delta$, where δ is threshold, the score of p_j will be saved into a array $S = \{s_1, s_2, s_3, \cdots\}$, $s_1 > s_2 > s_3 > \cdots$, and used in the following. Obviously, S implies user's profile.

- Step 2: Calculating an IR engine matching score to the topic.

As mention previously, the WebPages in S are user's favorite ones. So, we deem that an IR engine is more appropriate for the retrieval mission of t, the higher rank they in its result list. Based on this idea, the IR engine matching score is designed, the description is as following. Using t to carry out search mission in an IR engine waiting for the evaluation and get its search result list. Sum up the ordinal number of the favorite WebPages in the list, we can get the score. For an engine, the total IR matching score is computed with formula (2):

$$s_{IR}^t = \beta \cdot \sum 1/R_{p_j} \qquad (2)$$

Where s_{IR}^t is the matching score of engine IR, such as Google, with query term t. R_{p_j} is ordinal number of web page p_j in the list, $p_j \in S$. β is the weight, $1 \geq \beta > 0$. The matching scores of every available engine will be calculated and recorded (or updated) into a database in the form as table 1 in turn.

Table 1 The Matching Score of IR

	IR_1	IR_2	IR_3	...
N_{t_1}	$s_{IR_1}^{t_1}$	$s_{IR_2}^{t_1}$	$s_{IR_3}^{t_1}$	...
N_{t_2}	$s_{IR_1}^{t_2}$	$s_{IR_2}^{t_2}$	$s_{IR_3}^{t_2}$	...
N_{t_3}	$s_{IR_1}^{t_3}$	$s_{IR_2}^{t_3}$	$s_{IR_3}^{t_3}$	...
...	...	...	...	...

The row is available IR engine, and the column is the nodes with respect to query occurred in user's search histories. The engine, which has the maximal matching score in the column of node N_t, is named the matching engine of term t. For rapid calculation, table 1 may be transformed into index form in machine memory.

There exist two cases in the matching scores record, namely appending and updating. 1) Appending: the query term t doesn't appear in the table, we just insert record (N_t, s_{IR}^t) into the table. 2) Updating: (N_t, s_{IR}^t) has already existed in the table, we will update it with the average value by the formula (3):

$$s_{IR}^t = \sum (s_{IR}^{t\,*} + s_{IR}^t)/n \qquad (3)$$

$s_{IR}^{t\,*}$ is a new score which point to a triples same as s_{IR}^t, n is learning times.

The learning process goes constantly in the background. The more profile is, the more appropriate engine will be recommended in the following step.

2. Calculating matching score of IR engine with respect to a user query

In this modular, the most appropriate IR engine to a user's new query q will be recommended. The score calculations are mainly implemented according to the IR matching score in table 1, which has been gotten in the learning process above.

If q has been recorded in table 1, we may directly get the engine which has the maximal score. In case q has not been learned in table 1, an Ontology based similarity calculation is implemented so as to indirectly get the engine. Before the calculation, q also must be mapped to the corresponding node of WordNet. It is easy to carry out by comparing q with the glosses of node of WordNet. Here, we assume that q is mapped to node N_q.

On the Ontology, WordNet, if two nodes are closer, it means they are more relevant or similar Some WordNet-based relatedness measure has been testified in [4]. So the similarity between N_q and N_t in T may be calculated by formula (4):

$$sim(N_q, N_t) = -\log \frac{\min[len(N_q, N_t)]}{2D} \qquad (4)$$

Here, sim is the similarity of N_q and N_t, $len(N_q, N_t)$ is path length between N_q and N_t, $len(N_q, N_t)|_{N_q = N_t} = 1$, D is the depth of N_q and N_t on the Ontology. Our measure will take the matching IR engine of the Node, which has maximal similarity with N_q in the table 1, as the recommend IR for q.

3. Loading the most appropriate IR and search result list

The browser, equipped with our measure, will seamlessly load the recommended IR engine and implement the search. Then user may get a research result list in an interface which he is familiar with and the integrity is also protected. Obviously, different will be led to uncertain engine.

4   Evaluation Experiments and Results

In order to test the measure, we built up an application based on the measure to implement some experiment missions. Two famous IR engine, Google and Baidu, are selected to the test. The software is programmed with IDE VC++6.0, WinInet API, and Ted Pedersen's WordNet-Similarity tools [5]. We test the measure from two aspects, i.e. degree of satisfaction of users and performance.

In the satisfaction degree experiment, we test users' satisfaction to the measure. 20 undergraduates whose major respectively are business, science, education and medicine are samples. The degree is defined as: $Sat = Num_{right} / Num_{total}$. Sat is the satisfaction degree, Num_{right} is the number of the recommend IR which is just user want, and Num_{total} is the total number of test.

The test lasts 15 days and average satisfaction degree is given every day. The result is shown in Fig 2. From it, we can know that the overall tendency is continually growth. With users' favor information accumulating, the accuracy of measure also increases, and that is consistent with our design. But in the subsequent phase, the degree tends to be stable (43.8%). That maybe indicate our user profile method is still not good enough and some deep information should be dug out. In conclusion, the satisfaction degree growth tendency and range are basically rational.

Benefit form the design of index table, the calculation speed (including the recommendation and loading process) of our measure is relatively fast (average time is 1.6 sec.). Because the personalization of Web search results is a computationally-intensive procedure, it can not run in real-time now. Especial, if the query is out of WordNet, the calculation is very slow (average time is 10.2 sec.).

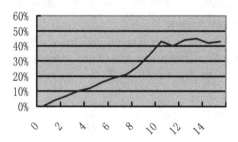

Fig. 2 The average of changing tendency of www.nwpu.edu.

5 Conclusions and Future Work

As our measure can dynamically select most appropriate IR engine for user query request, it can protect original engine integrity so as to improve satisfaction degree of personalizing search. IR is one of the most foundational services, so our method can be integrated into many applications to provide better services.

Undoubtedly, more work need to be successively done to improve it from two aspects, ie deep user profile and calculation performance, in next step. In order to work on semantic level, it is good idea to import the reasoning technique of Ontology into the learning process modular. Furthermore, the challenges of managing a large number of user profiles also need to be considered adequately.

Acknowledgments. This work is part of the research of "Semantic Information Resource Detecting Technique Based on Ontology". The research is supported by the province natural science foundation of Shaan'Xi under grant NO.2010JM8004 and 2010 doctor research starting funding of AFEU Telecommunication Engineering Institute.

References

[1] Pitkow, J., Schütze, H., Cass, T., Cooley, R., Turnbull, D., Edmonds, A., Adar, E., Breuel, T.: Personalized search. Commun. ACM 45(9), 50–55 (2002)
[2] Micarelli, A., Gasparetti, F., Sciarrone, F., Gauch, S.: Personalized Search on the World Wide Web. In: Brusilovsky, P., Kobsa, A., Nejdl, W. (eds.) Adaptive Web 2007. LNCS, vol. 4321, pp. 195–230. Springer, Heidelberg (2007)
[3] Fellbaum, C.: Wordnet: An Electronic Lexical Database. MIT Press, Cambridge (1998)
[4] Hirst, G., St-Onge, D.: Lexical chains as representations of context for the detection and correction of malapropisms. In: Fellbaum, C. (ed.) WordNet: An electronic lexical database, pp. 305–332. MIT Press (1998)
[5] http://search.cpan.org/dist/WordNet-Similarity/

A Study on QoS Routing Scheme for Tactical Ad-Hoc Network

Taehun Kang and Jaiyong Lee

Abstract. Recently developments in the field of telecommunications and network technology has been formed new paradigm shift in defense weapon system. The military also developing the Tactical Information Communication Network (TICN) system to meet the changing aspects of these wars. To equip its function from the battlefield, TICN system have to be received real-time information from the battlefield troop, and the key technologies that support these features is MANET (Mobile Ad-hoc Network) routing protocol. In this paper, basis on the tactical environment of small combat units, by studying on the MANET routing protocol that can be transmitted in real time & reliable information, suggest efficient routing protocol in tactical MANET that can be applicable to advanced digital battlefield.

1   Introduction

The Military has developing the Tactical information & communication network to replace current communication net-work system. To equip its function from the battlefield, TICN system have to be received real-time information from the battlefield troops. but, Current tactical information system has limitations on its functionality because of the information is not transmitted battlefield troops but battalion or higher. Therefore, To receive real-time information, must be introduced communication devices to the battlefield troops. and given the special circumstances of the battlefield, MANET technology will be most effective things to the battlefield troops for communicate of real-time information. In this paper, basis on the tactical environment of small combat units, by studying on the MANET

Taehun Kang
University of Yonsei Seoul Korea
e-mail: 04taehun@yonsei.ac.kr

Jaiyong Lee
University of Yonsei Seoul Korea

F.L. Gaol et al. (Eds.): Proc. of the 2011 2nd International Congress CACS, AISC 145, pp. 161–166.
springerlink.com   © Springer-Verlag Berlin Heidelberg 2012

routing protocol that can be transmitted in real time & reliable information, suggest efficient routing protocol in tactical MANET that can be applicable to advanced digital battlefield. [1][2]

2   Problem Statement

2.1   *Difference between Tactical Ad-Hoc and Common Ad-Hoc*

If, applies the common Ad-hoc routing protocols(Shown before) in the tactical Ad-hoc network, it may cause some problems. The reason is that there are some different characteristics between common Ad-hoc and tactical Ad-hoc. and the details are as follows.

First, the node speed of tactical Ad-hoc is more faster than common Ad-hoc. Second, In tactical Ad-hoc, For tactical information in real-time transmits the streaming data(video, audio) and graphic data as well as sensing data. it means that, in tactical Ad-hoc will take a long time to transmit data. Last, In tactical Ad-hoc, unlike a common Ad-hoc does not take into account the energy consumption.

Tactical environment, the communications device will be using only a few days and moreover, because of the mobility of nodes, energy source can be replaced at any time. Therefore, due to the characteristics of the afore-mentioned tactical Ad-hoc, Routing metric in tactical Ad-hoc should be applied another metric that unlike common Ad-hoc routing. In other words, in common Ad-hoc to minimize the energy consumption, such as a hop count or energy is important metrics of routing. but in tactical Ad-hoc, reliability or delay time will be more important metrics of routing.

2.2   *Connectivity Problem*

Most existing researches of Ad-hoc routing, node 1 selected node 2(Fig 1.-(a)) as a relay node due to the shortest path scheme that located in the outermost of the transmission range. But in this scheme, the node 2 is likely to goes out of transmission range due to the mobility and it can say that is more apt to cause link breaks. Whereas, node 1 selected node 3(Fig 1.-(b)) as a relay node that instead of node 2, even though node 3 has the mobility, which is likely to located within the transmission range of node 1.

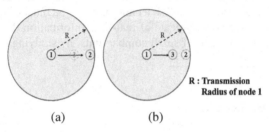

R : Transmission
Radius of node 1

(a)   (b)

Fig. 1 Selection of relay node

In Tactical Ad-hoc, which has high-speed node and take a long time to transmit data because of using the multimedia data to communicate, shortest path scheme for routing may causes some problem that link breaks during the data transmission due to the selected outermost node as a relay node.

2.3   Problem of Route Maintenance

Common Ad-hoc routing protocols uses error messages for route maintenance. When a node detects a broken link to the next hop, it first tries to locally repair the route by broadcasting Route Request(RREQ) packet to the destination, under some conditions in a particular time interval. If it gets Route Reply(RREP) from destination or intermediate route it makes the route active again without the knowledge of source. Otherwise, generates a Route Error(RERR) message that contains a list of unreachable destinations and sends it to related nodes. The source node should reestablish a new route discovery process to detect an alternative route to the same destination.[3] This scheme may causes additional time for route maintenance. and also, that is causing the problem of reliable data transmit due to the delay, data loss and jitter.

3   Proposed Scheme

3.1   Algorithm Overview

We proposed new algorithm that reflect the characteristics of tactical Ad-hoc and can reduce the additional delay time and to guaranteed the reliability.

The operations of proposed algorithm accordingly can be divided into the following two phases. The first phase is to calculate the probability of link-breakage that depends on the nodes position in one hop transmission range. The second phase is to calculate the expected delay time that depends on the probability which is calculated first phase. When finished above two phases, the destination node selects the routing path which can be maximize the reliability and minimize the delay time. This algorithm can be represents mathematically as follows.

$$\text{QoS Routing} = (\text{Max} \sum \text{Link}_{\text{lifetime}}, \text{Min} \sum \text{EDT}) ,$$ (1)

Where, represents the lifetime of the active link that is calculated by the probability of link-breakage. As long as the life-time of the link, guaranteed the reliability. and represents the expected delay time that is calculated by the probability of link-breakage and hop-count.

3.2   Link Breakage Probability

To calculate the probability of link-breakage according to the distance between nodes, we designed fig. 2. Fig. 2-(a) shows that the two node. The blue cycle is that the transmission rang of node 1. Node 2 may located in red dotted arrows(Fig. 2-(b)) after arbitrary time(Δt). and this is can be represented in the form of circle has a radius K (Fig. 2-(c)).

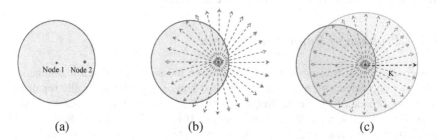

(a)   (b)   (c)

Fig. 2 Position of relay node

Where, K=Δt $\times$ V, V is the maximum speed of node 2. Therefore the probability of link breakage can be calculates as follows.

$$\text{Link breakage probability} = \frac{\int \text{Area}_{\text{outside of circle k}}}{\int \text{Area}_{\text{circle k}}} \tag{2}$$

Where, numerator means the outside area of red circle that does not overlap with the area of blue circle. and denominator means the total area of red circle. The equation (2) allows an easy calculation of the probability of link breakage by calculating the area of circle.

3.3   Analysis of Expected Delay Time

Let p the probability of link breakage in any link, then, the expected delay time can be calculated as follows.[4]

$$E[T_{\text{delay}}]_{\text{End to end}} = \sum_{i=1}^{j} \frac{RTT}{1-p_i} \tag{3}$$

Where, P_i is the probability of link breakage in i-th link and the RTT means Round trip time.

4   Simulation Result

In this section we conduct simulations to evaluate the performance of the devised protocol.

In case of one-hop, is good to selects the nearest node as a relay node for maximize the reliability and minimize the delay time. but if it extended to multi-hop, it's too long delay would be result in, because hop-count is increased.

Therefore, we conduct simulations in multi-hop to find the distance between nodes that maximize the reliability and minimize the delay time. Network distance is 600m. The transmission radius of each node is 200m. The distance between nodes varies from 60m to 200m. Fig. 3 illustrates the simulation design.

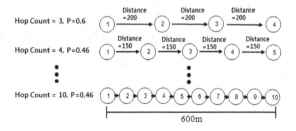

Fig. 3 Simulation design

Where, P is probability of link breakage that depends on the nodes distance. We can see the expected delay as follows

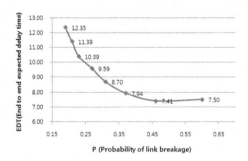

Fig. 4 Expected delay time depends on the nodes distance

Selects the node to send a three-hop case, the expected delay time is 7.50 and the reliability factor is 0.064 {(1-p)hop-count = (1-0.6)3 = 0.064} whereas, Selects the node to send a four-hop case, the expected delay time is 7.41 and the reliability factor is 0.085 {(1-p)hop-count = (1-0.46)3 = 0.085}.

It means that, it is good to selects more closer node as a relay node for maximize the reliability and minimize the delay time rather than using the short path scheme to select a outer nodes.

5   Conclusions and Future work

In this paper, in tactical Ad-hoc, to use the shortest path scheme is not al-ways efficient. Therefore we have proposed more efficient routing scheme, that is good to selects more closer node as a relay node for maximize the reliability and minimize the delay time rather than using the short path scheme to select a outer nodes.

In summary, to provide efficient QoS routing for tactical ad hoc networks, problems such as efficient position-aware information is need to be further investigated. and also, more simulation(by using NS-2)is needed to compared with common Ad-hoc routing scheme.

Acknowledgment. This research was supported by the MKE(The Ministry of Knowledge Economy), Korea, under the ITRC (Information Technology Research Center) support program supervised by the NIPA(National IT Industry Promotion Agency)" (NIPA-2011-C1090-1111-0006)

References

1. Jungsob, Hwag, Haehyun: Design methodology of Military technology for Network centric warfare. In: KIEES 19th, vol. 1, pp. 15–32 (July 2008)
2. Theofanis, P.: A Survey on Routing Techniques supporting Mobility in Sensor Networks. In: IEEE 5th ICMAS (2009)
3. Taddia, C.: An analytical model of the route acquisition process in AODV protocol. In: IEEE WCNC 2005, pp. 779–778 (2005)
4. Draves, R., Padhye, J., Zill, B.: Routing in Multi-Radio, Multi-Hop Wireless Mesh Networks. In: Proc. ACM Mobicom (September 2000)

Business Process Execution Based on Map Reduce Architecture

Chenni Wu and Wei Zhang

Abstract. Map Reduce is a programming model and software framework for writing applications that rapidly process vast amounts of data in parallel on large clusters of compute nodes. Users define a map function with key/value parameters to generate a set of intermediate key/value pairs and a reduce function that merges all intermediate values associated with the same intermediate key. At the same time, the business process execution of web services has the same context with map reduce. It can be integrated with map reduce which will be the basic infrastructure. Business process is automatically parallelized and executed on a large cluster of web services. This paper takes care of the details of partitioning the input data, definite map function, compare function, reduce function and analysis functions. On the other hand, map reduce functions are also responsible to realize statistical analysis to provide services' QoS data. We can finally know that this model can improve efficiency and is easy to realize the dynamic replacement of web services based on the generated QoS data.

Keywords: component, business process, web service, map reduce, QoS, service replacement.

1 Introduction

Map Reduce is a framework for processing huge datasets on certain kinds of distributable problems using a large number of computers (nodes). Map Reduce is widely used in varied fields, especially Google have implemented hundreds of special-purpose computations with large amounts of raw data, such as crawled documents, web request logs, etc [1]. Most of the cases , the input data is large and the computation have to be distributed across sets of computers , parallel programming would be a better method, Map Reduce will be a smart choice[2].

Chenni Wu · Wei Zhang
Software College of Northeastern University, Shenyang, China
e-mail: wcnbear@sina.com, weiz5535@gmail.com

F.L. Gaol et al. (Eds.): Proc. of the 2011 2nd International Congress CACS, AISC 145, pp. 167–172.
springerlink.com   © Springer-Verlag Berlin Heidelberg 2012

We know that web services could apply large results; therefore, we designed a new abstraction that allows us to express the simple computation we were trying to perform but hides the messy details of parallelization, fault-tolerance, data distribution and load balancing in a library. There are few web service executions based on Map Reduce, so we would use Map Reduce architecture to implement web services business process execution.

A booking process will be taken as an example. We would apply a series of map operations to user input and service execution result in order to compute a set of intermediate key/value pairs, and then apply a series of reduce operations to all the values with the same key to calculate response time, delay time, execution time and so on. Those QoS data will be stored in distributed databases and can be used by task node to do the service replacement.

2   Key Points

The main idea of this paper is a simple and powerful interface that enables business process execution automatic parallelization and distribution of large-scale computations, combined with an implementation of this interface that achieves high performance on large clusters of machines [3].

In the framework, map step and reduce step is necessary, during the map step, the master node in the main server takes the input and partitions it up into smaller sub- problem with key/value pairs, each sub-problem would be distributed to those work nodes in task server, the web services on the internet will take place of the sub-problem, a specified web service in a work node may do this again in turn, leading to a multiple-level tree structure [4]. The work node processes the booking requirements into much smaller problem, and passes the answer back to its master node. During the reduce step, the maser node would analyze these answers from the child-worked nodes, it will combine that results which shared the same key (may be just service name). Through a number of algorithms, those results are stored in distributed databases. Next time if we invoked the same type of web service, the QoS data stored in databases can be used to determine which service to choose and even get the best service.

3   Solution Overview

The Map and Reduce functions are both defined with respect to data structured in (key, value) pairs; Map takes one pair of data with a type in one data domain, and returns a list of values with the same key. In this paper, map function is responsible to execute one single service. User input the information and sends it on the internet; the map function will analyze the user input and map them into key/value pairs, and then invoke the web service with the input data. The result may be complex, but we can divide the result into multiple simple key-value pairs and transfer them to master node. The master node receives the result and invokes the next map or reduces function according to the business process blueprint. After that the reduce function will be triggered. A set of 'reducers' can perform the reduction phase – all that is required is that all outputs of the amp operation which share the same key are presented to the same reducer, between the map and reduce stages, the data is parallel-stored or exchanged between nodes. each output pair would be the next input to invoke the next web services, the next period will do this again in turn, until the smallest key/value pairs we called it atom.

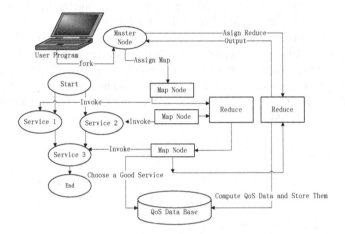

Fig. 1 This is the whole solution architecture. Finally the architecture will execute comparison function to combine the distributed data with the same key, such as using the "sum", "count" aggressive functions to compute the QoS data. The output writer writes the output of the reducer to stable storage.

3.1   Basic Concept

The architecture will trigger the developer self-programming program to each request identified by sessionId. The process will be started from which task node randomly and all task nodes are equivalent. Master node program in main server written by the user, will break information into many small pieces and produce a set of intermediate key/value pairs. The master node chooses one task node and distributes the request. The task which receives the request will invoke the linked web services. The web services will receive the parameter destination to invoke the associated services. In addition, the developer writes codes to fill in a map reduce specification object. According to results from web services, the system then invokes the Map Reduce function, passing it to the specification object.

A reduce worker who is assigned a reduce task reads the contents from the web services using remote procedure calls and sorts them by the intermediate keys. The sorting is needed by the different destinations, if the data storage is too large, an external sort is used. The reduce worker iterates over the sorted intermediate data and for each unique intermediate key encountered, it passes the key and the corresponding set of intermediate values to the user's Reduce function. The output of the reduce function is appended to a final output file for this reduce partition. At the same time, the attach information such as delay time, total time will be stored in the QoS database. When all map tasks and reduce tasks have been completed, the master wakes up the user program. At this point, the Map Reduce calls in the user program to return back to the user code. Next period, the output will be as input for the next circle, the map function is triggered once again; the map rules will be changed. Until the atom level, the resolving will be end.

Reduce function receives all the clients' results, then output the process result to client. At the same time, compute the QoS data into the QoS database. There is an executer in each map function. It is responsible to invoke the linked web service. The service specification (WSDL) and user input will be transferred to executer. The executer will choose the best service from QoS database according to history execution record.

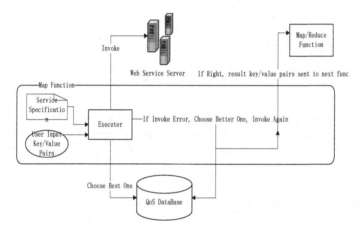

Fig. 2 Qos handler structure. If the execution process has error, executer will choose another best service from QoS database, otherwise it will divide the invocation result into key/value pairs and then transfer them to next map or reduce function.

3.2   *Data and Control Flow*

BPEL (Business Process Execution Language), short for Web Services Business Process Execution Language (WS-BPEL) is an OASIS standard executable language for specifying actions within business processes with web services [5]. Processes in Business Process Execution Language export and import information by using web service interfaces exclusively. From BPEL document, we can read out the web service definition, and then we can invoke web service.

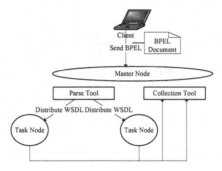

Fig. 3 The whole process of data and control flow structure. Master node is responsible to read control flow and data flow from BPEL document, and control the whole process to execute.

3.3 Experiment

Hadoop [6] is a framework of map/reduce; Axis is a service invocation framework; BPEL4J is responsible for parsing BPEL. With the above frameworks, we can construct the environment to execute business process.

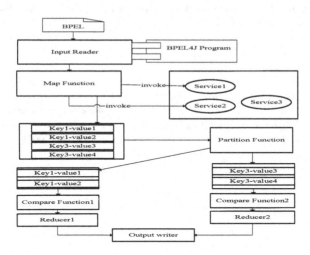

Fig. 4 It shows the data and control flow structure, combines the BPEL 4J and Hadoop.

Through the experiments, we can find that when a single client invokes the booking process, it seems a little slowly. But if there are 4 or more clients invoking the process at the same time, the speed raise up clearly. After invocation, we can find QoS data in database, which can be used for service dynamic replacing or QoS computing.

Table 1 Result table

Task Node Num	Result		
	Service num in process	Total time	QoS data num
1	3	12.5 sec	3
2	3	13.6 sec	3
3	3	10.2 sec	3
4	3	5.7 sec	3

The final result analysis table is just shown above in Table1.

4   Conclusions

This paper purposes a new method about business process execution based on map reduce. It combines web service and Map Reduce. This paper introduces how to implement the method and shows example results. It has high availability and makes business process execution easier and efficient. But this method is still not perfect; we will do our best to make the method better and better.

Acknowledgments. This research work is supported by "Natural Science Foundation of Liaoning Province" (20092006).

References

1. Lam, C.: Hadoop in Action, 3rd edn., pp. 68–73 (2010)
2. Cutting, D., White, T.: Hadoop:The Definitive Guide, 4th edn., pp. 64–69 (2009)
3. Google doc, MapReduce: Simplified Data Processing on Large Clusters (unpublish)
4. Liang, A.: Good at SOA, pp. 271–350. Electronic Industry Press (2007)
5. BPEL, http://zh.wikipedia.org/wiki/BPEL
6. Hadoop, http://zh.wikipedia.org/wiki/Hadoop

Developing Village Knowledge Sharing Capability in Supporting E-Village at Cipadu Village, Tangerang Region

Dian Anubhakti, Basuki Hari Prasetyo, Teddy Mantoro, and Nazir Harjanto

Abstract. Computer Network and Communication are tools, which is an indispensable in business life, since it will characterize economic competitiveness. This paper presents a study on knowledge sharing in groupware in the form of e-village web based communication. The development of a prototype of e-village will be discussed in order to support the national e-village. The Cipadu village, Larangan sub-district in Tangerang Region, Banten Province, Indonesia has been chosen as a pilot project because of its potential and resources. By developing e-village web-based communication, the knowledge holders and knowledge seekers, starting from e-village level, can communicate each other. This communication group will create new knowledge. The e-village knowledge management web application of Cipadu will support the web-based communication in storing, distributing, and accessing knowledge from other village or agencies. This paper proposes the development of Cipadu e-village which is knowledge management based for sharing knowledge. It is a hope that the communication on knowledge is in the form of web-based communication and in action becomes a knowledge sharing culzopture of this village.

Keywords: Computer Network, Data Communication, E-village, Knowledge sharing, web-based communication.

1 Introduction

In the globalization era, which will be marked with economic competitiveness, the use of computer networking and data communication become a character of economic competitiveness. Global competitiveness will very fast arise, but

Dian Anubhakti · Basuki Hari Prasetyo · Teddy Mantoro · Nazir Harjanto
Faculty of Information Technology, University of Budi Luhur
e-mail: {dian.anubhakti,basuki.hariprasetyo}@budiluhur.ac.id,
    teddy@ieee.org, nazirharjanto@yahoo.com

F.L. Gaol et al. (Eds.): Proc. of the 2011 2nd International Congress CACS, AISC 145, pp. 173–179.

generally in Indonesia, it is still in the early stages of this process and still tries to find a model on how to be a competitive country. The effective and efficient way for competitiveness is to develop network and communication. These efforts can be reached by using Information and Communication Technology (ICT) and knowledge management. In the era of knowledge-based economy, the rise of knowledge considered as an increase of competitiveness. Therefore, knowledge becomes an asset of a company/enterprise.

In supporting the Government efforts of developing economic competitiveness we did a survey in Cipadu. Cipadu is a famous village that produces materials for clothing and sells them to Cipulir market. This village is only 10 km form Cipulir market. But, the village people want to enhance their market to Tanah Abang market, the biggest clothing market in Jakarta. It means that they have to develop a communication tools and we propose a web-based communication for knowledge sharing means.

Unfortunately, Cipadu village do not have web-based or data communication, database and computer network facilities, even data and information still be stored manually. It means Cipadu needs e-village knowledge management for sharing knowledge. The problem is that the national e-village, as reference for Cipadu e-village, is not being built, which means there is no standardized data communication and network for each village in Indonesia yet.

Therefore, data Communication should be built in the form of web-based. Cipadu can't connect to the global market to introduce their product, especially for clothing product. Cipadu village people support the e-village of government programs. In order to open the market nationally or internationally, we propose to develop an e-commerce website for Cipadu product and service. This website is not only opening or accessing information, but also to support the village knowledge sharing. Their website advertises products and services, and may provide a means for taking orders. Small businesses could use the knowledge sharing network in selling their product and services.

The development of knowledge management must make provision for both direct human knowledge and indirect human knowledge, as mediated by machines, which extend and enhance the powers of mind or tacit knowledge. Therefore, knowledge management should be directed to make knowledge accessible and useful. This is one of the usefulness of village database, which data, information, and village knowledge easily accessed by everyone. Through externalization and internalization, in the Nonaka' Knowledge Cycle Model", network becomes important for the village people [1]. By developing simple network, the people easily access outside knowledge and know how to do something of value to the village. This knowledge should become skill knowledge. Skill of know-how in knowledge on doing something, become a value and important competencies [1]. By this, it means that valuable skill or knowledge should follow the rapidly changing business environment, in order to come to the market needs. The market needs also means the people of Cipadu needs and mostly means how to communicate with other villages through their web-based communication. They can get benefit from this communication, since they receive valuable new knowledge and also they have the opportunity to create knowledge from the information they receive through daily communication.

The rest of this paper describe the following: section 2 described related work, section 3 discussed framework for knowledge sharing center, section 4, discuss the working group and knowledge sharing and closed by conclusion in chapter 5.

2   Related Work

In order to develop the knowledge sharing tools for communication and enhancing their products, we studied the knowledge sharing approach using network and data communication tools. This paper describes how to develop knowledge sharing and then becomes a culture in this village and shows how to implement this knowledge sharing to become a village culture. The head of the village have to create a regular meeting with all groups of the village people. The meeting consists of the village young group, woman group, the craftsmen group, the farmer group, the trader group, the village NGO, the sport group, the counseling officers and the religion group, to discuss and sharing their knowledge [2]. Second, by developing blog as an Online Consultation Media, the blog can be used for consultation, for instance, for biotechnological agriculture online or chili commodity information. The communication conducted through Admin Blog (admin) and experts, by using synchronous technology (such as a chat online) or asynchronous technology (such as comment and email) [3]. Third, the development of e-village System in Pakistan can be a reference for Cipadu e-village. This benefit of ICT for rural people in Pakistan for uplifting their social, educational, healthcare and economic infrastructure [4] can be an example for the Cipadu village people.

3   Framework for Knowledge Sharing

The data communication in Cipadu village, focus on e-commerce web-based, as e-commerce conduct business on the Web. By using knowledge sharing through the online knowledge web application, village people have broad communications, especially for Cipadu village, other villages, markets, and other government agencies. Today, even the smallest computers can communicate directly with one another, with hundreds of computers on a company network, or with millions of other computers around the world – often via internet [5]. Therefore, in the web, it should also be included posting of knowledge in order to create a new knowledge.

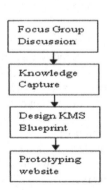

Fig. 1 Knowledge Sharing Process

The knowledge sharing process at Cipadu village, shown in Fig. 1, first the head of the village should develop Focus Discussion Group (FGD) which related to the capture of FGD result in order to design Knowledge Management System and develop prototype of knowledge sharing.

However International Agencies have already reviewed the benefit of Sri Lanka' e- village model [6] and Cipadu village people can learn from this model. The Sri Lanka' e-village model have six main steps, which are 1) Relationship building, 2) Installation of Equipment, 3) Training and Workshops, 4) Implementation of Educational Plan, 5) Other services to the Community and 6) Mesh Technology implementation.

Fig. 2 This knowledge based web service for knowledge sharing purposes (example: current available services, policy and market).

Fig. 3 The sub menu of those menu they need. such as knowledge sharing, knowledge transfer, knowledge implementation and knowledge creation.

In the case of our study, an online web-based communication in Cipadu will fill the gap between consumers and producers and find an easier way for consumers to communicate directly with sellers by using customer-to-provider or a business-to-business approach. In the global economies, business which run with web capability, make the CRM (customer relationship management) a much easier and accessible option. By using this type of web application, customers and sellers can watch when a piece of correspondence been submitted, what commitments were made to respond, when a response actually took place, who is handling the item. Fig. 2 shows the knowledge sharing for current available services, policy and market. Fig. 3 shows the sub menu of the Forum menu. The sub menu on

Knowledge sharing, Knowledge transfer, Knowledge implementation and Knowledge creation are means for communication and sharing their information within Cipadu village people or other people outside Cipadu village, especially in mapping the market demands. Cipadu village people used SMS Gateway in their web service to allow the communication using sms from a mobile phone, as mostly the Cipadu people have and use mobile phone rather than computer.

4 Discussion

The Cipadu village people need high quality human resources such as knowledge-based worker or knowledge-based human to formulae work performances and value-added, and as source of competitiveness. In the globalization era, "global village", give a challenge itself to Developing Countries network and human resources such as project designers and data interpreters who certainly are important knowledge sources [7]. On the other hand, computer or other systems generate substantial and significant knowledge contents.

The knowledge management must make provision for both, direct human knowledge and indirect human knowledge, as mediated by machines, which extend and enhance the powers of mind or tacit knowledge. Therefore, knowledge management is directed to the accessible and usefulness of enhancing their products into innovation. This is one of the usefulness of village database which data, information, and village knowledge easily accessed by everyone. Through externalization and internalization, in the Nonaka's Knowledge Cycle model, network becomes the core of the process [1]. By developing simple network, the people easily access outside knowledge and know how to develop value to the village. This knowledge could become skill knowledge. Skill knowledge worker knows how to do something of value [7]. This skill becomes the basis of core competencies for this village. By this valuable skill or knowledge, Cipadu citizen will follow the rapidly changing business environment, and then they will easily come to the market needs.

The market needs should become the people of Cipadu needs. They benefit it from this communication, since they receive valuable new information and they have the opportunity to create knowledge from the information they receive through daily communication. The building of web communications will focus on Groupware, which will help groups of people work together on creating valuable products and share information over a network. Groupware is a component of broad concept called *workgroup computing*, which includes network hardware and software that enables group members to communicate, manage projects, schedule meetings, and make group decisions [8]. A major feature of groupware is group scheduling, in which a group calendar is published in web-based communication to have a meeting. This meeting is a *knowledge sharing* which regularly conducted in the village and they should discuss in regard for finding solutions, with controversial issues, helping a group with many conflicts to work together, generating and prioritizing options, defining problems, reaching consensus, creating a performance record [8] and village potencies. This knowledge sharing could enhance village people welfare. This web-based communication is one of the tools for knowledge based to implement the Nonaka's knowledge cycles. Knowledge creation can be developed through this web-based communication, since they can develop a knowledge sharing between the knowledge holders and knowledge seekers [1]. The result of the village meeting or sharing knowledge will support the data and information by using the effective and efficient data communication. Data and information are used for supporting the development of new product or services, such as what will the market demand. Then, we expect that this village could become an entrepreneur village and their products or services become village competencies. Their knowledge sharing could be developed as knowledge culture in the village. When the innovation becomes a culture, The knowledge can be a source of innovation.

5 Conclusion

Cipadu village, as a famous village that produces materials for clothing, have a problem in introducing their products and services. The village has no knowledge based application including the data communication, database and computer network. Their data and information are still stored manually. We propose for Cipadu village to develop their e-village knowledge management web-based for sharing knowledge. Even though the national e-village, as reference for Cipadu e-village, is not being built yet, which means there is no standardized data communication and network for each village in Indonesia yet, Cipadu e-village can be started as a potential village leader in the future. This study presents the framework for knowledge sharing for Cipadu e-village uses a computer network and data communication such as sms gateway, in supporting e-village knowledge management, including:

- Knowledge transfer and Knowledge creation and innovation
- Web-based communication.
- e-commerce Cipadu website.

References

1. Nonaka, I., Takeuchi, H.: The Knowledge-Creating Company: How Japanese Companies Create the Dynamics of Innovation. Oxford University Press, Oxford (2007)
2. Harjanto, N.: Knowledge Sharing and Creating In Remote Areas Case Study, Pasir Region, East Kalimantan (2010)
3. Hasad, A., Nasution, N.H., et al.: Membangun Blog Sebagai Media Konsultasi Online Dalam Perspektif Manajemen Pengetahuan (2011)
4. Hussain, A., Qazi, S.: Development of E-Village in Pakistan (2010)
5. Shelly, G.B., Cashman, T.J., et al.: Discovering Computers, A gateway to Information, Web Enhanced Complete. Thomson Course Technology, Boston (2007)

6. The Information and Communication Technology Agency Of Sri Lanka, The Information and Communication Technology Agency, ICTA (2005)
7. Housel, T., Bell, A.: Measuring and Managing Knowledge. The McGraw-Hill Book Co., Singapore (2001)
8. Rumizen, M.C.: The complete IDIOT'S guide to Knowledge Management. CWL Publishing Enterprises, Madison (2002)

A State Oriented Buffer Control Mechanism for the Priority-Based Congestion Control Algorithm in WSNs

Jeongseok On, Yoonpil Sung, and Jaiyong Lee

Abstract. In the multiple-event WSNs, each event has different importance and most of the congestions are caused by the burst events with high importance. Therefore, the priority-based congestion control algorithm is required to WSNs. Most of the existing algorithms to control the congestions of the WSNs do not consider the priorities of the events. In addition, the existing algorithms rely on the rate control of source nodes using the backpressure mechanism but it is not efficient for the dynamic applications (ex. Tracking application). Therefore, we proposed a new algorithm that uses the queue state management and duty-cycle adjustment. This new algorithm tries to prevent the congestion previously and adjusts the duty-cycle to guarantee the reliability of the important events. The simulation results show that the proposed algorithm performs better than the existing algorithm in both of the static case and dynamic case.

Keywords: WSN, congestion control, priority-based, backpressure, queue state management.

1   Introduction

WSNs (Wireless Sensor Networks) [1] are event based systems that rely on the collective effort of several micro sensor nodes. Sensor nodes which are set up for various purposes are forming the network by themselves using auto self-configuration. WSN can be easily set up because sensor nodes use the wireless communication. Therefore, WSN would be very useful if the target place is not comfortable for people to install the network or when the temporal data collection

Jeongseok On · Yoonpil Sung · Jaiyong Lee
Department of Electrical & Electronic Engineering Yonsei University, 134,
Shinchon-Dong, Seodaemun-Gu, Seoul, Korea
e-mail: onchuck@yosnei.ac.kr

F.L. Gaol et al. (Eds.): Proc. of the 2011 2nd International Congress CACS, AISC 145, pp. 181–190.
springerlink.com   © Springer-Verlag Berlin Heidelberg 2012

is required. With these convenience and efficiency, WSN has very broad application area including a military, medical service, environment and home network etc. [1][2][3]. From these characteristics of the WSNs, a variety of new wireless communication protocols are developed and researched. Especially, the main research theme is about prolonging the network lifetime. Therefore, various energy efficient routing algorithms or MAC protocols are researched. However, the research themes of the WSN become very diverse because the WSN is used for a variety of applications. For example, the QoS of WSN, energy harvesting and heterogeneous sensor network are on the rise as new research themes. The congestion control problem of WSN is also one of the research themes. Congestion can occur in WSNs due to some reasons. First, the problems are on the topology of WSNs. The topology of WSN is not fully under control. In addition, because there are many source nodes and few sink nodes, a large amount of traffic might contend for one or few links or nodes, which become bottlenecks of the whole network. There are always congestions in bottlenecks. Second, in WSN, resources are limited and shared. Sensors have wireless shared media and they have to compete with neighbors to take the shared media. These competitions along the path to the sink are continued and make congestions. Third, recently, the application areas of the WSNs become wider due to advances in sensor and communication technology. New applications such as tracking or observation require the image or video data. These applications require high throughput and low latency. In addition, the applications with a variety of sensors which are deployed in wide area are increased. These development trends can make more congestion in the WSNs. Therefore, congestion control is positively required to commercialize WSN in many applications. Compared with wired networks, different types of the congestions occur in WSNs. In addition, WSN will exhibit its own phenomena when the congestions occur. Existing protocols to control the congestion of wired networks like TCP are not suitable for the WSN. [4]

2   Related Work

CODA [1] proposes the novel congestion detection and congestion control algorithm in WSN. [6] CODA detects congestion based on buffer occupancy as well as wireless channel load. It measures the channel loading and compares with a maximum throughput threshold applicable to CSMA. The formula (1) shows the maximum throughput of CSMA.

$$T_{max} \approx \frac{1}{\left(1 + 2\sqrt{\beta}\right)} (for \ \beta \leq 1), where \ \beta = \frac{\tau C}{L} \qquad (1)$$

CODA has two algorithms to control the congestions. First, 'open loop hop-by-hop backpressure' handles the local congestion by backpressure message and rate adjustment. Second, 'closed loop multi-source regulation' is for the persistent congestion. The mechanism is conducted by the sink node and the sink node transmits the ACK packet as the feedback. A transport protocol called ESRT is proposed in [7]. ESRT assumes the event-to-sink transmission rather than end-to-end transmission type which is generally considered in existing algorithms. That is, all sensor nodes in the event area become the source nodes, therefore basic transmission concept in [7] is the transmission between many source nodes and one sink node. In ESRT, congestion control is conducted by controlling the generating rate of source nodes. The sink node regulates the reporting frequency of source nodes to make the system operate in the optimal region. In ESRT, sink nodes dominate the congestion control. CCF [Congestion Control and Fairness] assumes the tree topology and it provides the fairness and the congestion control algorithm. [8] Congestion information, that is packet service time in CCF, is implicitly reported. CCF controls congestion in a hop-by-hop manner and each node uses exact rate adjustment based on its available service rate and the size of sub-tree. CCF guarantees simple fairness. However the rate adjustment in CCF relies only on packet service time which could lead to low utilization when some sensor nodes do not have enough traffic or there is a significant packet error rate.

3   Problem Statement

As we saw in related work, most of the existing algorithms do not consider the priorities of the events. However, we can see that the congestion of the WSN is caused by burst events with the high priority in research motivation. In addition, we can see that the existing algorithms based on the backpressure mechanism are not efficient to the dynamic applications. Therefore, we propose a new priority-based congestion control algorithm using the queue state management. The proposed algorithm should be efficient not only for the static applications but also for the dynamic applications

4   Algorithm Description

4.1   Overview

The proposed algorithm assumes a node with two queues, the RT [Real Time] queue and the NRT [Non Real Time] queue. All events are separated in two types, one is RT and the other is NRT according to their priority classes. The NRT queue is filled with lower priority events packets. The NRT events are for example, periodic events such a temperature, intensity of illumination or humidity. On the other hand, the RT queue is filled with high priority events. Those are events of emergency such a tracking, fire or earth quake monitoring etc. The proposed algorithm prevents the congestion by controlling the scheduling weights of the RT and NRT queue. If the congestion occurs in either the RT or NRT queue, the congestion control algorithm is triggered independently. In persistent congestion, the backpressure mechanism is triggered and it controls the source rates.

4.2   Node Model

Figure 1 shows the architecture of a sensor node. The total arrival rate into the sensor node is r_R, which is sum of the transmission rate from child node (r_t) and sensing rate by itself (r_S). The arrival packets enter either the RT queue or the NRT queue according to the priority of their events. We assume that there are n-priority classes. Packets from priority 1 to [n/2] enter the RT queue and packets from priority [n/2] +1 to n enter the NRT queue respectively. The rate of entrance in the RT queue is r_{R_RT} and the rate of entrance in the NRT queue is r_{R_NRT}. Packets in the RT or NRT queues are scheduled in MAC layer by their weight and relayed to the next hop node. The weight of the RT packet is W_{RT} and the weight of the NRT packet is W_{NRT}. The outgoing rate at the MAC layer is r_O.

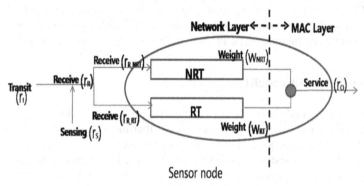

Fig. 1 Node Model

4.3   Queue States

There are three states of the queue according to the number of packets in each queue. The states of each queue are composed of non-congestion state, pre-congestion state and congestion state. The states of two queues (RT and NRT) are independent each other. N is the number of packets in the queue and there are two threshold of queue N_D and N_{th}. N_D is the desired number of packets. N_D is an application dependent value and the network works well without any problem if N is smaller than N_D. Nth is the threshold number of packets. It is identified as a congestion state when the N exceeds N_{th}.

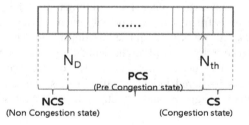

Fig. 2 Queue States

- Non-congestion state: ($N \leq N_D$): In this state, no congestion control algorithm is required.
- Pre-congestion state: ($N_D < N \leq N_{th}$): It is identified as a pre-congestion state when the N is between N_D and N_{th}. In this state, congestion could occur, if the preventive algorithm did not operate.
- Congestion state: ($N > N_{th}$): In this state, appropriate congestion control algorithm is required.

4.4 Algorithm in the Pre-congestion State

If the RT and NRT queues are in non-congestion state, no algorithm is executed. When the N is between the N_D and the N_{th} in either the RT or NRT queue, the adaptive weight adjustment algorithm is triggered. This algorithm is executed before the congestion occurs and it operates to prevent the congestion of the RT and NRT events.

Adaptive weight adjustment algorithm: If either RT queue or NRT queue enters the pre-congestion state, the adaptive weight adjustment algorithm would begin. This algorithm controls the scheduling weight according to the number of packets in each queue. That is, if the NRT queue is in non-congestion state and the RT queue is in pre-congestion state, the W_{RT} is raised and if the RT queue is in non-congestion state and the NRT queue is in pre-congestion state, the W_{NRT} is increased. The value NRT is the number of packets in the RT queue and N_{NRT} is the number of packets in NRT queue. Following formula is basic principal.

$$W_{RT} + W_{NRT} = 1, \quad \frac{W_{RT}}{N_{RT}} = \frac{W_{NRT}}{N_{NRT}} \tag{2}$$

The weights, W_{RT} and W_{NRT} are updated every transmission in pre-congestion state. However, if the RT queue becomes the congestion state, all these algorithms are ignored. The W_{RT} is set to 1 and all scheduling weight is dedicated for the RT queue. From this, we can see that the RT queue has priority to the N_{RT} queue.

4.5 Algorithm in the Congestion State

1. Congestion state of the NRT queue

In this case, the random discarding algorithm is executed in the NRT queue. For the NRT queue, the N_{th} value is the same as NRT queue size. Therefore NRT queue is congestion state when the NRT queue is full. When the NRT queue is in the congestion state, all packets in the queue have the dropping probability that is inverse proportional to the priority class. For example, if there are packets with priority 4, 5, 6, and 7 in NRT queue, each packet has the dropping probability P_{drop_4}, P_{drop_5}, P_{drop_6} and P_{drop_7} with $P_{drop\_4} < P_{drop\_5} < P_{drop\_6} < P_{drop\_7}$. The NRT queue receives the new packets after discarding packets based on those probabilities.

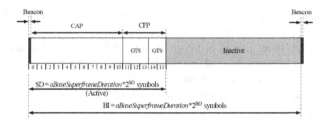

Fig. 3 Superframe structure

2. Congestion state of the RT queue

As mentioned in the overview, whole scheduling weight is dedicated for the RT queue by setting the W_{RT} to 1 when the RT queue is in the congestion state. The duty-cycle adjustment algorithm is triggered to support the throughput of the RT event. Duty-cycle is the proportion of time during which a component, device, or system is operated. Duty-cycle is determined based on the superframe structure. The superframe is the basic structure for the IEEE802.15.4 MAC standard and Figure 6 describes the structure of superframe.

Duty-cycle Adjustment algorithm: Duty-cycle is directly associated with the throughput. The portion of active period is approximately proportional to the throughput but it is inverse proportional to the energy efficiency. Therefore duty-cycle adjustment algorithm is executed only when the RT queue is in the congestion state. Only nodes along the route from the source nodes to the sink node increase the duty-cycle.

Duty-cycle is dependent on the SO (Superframe order) value. Therefore when a node becomes the congestion state, it increases its SO value and transmit SO information by piggybacking to the data packets to the parent node. This mechanism is continued to the sink node. Finally, nodes in the path from the source nodes to the sink node have increased duty-cycle. When the congestion state is over, all nodes in the path recover the original duty-cycle.

3. Persistent congestion state of the RT queue

Figure 4 is the structure of the basic data frame in the MAC layer. We add the 2 fields, priority and CN (Congestion Notification). The priority field can be from 1 to n and it is used to select the NRT queue or RT queue. The CN field is the notification of congestion. If the RT packets are in the congestion state node, the CN bit is set to 1. If the congestion state is continued for a long time, the throughput of the network is decreased and the energy efficiency is also decreased due to the increased duty-cycle. In this case, the backpressure mechanism is triggered and it controls the source rates. If the sink node receives the packet with the CN setting to 1, a timer in the sink node starts. The value t_{th} is the duration threshold and if the timer exceeds the t_{th} then, the sink node determines the persistent congestion state and transmits the backpressure message to the source nodes. When the CN bit becomes the 0, the persistent congestion state is over.

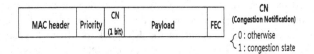

Fig. 4 Data frame

5   Performance Evaluation

In this part, we compared the proposed scheme with the algorithm using backpressure message (CCF) and no congestion control scheme. The simulation environments are the same as those in problem statement. We defined the reliability as Reliability R = (received packets by sink) / (generated packets by source nodes) and compared the performance based on the reliability. Total simulation time is 100 seconds. Start time of RT source nodes is 0 second and stop time is 15 second.

5.1   Simulation Environment

Simulation is done by MATLAB 7.0. Each node has RT queue and NRT queue. The size of each queue is 50 packets. Routing path from source nodes to the sink node is predetermined. N_D of the RT queue is 10 and N_{th} of the RT queue is 30. On the other hand, N_D of the NRT queue is 10 also and N_{th} of the NRT queue is 50. Topology is Figure 5.

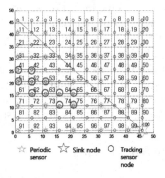

Fig. 5 Topology

5.2   Reliability Comparison

Simulation is conducted in two cases: static case and dynamic case. In this part, the static case means that the source nodes are not changed. For example, only four nodes 41, 42, 51 and 52 generate the packets from 0 second to 15 seconds in this simulation in figure 5. On the other hand, dynamic case means that source nodes are changed as time goes by. Tracking application can be an example. In this simulation, node 41, 42, 51, 52 generate the RT packets from 0 second to 5 second and node 52, 53, 62, 63 generate the RT packets from 5 second to 10 second and node 64, 65, 74, 75 generate the RT packets from 10 second to 15 second. The packet generation rate is 1 packet/minute for the NRT event and 4 packet/second for the RT event. Figure 6 indicates that the both of backpressure algorithm and the proposed algorithm increase the reliability of the RT event fairly in static case. However propose algorithm shows better performance about 10%. On the contrary, proposed algorithm shows that the reliability of the NRT events is lower than the other algorithms because the proposed algorithm gives the priority to the RT events. Figure 7 shows that the dynamic case. Figure 7 shows that the performance of the CCF is decreased in the RT event. In this case, the backpressure algorithm is not efficient to guarantee the reliability of RT events because the backpressure algorithm is not efficient when the source nodes are changed as time goes by. On the other hand, the proposed algorithm maintains the high reliability in the RT events.

5.3   Reliability According to the Rate of the RT Events

In this section, we measure the reliability variation as the change of the generation rate of RT events to investigate the performance according to the congestion level. Figure 8 shows the reliability of the RT event in the static case. The proposed algorithm and the backpressure algorithm show the similar phase until 6 packet/second but the proposed algorithm shows better performance when the rate is above the 6 packet/second. Figure 9 shows the reliability of the NRT event. When the rate is low, the proposed algorithm shows the lowest reliability because it gives the priority to the RT event. However, when the rate is above the 4 packet/second, the proposed algorithm and the CCF represent the stable reliability and the no algorithm's reliability is decreased rapidly. Next, Figure 10 shows the reliability of the RT event in the dynamic case. In this case, the proposed algorithm outperforms the other algorithm and shows the stable reliability. The CCF shows worse performance than the static case. Figure 11 represents the reliability of the NRT event in the dynamic case. In this case, the CCF shows the highest reliability.

5.4   Reliability of the NRT Events

In the previous part, we can see that the proposed algorithm is not very good for the reliability of the NRT events. Therefore, in this part, we measure the reliability of the NRT events based on the priority classes. The proposed algorithm uses the random discarding algorithm and this algorithm gives the priority to the high priority class. Figure 12 shows that the reliability of the NRT events. There are 3-priority classes and the packet generation rate of the NRT event is set to 5 packets/minute. The reliability of the class 1 is over the 90% although the reliability of the class 3 in lower than 60%. This result means that the proposed algorithm contributes the reliability of the NRT events guaranteeing the reliability of high priority class.

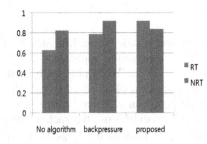

Fig. 6 Reliability in static case

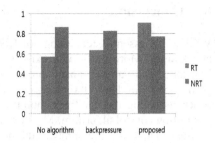

Fig. 7 Reliability in dynamic case

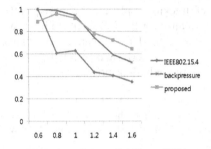

Fig. 8 Variation of the reliability of RT events in static case

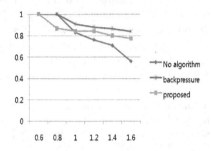

Fig. 9 Variation of the reliability of NRT events in static case

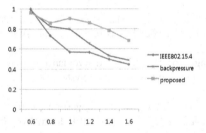

Fig. 10 Variation of the reliability of RT events in dynamic case

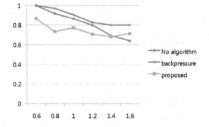

Fig. 11 Variation of the reliability of NRT events in dynamic case

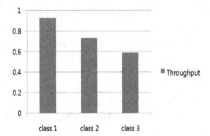

Fig. 12 Reliability of the NRT events

6   Conclusion

WSN is the essential technology of RFID/USN and it will be used in variety of applications. More traffic will exist on the WSN because the application area is broadened and various sensors are developed. To use the WSN practically, congestion problem due to the increased traffic must be controlled efficiently. In this paper, assuming the intelligent building, we proposed an efficient priority based congestion control algorithm. Most of the existed algorithms use the backpressure message. Therefore, we compare the proposed algorithm with no congestion control algorithm and the CCF algorithm which use the backpressure message. Simulation is done in two cases. One is the static case which means that the source nodes are not changed and the other is the dynamic case which means that the source nodes are changed as time goes by (cf. tracking application). Consequently, the proposed algorithm shows the best performance for the RT events in both cases. In particular, the proposed algorithm shows good performance in dynamic case. The proposed algorithm cannot show the best performance for the overall NRT reliability but it guarantees the reliability of the high priority class.

Acknowledgments. "This research was supported by the MKE (The Ministry of Knowledge Economy), Korea, under the ITRC (Information Technology Research Center) support program supervised by the NIPA (National IT Industry Promotion Agency)" (NIPA-2011-C1090-1111-0006).

References

[1] Ee, C.T., Bajcsy, R.: Congestion control and Fairness for Many-to-one routing in sensor networks. In: Proc. Second Int'l Conf. Embedded Networked Sensor Systems (ACM SenSys 2004) (November 2004)
[2] Wan, C.-Y., Eisenman, S.B.: CODA: Congestion Detection and Avoidance in Sensor Networks. In: Pro. First Int'l Conf. Embedded Networked Sensor System (ACM SenSys 2003) (2003)
[3] Wang, C., Sohraby, K., Daneshmand, M., Hu, Y.: Upstream Congestion Control in Wireless Sensor Networks Through Cross-Layer Optimization. IEEE Journal on Selected Areas in Communications 25(4)
[4] Sankarasubramaniam, Y., Akan, O.B.: ESRT:Event-to-Sink Reliable Transport in Wireless Sensor Networks. In: MobiHoc (2003)
[5] Patro, R.K., Raina, M., Ganapathy, V., Shanaiah, M., Thejaswi, C.: Analysis and improvement of contention access protocol in IEEE802.15.4 star network. Mobile Adhoc and Sensor Systems (2007)
[6] Callawa, E., et al.: Home networking with IEEE 802.15.4: A developing standard for low-rate wireless personal area networks. IEEE Communication Magazine 40(8), 70–77 (2000)
[7] Sung, Y., Youn, M.J., Lee, J.: A precautionary congestion control scheme in WSNs: Cross layer approach. In: ITC-CSCC, Shimonoseki, Japan (2008)
[8] Youn, M., Lee, J.: A Cross Layered Protocol Using Back-off Time in Wireless Sensor Networks. UKC (2007)

Survey on P2P Traffic Managements

Yingru Luo

Abstract. P2P applications have become more and more popular in the Internet. They have consumed most of the Internet bandwidth, and affected the performance of traditional Internet applications. Many researchers proposed various technologies to manage P2P traffic, such as P2P traffic control, P2P caching technologies. This paper introduced P2P technology and the factors that affecting P2P traffic distribution, and then surveyed the research efforts on P2P traffic management technologies.

1   Introduction

With the development of the Internet and Wide Band technology, peer to peer (P2P) network has rapidly developed for its unique technical advantages. The P2P traffic occupies more and more proportion of the network bandwidth. The large-scale applications of P2P have brought many challenges for the Internet service provider (ISP). The most important one is that the network bandwidth is been using up. To deal with this problem, many P2P management systems had attempted to control the internet. But P2P also have successfully avoided the control of P2P management system by means of encrypted communications, protocol disguising , encrypted target address, and so on [1].

Since P2P traffic is created between peers and the selection of peers is very flexible, it could to improve traffic distribution by improving the selection mechanism in P2P network. There are attempts to join the P2P cashing to reduce the P2P traffic of the major chains between the networks. Techniques such as P4P [2], Oracle [3], P2P caching [4] are the series of explorations in these areas.

2   Factors Affecting Traffic Distribution

It is different from the process of creating P2P traffic in different P2P systems. The communications between the parties opened file transfer, video chat, telephone and other communications activities according with demand in instant

Yingru Luo
Chinese People's Armed Police Forces Academy, LangFang, China
e-mail: luoyr99@21cn.com

F.L. Gaol et al. (Eds.): Proc. of the 2011 2nd International Congress CACS, AISC 145, pp. 191–196.
springerlink.com   © Springer-Verlag Berlin Heidelberg 2012

communication systems. In the file sharing and online video P2P systems, the node gives service demand. The traffic between nodes in P2P network is not a random distribution. In addition to be influenced by network environment, it is affected by the technical factors of the P2P system itself.

2.1   Strategy of Resource Blocking

In multiple-source data transmission system, resources are transmitted in blocks. When there is more than one node has a block of file, the download node may select. But it will download all the sub-block from only one upload node. A block's size is an important parameter of a P2P system. It is showed that, when a file is small, smaller block will reduce the resources download time, increase the activity of the P2P network [5]. For instance, if the sharing document's size is 5Mbyte, the most ideal block is 16kB. But when files is large, the block may not good as small as possible. For example, the most ideal block's size is 256kB for a 100Mbyte sharing file. The smaller blocks will slow down the performance of the system,

2.2   Blocks Selection Strategy

There are some strategies to determine the priority to download, Including Random block Selection method and Rarest Firs method. In BT systems, the blocks are downloaded by random or the first one is downloaded by random, the others by rarest first [6]. In PPlive systems, the peer will priority download video and audio data near the user, the other part of the movie will download by rarest first [7]. It had shown that rarest first strategy demonstrated a better system performance than random piece selection strategy [8].

2.3   Peers Provision and Selection Strategy

In a P2P system, the provision and select strategies of peers is sometime neglected. If there are many users, the peers of a P2P system will be widely distributed in the entire Internet. The selection will affect the P2P traffic distribution. BT system uses a choking algorithm to manage upload, so that each peer gets a higher download speed. In PPlive system, one peer will request 8-20 other nodes for download blocks at the same time. When the speed of data transmission between the nodes is not enough, the resource server of PPlive Company will provide the resources download service [9].

3   P2P Traffic Management

In order to challenging the network bandwidth consumption brought out by P2P traffic, ISPs tried to solve the problem by increased input and increased bandwidth. ISPs also want to find a suitable technical to limit P2P traffic. In accordance with the general P2P characteristics, the P2P traffic management research can be divided into three aspects: First, it blocks up the P2P traffic; Second, it is to set up cache equipment between domains and backbone links; Third, it is control the producing process of P2P traffic, leading more P2P traffic on the local network, and reducing the traffic between domains and backbone network.

3.1   P2P Traffic Identification and Blocking

The P2P traffic identification technology can be divided into three categories: techniques based on transport-layer port, Deep Packet Inspection(DPI) and techniques based on the characteristics of the traffic. The technique based on transport-layer port is the simplest and efficient method. ISPs can easily identify and restrict the P2P traffic. DPI detect P2P traffic according to different signatures in the payload of data packet [10]. It is generally recognized that DPI is a relatively mature technology. Many companies have released products for P2P traffic detection, such as Cisco's PDML and Juniper's netscreen-IDP, etc. But DPI could only detect P2P traffic based on well-known features of P2P applications, DPI algorithm is powerless to the new types of P2P traffic or encrypted P2P traffic, so it is not appropriate settled in high-speed backbone link.

Flow-based detection technique is identification of P2P traffic based on the characteristics of flow topology features and flow characteristics. For instance, detection based on P2P network topology feature is based on the diameter of P2P network is bigger and the node has both Server and Client features [11]; detection based on flow pattern is based on the extensive connections and used both UDP and TCP features to identify P2P traffic [12]; detection based on the statistical characteristics of flow is based on the demographic characteristics of the special characteristics using data mining or machine learning techniques to identify P2P traffic [13]. Currently, this type of flow identification and detection technology is still in research and exploration phase.

Each of the 3 types of P2P traffic identification has its own advantages and disadvantages. Table 1 presents the comparison of the three identifying method. Using traffic blocking to control the P2P traffic made the ISPs and P2P systems being in opposition.

Table 1 Comparison of P2P traffic identification

Technical basis	Advantages	Disadvantages
Port number	Easy to accomplish and deployment	Expiration if the port number is changed
Packet features	Higher accurate identification rate; relatively mature technologies	Lapse if not kowing the packet features Algorithm has highly spatial and temporal complexity
Flow characteristics	The versatility is good It can distinguish each kind of the P2P class, include encryption or emerging P2P current capacity	Indentification accurate rate is not high or algorithm is complex At the stage of exploration

3.2   P2P Caching

In P2P network, the traffic is often generated by same content, especially some hot resources, and formed the main traffic of the network. Therefore, people had attempted to apply the traditional cache technology in P2P.

The basic concept of P2P caching technology is cashing P2P content on the edge of the network [4]. It redirect the match content of P2P traffic to the cache server through monitoring the export network traffic of the metropolitan area network, and the cache server provides the user with content services. At present, the main purpose of this technology is to restrict the P2P traffic in the metropolitan area as far as possible, to the effect that relive the P2P traffic domination in the export of MAN and to ensure that broadband users' experience at the same time. P2P caching systems are usually composed by units of flow capture, protocol processing, cache inspection, content storage, transponder modules. Because there is a considerable uncertainty in P2P applications, Cache Sever need huge computing power and storage capacity.

Although P2P Caching equipments have been deployed in the network, it is not a good solution for P2P bandwidth consumption. First, P2P Caching equipments need to detect and identification P2P traffic, which need higher computing power, storage capacity, equipment bandwidth than traditional Web Caching. Second, P2P traffic is difficult to detect. Third, P2P caching equipment is difficult to adapt the rapid development. Forth, more and more P2P system used the manner of encrypted transmission, which makes cache technology useless no longer.

3.3   P2P Traffic Grooming Localization

Researchers are exploring technology of traffic grooming. We know that P2P can choose different peer to transmit data, but the underlying network architecture was ignored. It could not choose peers that make the network distance shorter and the cost cheaper for data transmission. Now, people try to take the underlying network of the P2P system into account.

The concept that makes the P2P traffic localization to improve system performance was put forward [14]. It was found that 70~90% traffic of inner peers is downloaded from external peers. The end-user's performance will be improved when the downloading process was changed to make it occur only between the inner peers.

P4P-A harmonize new architecture for P2P applications and ISPs was put forward [2]. In P4P system, the choice of peers joined the consideration of underlying network, made the traffic in the local network as much as possible. The experiments show that P4P technology can change P2P traffic distribution effectively and reduce the traffic, relieve backbone bandwidth pressure.

Similar to P4P, in Gnutella system, each ISP establish a server named "Oracle" to improve the neighbors relationship of peers [3]. Because Gnutella system relies on neighbor nodes in searching resources, the network location will have an important impact on the final distribution of traffic.

CDN (Content Delivery Network through) to improve P2P network structure was also introduced [15]. With the dynamic DNS redirection service provided by CDN, all the clients to visit a Web site connect to the mirror server. Thousands of the mirror servers distributed throughout the world. It may be considered that when two peers are redirected to visit with a mirror server, they are really very near.

Researchers had also taken into account the active measurement techniques in P2P system to improve peer selection strategy. The results show that the algorithm had raised the document copying performance of BT system and improve the performance of the underlying network.

Present research indicated that in the ordinary circumstances the P2P traffic localization can cause the P2P system and the ISP network achieves a win-win situation. The types of achievement can be divided into two categories: the first category is let the P2P peer to active measurement the underlying network topology and the running status. The other category is supplies these information by ISP to give the P2P system. Selecting the active measurement method, implements is quite simple, only need improve the concrete P2P protocol and to achieve the goal of traffic localization. Because of the P2P network's peer size is very large, and it is inevitable to face the following questions: 1) Massive peer's active measurement will have very tremendous influence on the underlying network; 2) The active measurement increased the peer computation and the memory burden and this will weaken the system performance instead. If the active measurement is not able to enhance the P2P performance evidently, the P2P system developer and operator are difficult to have the power to realize the improvement.

The intention to make ISP and P2P work coordinate is very good. But at present it is still at the experimental study stage. To realize in Internet's big area deployed and substitutes the P2P system significance, the most difficult problem is to establish the confidence between ISP and the P2P besides the technical factor. After many years of struggle, it is difficult for them to exchange their key information frankly.

Table 2 lists the various technology features of P2P traffic management.

Table 2 P2P traffic management technologycomparation

Technology type	Advantages	Disadvantages
Traffic blocking	Easy for implementation	Traffic is difficult to identify; Network quality of service reduce.
P2P Caching	Effectively reducing inter-domain traffic and meet P2P user demand	Traffic is difficult to identify; Higher computing power and storage capacity are requested.
Traffic localization (Transformation peer, Active measurement)	Benefit to both P2P and ISP; No extra equipment	Extra traffic is created and network performance is disturbanced.
Traffic localization (Add Server)	Benefit to both P2P and ISP; No extra traffic	Mutual confidence were deficient between P2P and ISP; Extra equipment invest were needed.

4 Conclusions

P2P traffic has occupied a great proportion of network traffic. Its influences to the network are growing larger and larger. Today, as an indispensable application of the Internet, P2P technology has a broad range of areas of application and a giant user groups. On the one hand, the Internet managers, such as ISPs, need to take measures to deal with the problem that the P2P traffic results in depletion of the Internet bandwidths. Therefore, they must divert the P2P traffic effectively without affecting P2P application service experience. To improve network performance and to reduce operating cost by changing P2P traffic distribution will be important directions of research in the future.

References

1. Thomas, K., Andre, B., Nevil, B., et al.: Is P2P dying or just hiding? In: GLOBECOM. IEEE Computer Society Press (2004)
2. Haiyong, X., Yang, R.Y., Arvind, K., et al.: P4P: Provider portal for (P2P) applications. In: Proc. of SIGCOMM (August 2008)
3. Aggarwal, V., Feldmann, A., Scheideler, C.: Can ISPs and P2P systems cooperate for improved performance? ACM SIGCOMM Computer Communications Review 37, 31–40 (2007)
4. Saleh, O., Hefeeda, M.: Modeling and caching of peer-to-peer traffic. In: Proc. of IEEE ICNP 2006, Santa Barbara, CA (2006)
5. Pawel, M., Nikitas, L., Arnaud, L., et al.: Small Is Not Always Beautiful. In: CoRR abs/ 0802, 1015 (2008)
6. Abram, H.: CSC523: Analysis of the P2P BitTorrent Protocol, http://presentation.abez.ca/BTB_paper.pdf
7. http://www.pplive.com/
8. Bharambe, A.R., Herley, C., Padmanabhan, V.N.: Analyzing and improving a bittorrent network's. performance mechanisms. In: Proc. IEEE Infocom 2006, Barcelona, Spain (April 2006)
9. Huang, Y., Fu, T.T.J., Chiu, D.M., et al.: Challenges, Design, and Analysis of a Large-Scale P2P VoD System. In: Proc. ACM SIGCOMM (2008)
10. Wang, R., Liu, Y., Yang, Y., et al.: Solving the App-Level Classification Problem of P2P Traffic Via Optimized Support Vector Machines. In: Proc. of ISDA (October 2006)
11. Onstantinou, F., Mavrommatis, P.: Identifying Known and Unknown Peer-to-Peer Traffic. In: IEEE International Symposium on Network Computing and Applications, pp. 93–102 (2006)
12. Thomas, K., Andre, B., Michalis, F.: Transport layer identification of p2p traffic. In: IMC 2004: Proceedings of the 4th ACM SIGCOMM Conference on Internet Measurement. ACM Press (2004)
13. Crotti, M., Dusi, M., Gringoli, F., et al.: Traffic Classification through Simple Statistical Fingerprinting. ACM SIGCOMM Computer Communication Review 37(1), 7–16 (2007)
14. Karagiannis, T., Rodriguez, P., Papagiannaki, K.: Should internet service providers fear peer-assisted content distribution? In: Proc.of IMC 2005, Berkeley, CA, USA (October 2005)
15. Choffnes, D., Bustamante, F.: Taming the Torrent: A practical approach to reduce cross-ISP traffic in P2P systems. In: Proceedings of ACM SIGCOMM (2008)

Solution of Collision Caused by Unidirectional Link in MANET on the Navy Ships

Jungseung Lee and Jaiyong Lee

Abstract. With the development of information and communication technology the future battlefield will be grow multidimensional integration war based on Network Centric Warfare(NCW). In this paper, to conducting NCW, reliable data transfer methods were studied through the resolution of collision caused by unidirectional link in MANET on Navy ship. In the proposed method, CTS is transfered one more hop by first CTS received nodes through in a way to transfer PCTS for near nodes. In addition, proposed method occurs a little overhead, but that speed loss level can be acceptable because the number of hops is limited by characteristic of Navy ship.

1 Introduction

According to South Korea Defense Security Forum which was held recently, the pattern of future war will be NCW-based multi-dimensional/concurrent war. Therefore, it was discussed that the way of prepare for this must be upgrade the current military force to network-based, reforming the military force structure. Now, it's time for improvement to network-based on the current state for each military force. Unlike other military force, Navy ship has undergone their operation at long distance that difficult to reach wire / wireless networks. But, Navy ship will be also has networks in order to perform NCW in the future. This means each Navy ship will communicate by network into themselves make nodes.

If Navy ship configure one group for perform the mission, it needs to be controlled for ensuring the reliability of communication such as Scheduling. But if communication being controlled by specific ship, it will be difficult to ensure survivability of network. Therefore, it will composed MANET(Mobile Ad hoc NETwork) in order to configure networks without depend on network controll from specific ship.

Jungseung Lee · Jaiyong Lee
Department of Electrical & Electronic Engineering Yonsei University, 134,
Shinchon-Dong, Seodaemun-Gu, Seoul, Korea

F.L. Gaol et al. (Eds.): Proc. of the 2011 2nd International Congress CACS, AISC 145, pp. 197–204.
springerlink.com   © Springer-Verlag Berlin Heidelberg 2012

In this paper, studied solution of collision caused by unidirectional link that occur due to characteristics of Navy ship.

2 Related Work

In a recent studies, there is no study about this subject-solution of collision caused by unidirectional link. So, direct method or algorithm to become good reference is limited. In this part, should be a brief introduction about the researches motivation of proposed method suggested in this paper.

2.1 Study on Partial Reverse Route for Uni-directional Link in Ad-Hoc Network

- Make neighbor information table(NIT) using HELLO message
- When receive HELLO message, node renewal or create their NIT
- Cumulative path will be update by limiting three-hop
- If NIT in the incoming HELLO message has not been include in the own node's, it would be their opponents did not receive their HELLO message, then consider Unidirectional Link.

In this paper, main idea is make neighbor information table using hello message. With this method, we can know   existance of unidirectional link and can handling data transmission. But this approach resolution in terms of routing. It can't be implemented directly in MACs.

2.2 A Method to Reduce Collision by Interference in MANETs

- The power that required to transfer is bigger than the power that required to interference.
- So, collision can occur even if to start transmission from the node does not belong to transmission range.
- Solution

  – RTS-CTS-STS-DATA-ACK
  – CTS received node : Send STS packet
  – STS received node : calculate the interference effect(SIR) in the data receiving node when they start the transfer
  – If the effect is larger, transmission shall delay

$$\frac{\text{received power by sending node}}{\text{Interference power in the neighbor nodes}} = \frac{P_{rts}}{P_{cts}}$$

In this paper, interference range almost twice time then transmission range, so collision can occur even if to start transmission from the node does not belong to transmission range. The solution is using STS packet. When nodes received CTS, they send STS packet. STS received nodes calculate the SIR in the data receiving node when they start the transfer. If effect is larger, transmission shall delay. The solution focused just interference and CTS. However, passing STS is the reference element.

3   Problem Statement

As we know, we have hidden node problem in MANET, we applied CSMA/CA approach to the resolution. And we can prevent this collision problem with RTS(Request To Send)/CTS(Clear To Send). But when unidirectional link occurs, that node cannot receive RTS/CTS. Then cannot avoid collision. Unidirectional link means one way link. Unidirectional link in multi-hop network just like ad-hoc network caused by asymmetric transmission range, transmit power difference and node movement, and so on.

3.1   Occurance of Unidirectional Link and Caused of Collision in Accordance

As shown in figure 1 above, in the node B can transfered to the node A, but the transmission range of node A is formed asymmetrically, so transfer to node B is impossible.

Node C's size of the transmit power can be transfered to node D, but node D's transmission power is not enough, so transfer to node C is impossible.

This case is the situations of Unidirectional Link occurs, at this point, if node A/D to send the RTS or CTS to node B/C, because node B/C did not receive it, node B/C can transmit to node A/D during in the transmission of node A/D, and in this case collision occurs.

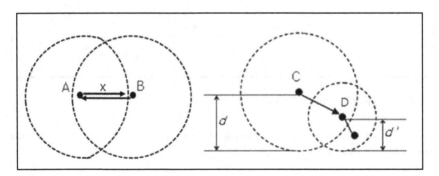

Fig. 1 Situation of Unidirectional Link occurs.

3.2   Collision Occurs Caused by Unidirectional Link in Navy Ships

Because each Navy ship has different missions to perform by ship type&class, so their data throughput is different between each other. And for security, because the range of radio transmission is minimizes within level of service, transmission range is different. Also due to the limitations of the location where AP installed (because in the navy ship radar positioned top place), AP will undergo interference by near structure. after all, transmission power will be reduced on specific angle. So, every ship has loss section in a particular direction. Due to these characteristics, unidirectional-link is expected in MANET on Navy ships, this will cause collision.

As shown in figure 2 below, in the state of ship's maneuver group formation, path from node C to node B is unidirectional link.

- At this time, node A broadcast RTS to node B for data transmission, and node C cannot receive it.
- Node B broadcast CTS, and node C cannot receive it.
- Node A send data to node B.
- In the process, node C broadcast RTS for data transmission, collision occurs in node B.

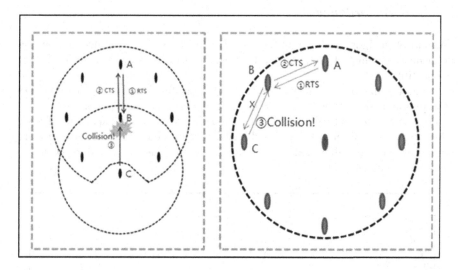

Fig. 2 Situation of Unidirectional Link occurs.

This situation, no matter any state of ship formation, occurs if the following conditions are satisfied.

- Node A is located outside of communication radius from node B.
- Node C is located outside of communication radius from node B.
- Node B in unidirectional link from node A.

If this condition occurs, collision will occurs with a very large probability. Because Navy ship maneuvering slowly and if formation formed at once, it would be less change.

Data loss due to collision caused by unidirectional link can be caused huge impact on the military operations.

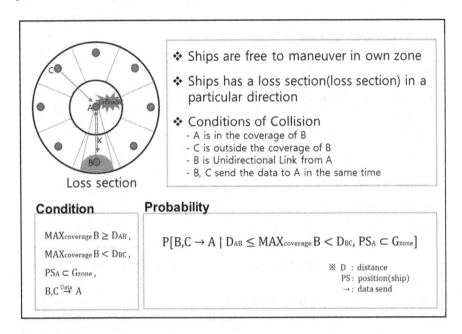

Fig. 3 Conditions of collision.

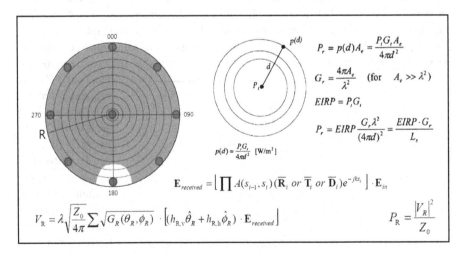

Fig. 4 Reference of loss section.

4   Proposed Scheme and Simulation Result

In this paper, propose the method that pass CTS once more to nearby nodes from the node that first received CTS, to avoid sending from node did not receive CTS by unidirectional link.

4.1   Proposed Scheme

- Solution.

1. Set PCTS(Pass CTS) for transfer CTS information to near node
2. PCTS contains NAV (PCTS) and Random value.
3. It work when receiving CTS, send PCTS to one hop

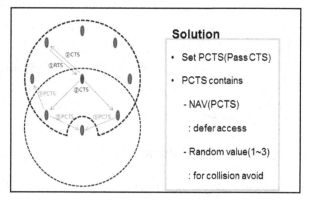

Fig. 5 The process of passing CTS to Unidirectional Link.

Above in figure 5, node B broadcast RTS for transmission, and the target node that received RTS will broadcast CTS. After then, the nodes that receives CTS are broadcast PCTS in order to transmit CTS one more hop for nodes didn't received CTS caused by unidirectional-link. This PCTS includes NAV(PCTS) which in the same role for existing CA operation. It would be delays transmission of the receiving nodes during PCTS defer access time. Also for the prevention of collision caused by broadcast PCTS at the same time, transport including random value of 1~3. Values of 1~3 is selected randomly from sending nodes, then receiving nodes received PCTS a sequential order.

When you add such the PCT delivery process, can be prevented collision caused by unidirectional link that occurs during CTS transmission.

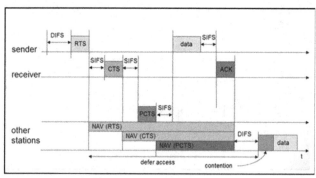

Fig. 6 CA operation.

4.2   Simulation Result

1. Before applying proposed method.

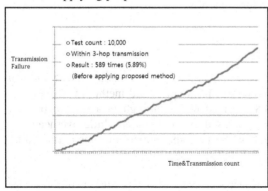

- Test value(conditions)

- Number of ship : 9 (fixed)
- Formation : circle formation (fixed)
- The distance between each ship is fluid, but the minimum & maximum distance is limited
- Every ship has their own zone, free to maneuver in own zone.

Fig. 7 Simularion Result(Before).

- Result : Transmission failure 5.89%

2. After applying proposed method.

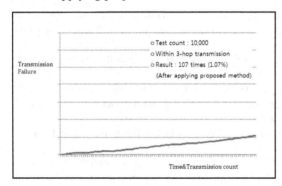

- Test value(conditions)
- Number of ship : 9 (fixed)
- Formation : circle formation (fixed)
- The distance between each ship is fluid, but the minimum & maximum distance is limited
- Every ship has their own zone, free to maneuver in own zone.

Fig. 8 Simulation Result(After).

- Result : Transmission failure 1.07%
  ⟹ Improve 4.82%

4.3   Alalysis

Of course, this CTS transfer process undergoes, overhead gets bigger, it can slow down the transfer. However, Data Communication at the situation in military operations, reliability and survivability is the best important element than Speed and efficiency. So, even at the risk of a little slow down, the guarantee of Data reliability is required. In addition, consideration of Navy ship maneuver group formations, because far from source to destination is within 3~5 hops, this amount of overhead is sufficiently can afford for reliability.

5 Conclusion

In this paper, proposed the solution of collision caused by unidirectional link that occur due to characteristics of Navy ship. In the proposed method, CTS is transfered one more hop by first CTS received nodes through in a way to transfer PCTS for near nodes. In addition, proposed method occurs a little overhead, but that speed loss level can be acceptable because the number of hops is limited by characteristic of Navy ship.

Acknowledgments. "This research was supported by the MKE(The Ministry of Knowledge Economy), Korea, under the ITRC(Information Technology Research Center) support program supervised by the NIPA(National IT Industry Promotion Agency)" (NIPA-2011-C1090-1111-0006).

References

[1] Venugopalan, R., Moss, D.: Statistical Analysis of Connectivity in Unidirectional Ad Hoc Networks. In: Proceedings of the International Conference on Parallel Processing Workshops (2002)
[2] Choi, M., Kim, H.: Extension of DYMO for using unidirectional link. In: Korea Computer Congress (2009)
[3] Kim, H., Lee, K., Lee, J., Kim, B.: Design of Adaptive DCF algorithm for TCP Performance Enhancement in IEEE 802.11 based Mobile Ad-hoc Networks. Institute of Electronic Engineering (October 2006)
[4] Park, S.: Routing Protocol for Ad-hoc Networks with Unidirectional Links. Information Security Institute (September 2007)
[5] Hwang, A., Lee, J., Kim, B.: IEEE 802.11 DCF medium access control algorithms to improve performance of the design and performance analysis. Journal of Electrical Engineering (October 2005)

Fast and Memory Efficient Conflict Detection for Multidimensional Packet Filters

Chun-Liang Lee, Guan-Yu Lin, and Yaw-Chung Chen

Abstract. Packet classification plays an important role in supporting advanced network services such as Virtual Private Networks (VPNs), quality-of-service (QoS), and policy-based routing. Routers classify incoming packets into different categories according to pre-defined rules, which are called packet filters. If two or more filters overlap, a conflict may occur and leads to ambiguity in packet classification. In this paper, we propose an algorithm which can efficiently detect and resolve filter conflicts. The proposed algorithm can handle filters with more than two fields, which is more general than algorithms designed for two-dimensional filters. We use the synthetic filter databases generated by ClassBench to evaluate the proposed algorithm. Compared with the bit-vector algorithm, simulation results show that the proposed algorithm can reduce the detection times by over 84% for 10 out of 12 filter databases, and only uses less than 26% of memory space.

1 Introduction

For providing advanced network services such as firewalls, policy-based routing, and quality-of-service (QoS), routers have to classify incoming packets into different flows. The rules used to classify packets are called packet filters or simply filters. A filter $F = (f[1], f[2], .., f[k])$ is called k-dimensional if the filter consists of k fields, where each field can be a variable length prefix, a range, an exact value or wildcard. The most common fields are the IP source address (SA), IP destination address (DA),

Chun-Liang Lee
Dept. of Computer Science and Information Engineering, Chang Gung University
Taoyuan 333, Taiwan
e-mail: `cllee@mail.cgu.edu.tw`

Guan-Yu Lin · Yaw-Chung Chen
Dept. of Computer Science, National Chiao Tung University
Hsinchu 300, Taiwan
e-mail: `{guanyu,ycchen}@cs.nctu.edu.tw`

F.L. Gaol et al. (Eds.): Proc. of the 2011 2nd International Congress CACS, AISC 145, pp. 205–211.
springerlink.com   © Springer-Verlag Berlin Heidelberg 2012

port numbers of source/destination applications (SP/DP), and protocol type (PT). We say that a packet P matches a particular filter F if for all i, the i-th field of the header satisfies $f[i]$. Each filter has an associated action. For example, suppose that a filter is of the format (SA, DA, SP, DP, PT). The filter $F1 = (*, 163.25.*.*, *, 23, TCP)$ specifies a rule for traffic flow sent from any host to subnet 163.25 using TCP destination port 23, which is used for incoming Telnet. The associated action of $F1$ will be set to "Reject" if Telnet connections are not allowed to enter subnet 163.25.

It is possible that a packet matches multiple filters. In this case, an ambiguity may arise if the actions of matching filters conflict. For example, suppose that we have a filter $F2 = (140.113.*.*, *, *, *, TCP)$ and its associated action is "Accept". We say that $F1$ and $F2$ are in conflict since there exists a packet, say $P = (140.113.1.1, 163.25.1.1, 5000, 23, TCP)$, which matches both filters. Since the actions of $F1$ and $F2$ conflict, this will cause a security problem if packets that should be blocked are accepted to pass through. To solve the problem of filter conflict, three possible solutions are listed in [1]:

1. The first matching filter in the filter database is selected.
2. Assign a priority to each filter. The matching filter with the highest priority is selected.
3. Assign a priority to each field. The matching filter with the most specific matching field with the highest priority is selected.

However, none of the above solutions can fully solve the ambiguity problem caused by filter conflict. A possible way to get around of this difficulty is to use resolve filters [1]. A resolve filter is a filter that matches the packets in the overlap region of two conflicting filters. However, finding the minimum number of resolve filters is an NP-hard problem [1, 2]. Therefore, it is a challenging work to design an algorithm that can efficiently detect and resolve filter conflicts. In this paper, we propose an efficient algorithm to detect filter conflicts using the tuple space search [3]. Although the tuple space search were designed to deal with two-dimensional (2-D) filters, the proposed algorithm can be used in filters with more than two fields.

2   Related Work

The simplest way to detect all filter conflicts is to compare every pair of filters. However, this approach takes $O(n^2)$ time to detect all conflicts, where n is the number of filters. For large filter databases, it cannot provide satisfactory performance in terms of detection time. Hari et al. [1] propose the use of grid of tries [4] to detect all conflicts in 2-D prefix filters. Baboescu and Varghese [5] propose several conflict detection algorithms by using the bit vector scheme [6] and the aggregated bit vector scheme [7]. Lu and Sahni [2] propose a plane-sweep algorithm that improves the performance of both time and space. The key idea behind the plane-sweep algorithm is to treat each 2-D filter as a rectangle with four line segments in the space, and then find all overlap regions (i.e., conflicts) through finding orthogonal line segment intersections.

Since the algorithm presented in this paper uses the tuple space [3] to store filters and detect conflicts, the following briefly reviews the tuple space search. The basic idea behind the tuple space search is that the number of distinct prefix lengths of filters is much less than the number of filters. The filters with specific prefix lengths are grouped into a tuple. Since each tuple has a specific length for each field, these bit strings can be concatenated to form a hash key, which can be used to perform the tuple lookup. For an incoming packet, the matching filters can be found by probing all tuples. For example, filter $F = (11*, 110*)$ belongs to tuple $T_{2,3}$, while filter $G = (110*, 0011*)$ belongs to tuple $T_{3,4}$. Given a packet $P = (11100000, 11011111)$, if we want to probe tuple $T_{2,3}$, two and three bits will be extracted from two fields respectively to construct the hash key (i.e., 11110). Similarly, the hash key 1111101 will be constructed for tuple $T_{3,4}$. It is obvious that even a linear search in the tuple space represents a considerable improvement over a linear search of the filters.

To improve the performance, a tuple-based algorithm called rectangle search was proposed in [3]. Through the use of markers and pre-computation, which were introduced in [8], the number of tuples needed to be probed can be greatly reduced. Let filter $F = (f[1], f[2])$ with the length combination (i, j) belong to tuple $T_{i,j}$. For each tuple left to $T_{i,j}$, say $T_{i,j'}$, $0 \leq j' < j$, a marker will be generated by eliminating the bits behind the j'th bits.

3 Proposed Conflict Detection Algorithm

In the proposed algorithm, the process of conflict detection is divided into two parts. The first is to deal with the source and the destination IP addresses, and the second is to deal with other fields. Each filter is first inserted into the corresponding tuples according to the lengths of the source and the destination IP address prefixes. Filters with the same source and destination IP address prefixes are chained by pointers. Thus, the proposed algorithm can uses the tuple space to reduce the number of possible conflicting filters, and then compare the other fields to find the real conflicting ones.

Suppose that the length of both the source and destination IP addresses is W bits. Each filter is mapped to one of $(W + 1)^2$ tuples. The simplest way to detect all possible conflicts is to lookup all tuples for each filter. Therefore, the number of tuples required to lookup for a specific filter is simply $(W + 1)^2 - 1$, which is obviously not a feasible way. We need more insightful observations on the tuple space to reduce the number of tuples to lookup. Two filters F and G are in conflict if and only if all of the following three conditions hold [2]:

1. There is at least one packet that is matched by both F and G.
2. There is at least one packet that is matched by F but not by G.
3. There is at least one packet that is matched by G but not by F.

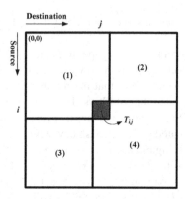

Fig. 1 Categories of Tuples for $T_{i,j}$.

For a filter F in tuple $T_{i,j}$, all tuples excluding tuple $T_{i,j}$ can be partitioned into three disjoint sets [3]:

1. Shorter tuples: A tuple $T_{i',j'}$ belongs to this set if and only if $i' \leq i$ and $j' \leq j$.
2. Longer tuples: A tuple $T_{i',j'}$ belongs to this set if and only if $i' \geq i$ and $j' \geq j$.
3. Incomparable tuples: A tuple belongs to this set if and only if it neither belongs to the shorter tuples nor the longer tuples.

These three kinds of tuples are shown in Fig. 1, region (1) contains the shorter tuples, and region (4) contains the longer tuples. Regions (2) and (3) contain the incomparable tuples. Note that tuple $T_{i,j}$ are excluded from these regions. Otherwise, these regions are not disjoint.

For a filter F, conflicting filters may exists in any tuple. However, for the incomparable tuples, only those tuples in region (2) have to probe since we only need to generate one resolve filter for two conflicting filters. Suppose that there is a conflicting filter G in region (3). Since we do not probe the tuples in region (3), the conflict between filter F and G cannot be detected when processing filter F. However, the conflict can be detected when we process filter G because filter F is in region (2) of filter G. Similarly, we only need to probe the shorter tuples (i.e., region (1)), and omit the longer tuples. As a result, the number of tuples needs to lookup is reduced by nearly a half, which is indicated by the bold line in Fig. 2.

To further reduce the number of tuples needed to lookup, we can use the markers mentioned in Section 2. Suppose that there is a filter G that is in conflict with filter F. Filter G generates a marker for each tuple on the left of $T_{i',j'}$, which is indicated by light gray color in Fig. 2. Through the help of markers, this conflict can be found by probing the left-most tuple (i.e., $T_{i',0}$). As markers only indicate there are possible conflicting filters in the tuples on the right of the same row, we need a way to efficiently find those filters and generate resolve filter if necessary, which will be addressed later. Therefore, for each filter F in tuple $T_{i,j}$, we only need to lookup $i+1$ tuples (i.e., $T_{0,0}, T_{1,0}, ..., T_{i,0}$). The number of tuples to lookup has been reduced from $O(W^2)$ to $O(W)$.

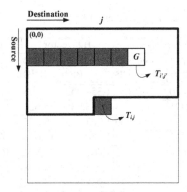

Fig. 2 Reducing the number of tuples needed to lookup.

We have described the key idea behind the proposed algorithm. However, with the markers presented in [3], if we use a filter to probe a tuple and find a matching marker, we still have to probe all tuples on the right side of the same row for finding the conflicting filters. To efficiently find the conflicting filters, we add a new field to the marker, which is called *marker pointer*, and it points to the filter that generates it. Note that two filters may generate identical markers in a tuple, but only one of them can be stored in the tuple. We can categorize the settings of marker pointer into three types:

1. The first type is simple. If the marker M generated by filter F does not find any duplicate marker or filter in the tuple, the marker pointer of M will point to F.
2. If the marker M generated by filter F finds a duplicate filter F_D in the tuple, M will not be inserted to the tuple. In addition, the filter stops generating markers in all remaining tuples. However, the *filter pointer* of G will point to F. Filter pointer is also a newly introduced field. It is similar to the marker pointer while it is only for filters.
3. If the marker M generated by filter F finds a duplicate marker M_D in the tuple, M will not be inserted to the tuple. Following the marker pointer of M_D, we can find the filter, say G, that generates it. The filter pointer of G will point to F.

To detect all conflicts, tuples are processed from left to right and from top to bottom. For a tuple to process, all filters in the tuple are retrieved one by one, and then the retrieved filter is used to probe the tuples as mentioned above. If there is a match in a tuple, which means a conflict is detected, resolve filters can be generated by following the marker pointer to find the filter or the filter pointer for the chained filters. Note that although there are no duplicate filters (here we only consider SA and DA fields) in a tuple, it is possible that more than one filters have identical SA and DA fields. According to the above ways to set marker pointers and filter pointers, filters with identical SA and DA fields will be chained using the marker pointers. Thus, the proposed algorithm can detect all conflicts between multi-dimensional filters. Also note that the proposed algorithm definitely can be used for 2-D filters. However, since 2-D filters exhibit more properties than multi-dimensional filters,

Table 1 Performance Evaluation for Synthetic Databases.

Databases	Statistics		Bit Vector		Proposed Algorithm	
	Number of Filters	Number of Conflicts	Time (us)	Space (KB)	Time (us)	Space (KB)
ACL1	29,870	379,155	1,356,453	5,198	15,195	698
ACL2	29,971	3,987,104	1,238,935	5,625	184,455	861
ACL3	29,959	698,700	1,267,190	8,095	111,738	868
ACL4	29,942	479,867	1,358,589	13,083	167,294	1,120
ACL5	29,880	22	1,409,527	35,854	109,605	1,190
FW1	29,999	4,211,145	1,057,411	4,428	127,285	692
FW2	29,982	572,541	934,864	2,612	11,387	669
FW3	29,984	4,781,576	929,478	3,736	129,743	678
FW4	29,998	2,245,618	1,254,447	29,786	193,423	710
FW5	29,905	7,915,689	1,032,359	50,264	370,483	811
IPC1	29,932	1,365,774	1,202,888	4,902	73,077	756
IPC2	29,975	7,004,835	1,541,960	296,832	1,278,961	2,642

the proposed algorithm can be modified to provide better performance in detection time. Due to the space limit, we focus on multi-dimensional filters, which is more general than 2-D filters.

4   Experimental Results

In this section, we compare the performance of the proposed algorithm with that of the *BitVector* algorithm [5]. The *BitVector* algorithm incorporated path compression and aggregation as suggested in [5], and the aggregation size was set to 32. Both algorithms were implemented in C++, and benchmarked on a 3.2 GHz AMD Phenom II X4 PC that has 4 GB of memory. The tested filter databases were synthesized by a public tool, ClasssBench [9]. For each parameter file provided by ClassBench, we generated a filter set with 30K filters. Since a filter set generated by ClassBench may contain duplicate filters, the number of filters for each seed file is less than 30K after removing duplicate filters. Fields specified by range (i.e., source port and destination port numbers) are converted to prefixes using techniques shown in [3, 4]. The memory requirement does not include the memory required to store the original filter set. For the proposed algorithm, the memory used to implement the hash function is included.

The memory and time requirements for the *BitVector* algorithm and the proposed algorithm are given in Table 1. The improvement on time is significant. The smallest improvement is for IPC2. The time required to detect and resolve all conflicts is reduced by 17.1%. The largest improvement is for ACL1. The required time is reduced by 98.9%. One thing worth noting is the number of conflicts of ACL1 is around 379K, which is the second smallest among all seed files. As for IPC2, the number conflicts is around 7M, which is the largest. In short, the proposed algorithm can reduce the times by over 80% for 10 out of 12 seed files. As for the memory requirements, the proposed algorithm requires much smaller memory space than the *BitVecor* algorithm. In addition, the memory requirements for the proposed algorithm are more stable than the *BitVector* algorithm.

5   Conclusion

Packet classification is essential for providing advanced services in routers. If filter conflicts exist, it may lead to ambiguity in packet classification. This paper proposes a conflict detection algorithm that can efficiently detect and resolve all conflicts. With the insightful observations about the tuple space, the proposed algorithm can reduce a significant number of tuples to lookup. Compared with the *BitVector* algorithm, the experimental results show that the proposed algorithm can reduce the detection time by 17.1% to 98.9%. More important, the required storage can be significantly reduced by 74.4% to 99.1%.

Acknowledgment. This work was supported in part by the National Science Council under Grant No. NSC 96-2221-E-182-013 and NSC 99-2221-E-182-053.

References

1. Hari, A., Suri, S., Parulkar, G.: Detecting and Resolving Packet Filter Conflicts. In: Proceedings of INFOCOM, pp. 1203–1212 (2000)
2. Lu, H., Sahni, S.: Conflict detection and resolution in two dimensional prefix router tables. IEEE/ACM Transactions on Networking 13(6), 1353–1363 (2005)
3. Varghese, G., Suri, S., Varghese, G.: Packet classification using tuple space search. In: Proceedings of ACM SIGCOMM, pp. 135–146 (1999)
4. Srinivasan, V., Varghese, G., Suri, S., Waldvogel, M.: Fast and scalable layer four switching. In: Proceedings of ACM SIGCOMM, pp. 191–202 (1998)
5. Baboescu, F., Varghese, G.: Fast and Scalable Conflict Detection for Packet Classifier. Computer Networks 42(6), 717–735 (2003)
6. Lakshman, T.V., Stiliadis, D.: High-Speed Policy-based Packet Forwarding Using Efficient Multi-dimensional Range Matching. In: Proceedings of ACM SIGCOMM, pp. 203–214 (1998)
7. Baboescu, F., Varghese, G.: Scalable Packet Classification. In: Proceedings of ACM SIGCOMM, pp. 199–210 (2001)
8. Waldvogel, M., Varghese, G., Turner, J., Plattner, B.: Scalable High Speed IP Routing Lookups. In: Proceedings of ACM SIGCOMM, pp. 25–36 (1997)
9. Taylor, D.E., Turner, J.S.: Classbench: a packet classification benchmark. IEEE/ACM Transactions on Networking 15(3), 499–511 (2007)

Hierarchical Codes in Bandwidth-Limited Distributed Storage Systems

Zhen Huang, Lixia Liu, Yuxing Peng, and Shoufu Xue

Abstract. Distributed storage systems store data on the "unreliable" network peers that can leave the system at any moment and their network bandwidth is limited. In this case, the only way to assure reliability of the data is to add redundancy using either replication or erasure codes. As a generalization of replication, erasure codes require less storage space with the same reliability as replication. Recently, a near-optimal erasure code named Hierarchical Codes, has been proposed that can significantly reduce the repair traffic by reducing the number of nodes participating in repair, which is referred to as repair degree d. To overcome the complexity of *reintegration* and efficiently control the reliability of Hierarchical Codes, we refine two concepts called *location* and *relocation*, then we propose an integrated maintenance scheme, which allow us to tune the code construction.

1 Introduction

As the development of distributed storage systems, the comparison between "classical" erasure code and replication has been a hot topic in the research community. The main drawback of erasure code is in the phase of *maintenance*. To *maintain* data reliability, lots of *repair traffic* are introduced as using erasure code [1, 2]. Even worse, in a bandwidth-limited environment, *repair time* will be increased significantly as the increasing of *repair degree* [3] which further degrades data reliability.

Hierarchical Code explores a field that the repair traffic is significantly reduced by reducing the *repair degree* [4]. This motivates us to evaluate its performance in the

Zhen Huang · Lixia Liu · Yuxing Peng
National Laboratory of Parallel and Distributed Processing,
National University of Defense Technology
e-mail: {huangzhen,liulixia,pengyuxing}@nudt.cn

Shoufu Xue
Air Defence Command College
e-mail: xsf715e@sina.com

F.L. Gaol et al. (Eds.): Proc. of the 2011 2nd International Congress CACS, AISC 145, pp. 213–218.
springerlink.com   © Springer-Verlag Berlin Heidelberg 2012

bandwidth-limited distributed storage system. Although several experiments have been carried out in [4], some important properties are not included in the maintenance strategy: i.e. *reintegrate* the blocks that have been *timeout* and where to *locate* the repaired block in the code construction. Then, we refine the concept of *location* and *relocation*, prove propositions on them and propose an integrated maintenance scheme for Hierarchical Codes.

To the best of our knowledge, we are the first study to consider the practical scenario that the bandwidth is limited and failures occur in the process of repair [5]. For these failures, system has to inform other available peers to continue the repair. We simplify the cost of detection and information as a research work. To gain insight into practical systems, we use the real traces to simulate the distributed storage system and carry out the experiments with several coding schemes to find out their real performance.

By several detailed experiments, we can find that Hierarchical Codes perform best as compared with the other famous erasure codes: Regenerating Codes [6] and Random Linear Codes [7]. Two important results are listed:

- *The number of transfers* for Hierarchical Codes are much less than other codes.
- *The number of failures* on Hierarchical Codes are much less than other codes.

The paper is organized as follows: in section 2 we introduce the maintenance traffic problem, *Random Linear Combination* for repair. In section 3 we analyze the properties Hierarchical Codes, deduce our propositions on *location* and *relocation*. Finally in section 4, we do the experiments by simulate two real traces.

2 Background and Related Work

A key issue in distributed storage systems is data maintenance: storage peers are leaving frequently from the system and the data stored on these peers need to be repaired to maintain data reliability, which introduces network traffic and the available network bandwidth become the bottleneck resource [2] that limits the total amount of data that can be stored in the system. The maintenance traffic will be especially high in dynamic distributed environments with significant peer churn.

In response to the problem, *Regenerating Codes* and *Hierarchical Codes* are proposed. Although *Regenerating Codes* can reduce the repair bandwidth, more than k nodes are involved in the repair process which can also cause a high failure probability for the repair process [6]. Hierarchical Codes [4] allow the failed block be repaired by few blocks but not the total k blocks, which means that we can use few nodes to repair. In this paper, we will evaluate the performance of these codes in a representatively bandwidth-limited environment.

3 Analysis of the Hierarchical Codes

3.1 Location of the Block in the Code Construction

According to the Proposition 3 in [4] for Fig. 1(b), p_8 can also be repaired in *Group* $G_{4,1}$ by selecting $d_1 = 4$ blocks p_1, p_7, p_4 and p_5. So, if p_3 is also repaired by the p_1, p_7, p_4 and p_5, each of the repaired block p_3' and p_8' is a *random linear combination* of the 4 blocks. This implies that p_3' and p_8' can act as the same function after repair. So we can locate p_3' and p_8' in the same place of the code construction. **Location** denotes the *minimum* group where the block is located. For instance in Fig. 1(b), the location of p_3 and p_8 is respectively $G_{2,1}$ and $G_{4,1}$.

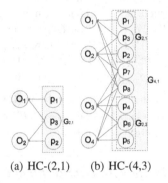

(a) HC-(2,1)   (b) HC-(4,3)

Fig. 1 Samples of Code Graphs for Hierarchical Codes.

Now let us propose a proposition for the *location* of repair:

Proposition 1. *Consider an Information Flow Graph of a hierarchical code at time t. Denote as $R(p)$ the set of blocks in P_{t-1} that have been combined to generate repaired block p. Denote as $L(p)_t$ the location of block p at time t.*

If $\exists G_{d,i}$ that fulfills the condition (C1):

$$\exists\, G_{d,i} \in G(p) : R(p) \subseteq G_{d,i}, |R(p)| = d \qquad (C1)$$

Then $L(p)_t$ can be $G_{d,i}$ that reconstruction property is preserved at time step t.

Proof: Since $G_{d,i}$ fulfills the condition (C1), $|R(p)| = d$. Suppose $R(p) = \{r_1,...,r_d\}$, $\forall p' \notin R(p)$ and $L(p')_t = G_{d,i}$, it can be repaired by $\{r_1,...,r_d\}$ using Proposition 3 [4]. Due to the fact that the blocks p and p' are all randomly linear combination of these d block, we place p in the same location as p'. So $L(p)_t = L(p')_t = G_{d,i}$, proposition 1 holds on.

Moreover, can we change the location of blocks? **Relocation** denotes a change of the block's location. To answer this question, we propose a proposition for the *relocation*:

Proposition 2. *Consider an Information Flow Graph of a hierarchical code at time step t. Consider any block p relocated at time step t. Denote as $L(p)_t$ the location of block p at time t.*

If $\forall\, t$ and $\forall\, p$,

$$\forall\, L(p)_t \subseteq L(p)_{t-1} \tag{C2}$$

Then the reconstruction property of the code graph is preserved.

Proof: To select any P^k, a set of k blocks in $HC - (k,h)$, which fulfills condition (C3) at time step t:

$$|G_{d,i} \cap P^k| \le d \;\; \forall G_{d,i} \text{ belonging to the code} \tag{C3}$$

Now Let us consider the alternative cases:

- If $p \notin P^k$, then P^k fulfill condition (C3) at $(t-1)$.
- If $p \in P^k$, respectively denote as $G(p)$, $G(\bar p)$ the hierarchy of groups that contains or doesn't contain the block p. (i) $\forall\, G_{d,i} \in G(\bar p)$, where $|G_{d,i} \cap P^k| \le d$ at t, then $|G_{d,i} \cap P^k| \le d$ at $(t-1)$. (ii) $\forall\, G_{d,i} \in G(p)$ where $L(p)_t \subseteq G_{d,i} \subset L(p)_{t-1}$, since $|G_{d,i} \cap P^k| \le d$ at t, $|G_{d,i} \cap P^k| \le (d-1) < d$ at $(t-1)$. (iii) $\forall\, G_{d,i} \in G(p)$, where $L(p)_{t-1} \subseteq G_{d,i}$ or $G_{d,i} \subset L(p)_t$, $|G_{d,i} \cap P^k| \le d$ at t, then $|G_{d,i} \cap P^k| \le d$ at $(t-1)$.

In doing so, we can conclude that P^k fulfills condition (C3) at time step $(t-1)$. According to the Proposition 2 of [4], at time step $(t-1)$, P^k are sufficient to reconstruct the original blocks. So Proposition 2 holds on.

3.2   Maintenance Scheme with Hierarchical Codes

In distributed storage system, we refine three maintenance operations for the data stored by Hierarchical Codes: *upgraded repair*, *downgraded locate* and *reintegrate*.

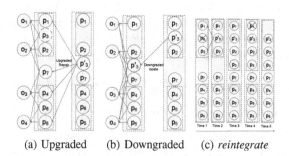

(a) Upgraded   (b) Downgraded   (c) *reintegrate*

Fig. 2 Maintenance scheme with Hierarchical Codes-(4,3)

1. **upgraded repair:** According to the proposition 1, the location of repaired block can be upgraded as compared to the lost block. As depicted in Fig. 2(a), p'_3 is repaired by p_1, p_7, p_4, p_5 and then it is upgraded to level 1.

2. **downgraded locate:** According to the proposition 2, the location of block can be downgraded. As depicted in Fig. 2(b), we downgrade the location of p'_3. But the operation degrades the data reliability. To maintain the reliability, we propose a rule to the *relocation: p′ denotes the repaired block for the lost block p, L(p′) can at most downgrade to L(p)*.
3. **reintegrate:** reintegrate can be regarded as a *costless* repair. The only problem is where to locate this block in the code construction. Since we apply the *random linear combination* as the way to repair, by which we locate the lost block in the group where it left before. An example is depicted in Fig. 2(c).

4 Evaluation

The experiments are based on an event-driven simulator, which simulates a distributed storage system. Peers' behavior is from the real availability of traces [8]. We aim to find out the real performance of Hierarchical Codes as compared to other codes. The results are listed as follows:

1. *compare the number of started transfers (N(st)):* The results on $N(st)$ are depicted in Fig. 3(a) and Fig. 3(b). As we know from [4], *HC* are involved with much more repairs to assure the reliability. However, for one generic repair, the number of transfers depends on the repair degree, which leads to a lowest $N(st)$ for *HC*. Conversely for *RG*, the repair degree is the largest so that $N(st)$ is the largest.

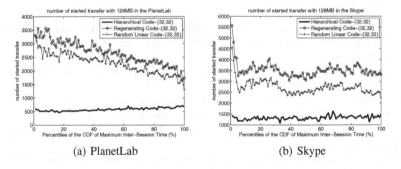

| (a) PlanetLab | (b) Skype |

Fig. 3 Number of started transfers $N(st)$ in the traces

2. *compare repair's failure (R_{FR}):* The results on R_{FR} are depicted in Fig. 4(a) and Fig. 4(b). *HC* perform best in all the traces since *HC* normally use a small repair degree $d < k$, which leads to a lowest probability to fail for one generic repair. This result is very important to the deployment of coding scheme for practical system. Since a large probability to fail will cause a bad reliability to successfully repair and thus data reliability will be further degraded.

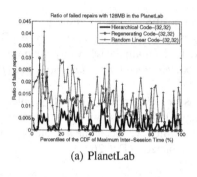

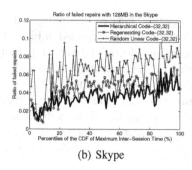

(a) PlanetLab   (b) Skype

Fig. 4 Ratio of failed repairs R_{FR} in the traces

5 Conclusions and Future work

Our experiments validate our claims that Hierarchical Code perform best in the bandwidth-limited distributed storage system. The next step is to deploy the scheme to a practical system.

Acknowledgements. The authors work is supported by the National Basic Research Program of China under Grant No.2011CB302601.

References

1. Lin, W.K., Chiu, D.M., Lee, Y.B.: In: P2P 2004: Proceedings of the Fourth International Conference on Peer-to-Peer Computing, pp. 90–97. IEEE Computer Society, Washington, DC, USA (2004)
2. Rodrigues, R., Liskov, B.: In: Peer-to-Peer Systems IV 4th International Workshop IPTPS, Ithaca, New York (2005)
3. Pamies-Juarez, L., Garcia-Lopez, P., Sanchez-Artigas, M.: In: Proceedings of the 10th IEEE International Conference on Peer-to-Peer Computing, P2P (2010)
4. Duminuco, A., Biersack, E.W.: Journal of Peer-to-Peer Networks and Applications 2 (March 2009)
5. Monteiro, J.G.: Modeling and analysis of reliable peer-to-peer storage systems. Ph.D. thesis, INRIA (2010)
6. Dimakis, A.G., Godfrey, B., Wu, Y., Wainwright, M.J., Ramchandran, K.: Computer Research Repository (CoRR) (2008); arXiv:0803.0632v1 http://arxiv.org/abs/0803.0632
7. Acedacnski, S., Deb, S., Medard, M., Koetter, R.: NETCOD (2005)
8. Godfrey, B.: Repository of Availability Traces (2006), http://www.cs.berkeley.edu/~pbg/availability/

Middleware-Based Distributed Data Acquisition and Control in Smart Home Networks

Yaohui Wu and Pengfei Shao

Abstract. For each of the networking techniques widely adopted in the field of home appliance control has its strength and weakness, it is more desirable to use multiple network technologies at the same time. Multi-platform smart home system would be an objective existence. In general point to point method, with the increasing of underlying hardware units and upper applications, the complexity of the central control unit will increase linearly. By using a universal home control middleware adapter, we decouple the application layer from the hardware layer, and make it possible to enable distributed data acquisition and control in a multi-platform smart home system. Many validated methods efficient to implement distributed data acquisition and control are discussed under this architecture, like using a timer or special thread with the unified API of the middleware adapter, redefinition callback or delegation functions for one type of the application interface, or using middleware integrated database to directly save the collected data. For each of these discussed methods, higher quality and usability both of devices and user interfaces, higher performance, lower production costs and loose-coupled hardware networks of different technologies are concerned.

1   Introduction

As [1] suggested, a smart home is understood as an integration system, which takes advantage of a range of techniques such as computers, network communication as well as synthesized wiring to connect all indoor subsystems that attach to home appliances and household electrical devices as a whole. In this way, smart home techniques enable households to effectively centralize the

Yaohui Wu · Pengfei Shao
Department of Electronic & Information Engineering,
Zhejiang Wanli University, P.R. China 315100
e-mail: yaohuinb@yahoo.com.cn, cbb_bb@126.com

F.L. Gaol et al. (Eds.): Proc. of the 2011 2nd International Congress CACS, AISC 145, pp. 219–226.
springerlink.com   © Springer-Verlag Berlin Heidelberg 2012

management and services, provide them with functions for internal information exchange and help to keep in instant contact with the outside world. In terms of convenience, they help people in optimizing their living style, rearranging the day-to-day schedule, securing a high quality of living condition and in turn enable people to reduce bills from a variety of energy consumptions in a house [2].

As more and more appliances are used, intelligent home controls are enriched and requirements of the future smart home system should contain some kind of adaptability and pro-activity. The home control network infrastructure is the basic skeleton for the construction of a smart home in which it unifies home appliances. With the lack of standardized or completely missing communication interfaces in home electronics, there is no perfect solution to address every aspect of smart homes based on the existing technologies. It is in practice more desirable to use multiple network technologies instead of one single home network technology. In the future, with more and more complex home applications, it is very important to support distributed data acquisition and control in a multi-platform smart home system. This topic focuses on how to implement distributed data acquisition and control in such cases to lead to improved and enhanced smart-home utilization.

2   General Multi-platform Smart Home System

For the technologies of home control networking popular in the domain of smart homes, home fieldbus technologies and short-range low-rate wireless network technologies have attracted more attention. The home fieldbus technologies mainly include HBS, EIB and power line communication (PLC) based technologies, such as X-10, INSTEON, HomePlug, CEBus, LonWorks and PLC-BUS [3]. The representative protocols of short-range low-rate wireless network technologies include infrared, wireless Rf, Bluetooth, 802.15.4/ZigBee and Z-wave [4][5]. Each technique has its strengths and weakness, there is no perfect solution to address every aspect of the smart homes. It is in practice more desirable to use multiple network technologies instead of one single network technology. For example, considering the aspect of power conservation, openness of protocol stack, interoperability based on layering and cost-effectiveness, a backbone network of HomePlug C&C plus ZigBee seems more promising in a smart home. Combining X-10 with ZigBee is more practical when installing networks and devices in older buildings. However, the lack of standardized or completely missing communication interfaces between the existing home control network technologies results in the central control unit must support most of the above and self-defined home networking standards [6].

For the control applications, it always contains home electronics control, light control, door and window control, etc, complex scenarios leading to improved and enhanced smart-home utilization. It at least needs the application interfaces (API) of user interface (UI), devices and equipment interfaces, and interfaces for application services to support these home applications.

With all the elements of the above two aspects, it is natural to get the general smart home framework shown as figure 1.

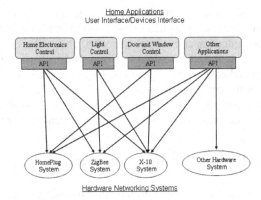

Fig. 1 General Multi-Platform Smart Home System

3 General Data Acquisition and Control

In general, all of the home applications directly access to the underlying hardware network through the central control unit, perceiving the state of the home through sensors and acting upon the environment. Each application works point-to-point with specific hardware control unit. In order to ensure that each application can detect data in real time, system software would create a thread for each application. Each thread only checks its respective sensors' status and catches the required data, then processes directly or storages data to a centralized database for further processing. It does not send operation commands to its respective actor devices. Control operations are carried out by the system application. The system application accepts user's requests or reacts with some internal requests triggered by pre-programmed complex scenarios, and sends the appropriate operating commands to the underlying sensor control unit. Figure 2 shows these relations.

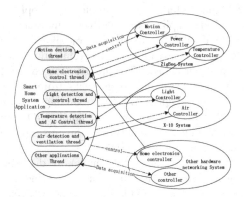

Fig. 2 Point-to-Point Data Acquisition and Control

In a general multi-platform smart home environment with multiple home applications, it needs the system application and central control unit to complete most of the tasks of data acquisition and control operation. This will occupy a lot of system resources to run multiple threads and do centralized data processing. Overload will be a problem of the central control unit. System performance can be easily influenced by other applications or the change of the environment. In a word, this general system is networking but not distributed, it must create corresponding control application for the underlying hardware, and data communications between them are point to point.

4 Distributed Data Acquisition and Control

We proposed a universal smart home architecture by using a middleware adapter, shown as figure 3.

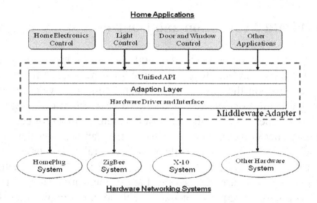

Fig. 3 A Universal Multi-Platform Smart Home System

The middleware adapter offers an abstraction of the application level from the hardware specific and decoupling the hardware interface modules from the application level. One doesn't need to concern about the differences of the underlying hardware networks any more in this architecture. All the I/O operations will be implemented asynchronously. The applications no longer need to directly access the underlying hardware, but to get data from the unified application interfaces of the middleware adapter. The corresponding hardware driver instances created by the middleware adapter will complete I/O operations and data acquisition of the underlying sensor networks. So this smart home architecture enables distributed data acquisition and control.

In the following discussion of the implementation of distributed data acquisition and control, higher quality and usability of both devices and user interfaces, higher performance, lower production costs and loose coupling between hardware networks of different technologies are concerned.

4.1   Using Timer or Thread with Unified API

As the middleware adapter mainly deals with data communications both on down link and up link, underlying data collection can be finished by the adaptation layer. The system application creates a special timer or thread to periodically read real-time hardware status from the unified application interface of the adaptation layer. Then it will either save the data to a networking database or process them directly, displaying status information, prompting alarms, performing control operations after prediction or judging. For example, the temperature monitoring of a room and courtyard can be stored for a long-term analysis, while the brightness changes of a room need not to be stored, it only works when residents are at home and it is too dark to turn on a light. The application thread created here is different from the data acquisition thread used in the general smart home system. This thread is used to separate data collection from the whole application system and to get all of the gathered data from the unified application interface of the middleware adapter via a dedicated single task. But each of the threads used in the general smart home is only responsible for the data acquisition of its related hardware system.

Control operations are independent from data acquisition. Home applications send commands to the middleware adapter via the unified application interfaces and throw them to a specific hardware driver or instance which acts upon the environment through its bundled controllers.

4.2   Using Callback or Delegation Functions

In practical smart home environment, it is always according to the change of monitoring status to trigger corresponding actions, but some monitoring sensors change their status very little or not during a period of time. So collecting and saving these unchanged status periodically would be a waste of processor resources and database spaces. It is better to get and save status data only when the underlying sensor devices change their status. However, the system application has trouble in completing this work because the underlying data acquisition is now implemented not by itself but by the hardware driver instance of the universal adaptation layer. The system application is difficult to find real-time status change of the monitoring sensors. It is more suitable for the middleware adapter to find the change of sensors' status and finish data acquisition.

Callback or delegation functions can be designed upon the unified application interface of the middleware adapter to provide special data access capability. The system application calls the callback function or redefines the delegation function in its implementation to achieve synchronous data acquisition completed by the bundled hardware interface instance. There is no need to set up a callback or delegation function for each monitoring object. For the same type of application interface, a unified callback or delegation function is enough. For example, 'IntStatusChange' can be designed for all the 'int' type data access, such as switched on/off operations, digital measurement, and 'RealStatusChange' can be designed for all the 'float' type data access, such as temperature and voltage detection. The system application gets the same type of changed status data of all the connected hardware devices via the same callback or delegation function and does further data processing, figure 4 shows the 'int' type data transmission. Developers should distinguish status data of different home applications in the re-implementation of the callback or delegation function.

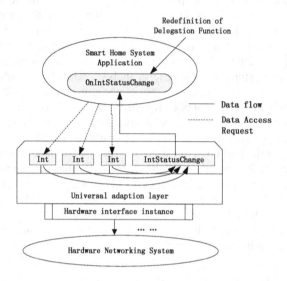

Fig. 4 Using 'Int' Type Delegation Function

As the middleware adapter running in each hardware system finishes underlying data acquisition of its related hardware devices and automatically triggers data transmission to the system application, while each type of the hardware system connected to the universal middleware adapter is loose-coupling, distributed data acquisition is implemented in the whole smart home system. By this way, it also simplifies the structure of the adaptation layer, and reduces system processing burden.

4.3   Using Middleware Integrated Database

In case of using callback or delegation functions to get real-time changed data, the system application has to read the collected data from all of the application interface units in the end. The concentrated data access between the system application and middleware adapter may cause large amount of burst data transmission on the network. Both of them may be involved in the simultaneously data acquisition and control of the multiple different types of hardware devices, especially in a large scale of smart home environment. So why not let the middleware adapter save the collected data to the database directly after discrete underlying data acquisition? The system application can get requested data from the networking database in real-time or when it needs. For this purpose, a database system is abstracted as a virtual hardware device and integrated into the middleware adapter, working as one type of hardware systems for distributed data storage and processing. With the integrated database, the middleware adapter completes data acquisition and saves the data directly to the database via its virtual database interface in each different networking technology based home hardware subsystem. The system application reads the data from the database for further analysis and treatment. Figure 5 shows this data transmission logic. Finally, dada acquisition and storage is completely separated from data analysis and processing, various parts of the whole system can do their works at the same time and provide timely access services for higher level. With the middleware integrated database, the structure of the smart home system is also optimized, and processing costs of various parts of the system are reduced.

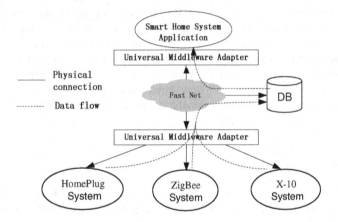

Fig. 5 Using Middleware Integrated Database

As each of the underlying hardware subsystems based on different networking technology has its own middleware adapter unit to finish its own data acquisition and storage, to receive operation commands and act upon the intelligent environment, there is almost no communications occurred between the multiple hardware subsystems. They are loose-coupled and almost irrelevant. With the function of the middleware integrated database, the system application can be totally decoupled from the multiple specific hardware subsystems. The system application doesn't need to care about how many network techniques used in the whole system, and how the hardware controllers and devices are networked. It only needs to get the required data and do further data processing. The system application gets required data from the networking database or by directly accessing the unified application interface of the middleware adapter. Finally, it is proposed that most of the connected devices (equipments and UI devices) have a built-in local database to store current status, streaming data, scheduled tasks and a list of commands in the queue.

5 Conclusion

As we can see, distributed data acquisition and control is very important in multi-platform smart home systems. It simplifies the system architecture and provides high quality and performance by separating underlying data acquisition from upper data processing. It makes the smart home system work in its full potential.

The given many methods of distributed data acquisition and control in this paper are based on the universal smart home architecture. It uses a middleware adapter to make itself the capability for distributed applications. It is easy to create a timer or special thread to timely get the collected data from the unified application interface of the middleware adapter, but it is difficult to only care of the changed status. By redefining the callback or delegation functions of the unified application interface, it is very efficient to get the changed data of the same type of sensors in one task and to do further distinguished data processing in the system application, but the concentrated data communications may bring burst data traffic and higher system load. So it is recommended to integrate the networking database system to the middleware adapter to complete the underlying data acquisition. By using the integrated database, each different networking technology based hardware subsystem completes its own data acquisition and storage, while receiving control commands from the system application and acting upon the intelligent environment.

References

1. Ricquebourg, V., Menga, D., et al.: The Smart Home Concept: our immediate future. In: 1ST IEEE International Conference on ELearning in Industrial Electronics, pp. 23–28 (December 2006)
2. Cheng, J., Kunz, T.: A survey on Smart Home Networking. Carleton University, Systems and Computer Engineering, Technical Report SCE-09-10 (September 2009)
3. Chunduru, V.: Effects of Power Lines on Performance of Home Control System. In: International Conference on Power Electronics, Drives and Energy Systems, pp. 1–6 (December 2006)
4. Kim, B.-K., Hong, S.-K., et al.: The Study of Applying Sensor Networks to a Smart Home. In: Fourth International Conference on Networked Computing and Advanced Information Management, vol. 1, pp. 676–681 (September 2008)
5. Ferrari, G.: Wireless Sensor Networks: Performance Analysis in Indoor Scenarios. EURASIP Journal on Wireless Communications and Networking 2007, 14 pages (2007) ID 81864
6. Bregman, D., Korman, A.: A Universal Implementation Model of the Smart Home. International Journal of Smart Home 3(3), 15–30 (2009)

Improving the Efficiency and Accuracy of SIFT Image Matching

Daw-Tung Lin and Chin-Hui Hsu

Abstract. Developing an accurate mechanism of correspondence and increasing matching stability are crucial tasks in many computer vision applications. This work improves the accuracy and efficiency in image matching via a novel method. The Modifiable Area Harmony Dominating Rectification (MHDR) method is proposed to eliminate mismatched key-point couples automatically and protect matching couples. The matching performance of the proposed scheme was evaluated on a test image database and via the transformation of the shearing effect and thin-plate splines. Compared with other methods, including the Exhaustive Search, Best Bin Search, and Sliding Midpoint Splitting, the proposed method had promising results in improving the accuracy and efficiency of the SIFT image matching.

1 Introduction

Image matching plays an important role in image analysis. The objective of image matching is to find point correspondences between two images in a similar scene or the same object. Scale Invariant Feature Transform (SIFT) extracts distinctive invariant features from images [9]. The SIFT algorithm is applied to two images to obtain key-points descriptors. Key-points from the two images are then compared via their descriptors. Most similar key-point couples from two images are found and deemed matching couples. To accelerate matching, Fridedman *et al.* proposed the k-d tree method, which can identify the exact nearest neighbors of points in space [5]. Nevertheless, the k-d tree algorithm is slower than an exhaustive search for feature space with more than 10 dimensions. Therefore, Lowe developed the best-bin-first (BBF) approach [3]. The BBF approach is approximate in the sense that it returns the closest neighbor with the highest probability. The BBF algorithm is computationally expensive. Arya and Mount developed the approximate nearest neighbors

Daw-Tung Lin · Chin-Hui Hsu
Department of Computer Science and Information Engineering, National Taipei University,
No. 151, University Road, Sanshia, New Taipei City, Taiwan
e-mail: dalton@mail.ntpu.edu.tw

F.L. Gaol et al. (Eds.): Proc. of the 2011 2nd International Congress CACS, AISC 145, pp. 227–233.
springerlink.com   © Springer-Verlag Berlin Heidelberg 2012

method and improved the outcome of searching [2]. Maneewongvatana and Mount demonstrated that the chosen of splitting rule affects the shape of cells and the structure of the resulting tree [10]. This method achieves efficient search times for the approximate nearest neighbor. Recent advances of the Locality Sensitive Hashing (LSH) technique also improved the efficiency of nearest neighbor search in high dimensional data space [6]. Weisheng and Youling developed a method for locating key-point couples [12]. The harmony is defined as a measurement of how a matching couple fits to the local area structure. A correctly matched pair always has high harmony, while an incorrectly matched pair always has very low harmony. This method eliminates mismatched key-point couples, and has short rectification time. However, when the number of key-point couples is massive, rectification time becomes slow. Thereby, the rectification method was improved, to keep rectification time stable. This work added the scale acquired from the first step of the SIFT method to examine the key-point couples. This step makes the verification process computationally efficient.

2 Modifiable Area Harmony Dominating Rectification Method

Following the three main steps for searching for discrete image point correspondences as presented in our previous study [7], key points from any two similar images are detected individually; these key-points are then converted into descriptors. In this step, the SIFT detector lays the foundation for image matching. These key-points are highly distinctive, and the descriptor outperforms existing feature descriptors. Finally, this work rectifies the incorrect match couples method using the proposed Modifiable Area Harmony Dominating Rectification (MHRD) method. Figure 1 shows a flowchart of the matching procedure with MHRD function block drawn in bold lines. Regardless of whether an error exists in key-point matching,

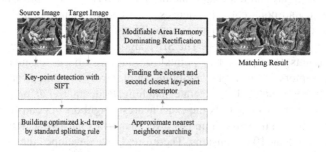

Fig. 1 Image matching procedure.

incorrect matching inevitably occurs with all algorithms. Weisheng and Youling developed a Area Harmony Dominating Rectification (AHDR) method for verifying that key-point couples are extracted by the SIFT algorithm from two relative images [12]. Harmony is defined as a measure of how a key-point couple fits the local area structure. The AHDR algorithm of matching point checking consists of three

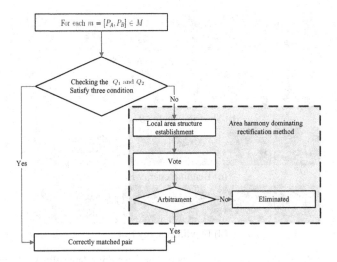

Fig. 2 Flow-chart of the proposed MHRD method.

major stages: (1) local area structure establishment, (2) vote, and (3) arbitrament. The AHDR method is both simple and fast. However, when the number of key-point couples is large, rectification is slow. Therefore, this work utilized the scale acquired from the DoG. Thus, nearby key-point couples are checked to determine if the distance is below a scale $k\sigma$: $disance(c,n) = \sqrt{(c_x - n_x)^2 + (c_y - n_y)^2} < k\sigma$, where (c_x, c_y) and (n_x, n_y) represent the key-point couples c and nearby key-point couples n, respectively. If the matched key-point couple is correct, each key-point has the same scale.

This work defines parameter Q_1 and Q_2 are appropriate quantities of nearby key-points in images I_A and I_B, respectively. This work then assesses whether Q_1 and Q_2 satisfy the following three conditions:

1. $Q_1 > 1$ and $Q_2 > 1$.
2. $Q_1 < N$ and $Q_2 < N$, where N denotes the number of total key-point couples.
3. $Q_1 = Q_2$.

If a key-point couple satisfies these three conditions, then it is a correctly matched pairs; otherwise, it must be checked by the standard area harmony dominating rectification method. Figure 2 shows the flow chart of the proposed process. The MHDR method is not only faster than the standard AHDR method, but also retains the accuracy of key-point couples matching

3 Experimental Results

The test image database was obtained from [1]. Five changes exist in imaging conditions, including viewpoint changes, scale changes, image blurring, JPEG compression, and illumination. The proposed method was also tested with 2D protein

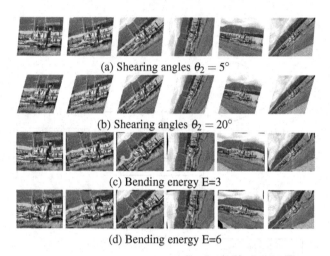

(a) Shearing angles $\theta_2 = 5°$

(b) Shearing angles $\theta_2 = 20°$

(c) Bending energy E=3

(d) Bending energy E=6

Fig. 3 (a), (b): Shearing effect on image "boat"; (c), (d): thin-plate splines effect on image "boat".

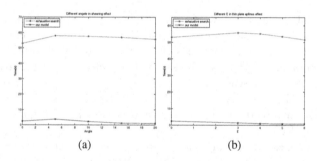

(a)   (b)

Fig. 4 The computational time performance of the proposed method compared with that of exhaustive search with (a) the shearing effect, and (b) thin-plate spline effect.

gel electrophoresis (2DEG) images used in [8]. The test images were also altered using the shearing effect and thin-plate spline effect from the test images [8, 4]. Figure 3(a) and (b) demonstrate the example images with different shearing angles ($\theta_2 = 5° \sim 20°$). Figure 3(c) and (d) illustrate the deformation resulting from a thin-plate spline transformation with E=3 and 6 on image "boat". Two-hundred and sixteen affine transformation images with various shear angles ($\theta_2 = 5° \sim 20°$) and 216 shape deformation images with various degree of thin plate spline transform ($E = 3 \sim 6$) were tested. Performance is based on image recall rate and precision of key-point couples matching.

3.1 Comparison of the Proposed Method and Exhaustive Search

In this work, the source image was compared with the other test images, including the original test images, the shearing effect, and thin-plate spline effect. Additionally, the exhaustive search method and the proposed model were compared. The proposed method was more efficient in term of computation time than that of the exhaustive search for both shearing and thin-plate spline effects. Figure 4 shows the simulation results. The recall rate of the proposed method outperformed the exhaustive search. Tables 1 shows the performance comparison of recall rate with the shearing effect and thin-plate spline effect. Tables 2 shows the performance comparison of precision rate with the shearing effect and thin-plate spline effect. In summary, the proposed model model was more efficient and accurate than exhaustive search.

Table 1 Recall rate comparison of the proposed method and exhaustive search method with the shearing effect and thin-plate spline effect.

Method	Shearing Effect (θ_2 value)					Thin-plate Spline Effect (E value)				
	0°	5°	10°	15°	20°	0	3	4	5	6
Proposed method	98.2%	98.2%	98.2%	98.2%	98.2%	98.2%	100%	100%	100%	98.2%
Exhaustive Search	96.3%	96.3%	96.3%	96.3%	96.3%	96.3%	98.2%	98.2%	94.4%	96.3%

Table 2 Precision rate comparison of the proposed method and exhaustive search method with the shearing effect and thin-plate spline effect.

Method	Shearing Effect (θ_2 value)					Thin-plate Spline Effect (E value)				
	0°	5°	10°	15°	20°	0	3	4	5	6
Proposed method	93.9%	93.9%	94%	92.9%	90.5%	93.9%	95.2%	94.4%	94.0%	93.6%
Exhaustive Search	92.3%	92.2%	91.6%	92.5%	87.4%	92.3%	92.9%	93.4%	86.9%	87.8%

3.2   Comparison of the Proposed Method with Other Methods

Furthermore, this work compared the performance with the Best Bin First (BBF) approach with a radius of 250, and construct a k-d tree by sliding midpoint splitting with an error bound of 30. The proposed model is better than exhaustive search and BBF in computation time, recall rate, and precision rate (Table 3). Sliding midpoint splitting was faster than the proposed method, however the recall and precision were better with the proposed approach. To demonstrate the advantage of the proposed method, false positive rate was further evaluated and compared with other methods using the same test images as in [11]. The proposed method is more precise than other methods (Table 4).

Table 3 Performance comparison of the proposed method and other methods.

Method	Time	Recall	Precision
Proposed Method	1.69 s	98.70 %	93.64 %
AHDR	2.56 s	95.0 %	93.64 %
Exhaustive Search	55.02 s	96.48 %	90.93 %
Best Bin First	31.28 s	92.76 %	94.38 %
Sliding Midpoint Splitting	0.26 s	98.15 %	91.36 %

Table 4 False positive (FP) rate of the proposed model compared with that of other methods. The experiments were based on the same test images from [11].

Method	FP	Method	FP
Proposed method	0.34	Cross correlation	0.72
GLOH	0.52	Steerable filters	0.78
SIFT	0.56	Spin images	0.84
Shape context	0.59	Differential invariants	0.87
PCA-SIFT	0.65	Complex filter	0.89
Moments	0.67		

4 Conclusion

This work improves the efficiency and accuracy in image matching via a novel MHDR method. The SIFT algorithm was used to detect key-points and transform detected points into descriptors. To improve the efficiency, this work proposed the MHDR algorithm to eliminate mismatched key-point couples. This rectification method is based on harmony, which is a measure of how much a matching couple fits the local area structure. This method eliminates key-point couples with low harmony and retains key-point couples with high harmony. The rectification result attain good matching accuracy. When the number of key-point couples was huge, the computational cost increased. To solve this problem, this work added scale information for the rectification method. This keeps computational time stable, and retains highly accurate key-point couples. Performance of the proposed method has been assessed on images with different effects. This work demonstrated that the proposed method outperforms other methods.

References

1. http://www.robots.ox.ac.uk/vgg/research/affine/.html
2. Arya, S., Mount, D.M., Netanyahu, N.S., Silverman, R., Wu, A.Y.: An optimal algorithm for approximate nearest neighbor searching fixed dimensions. Journal of the ACM (JACM) 45(6), 891–923 (1998)
3. Beis, J.S., Lowe, D.G.: Shape indexing using approximate nearest-neighbour search inhigh-dimensional spaces. In: Proceedings of IEEE Computer Society Conference on Computer Vision and Pattern Recognition, pp. 1000–1006 (1997)
4. Donato, G., Belongie, S.: Approximation Methods for Thin Plate Spline Mappings and Principal Warps. In: Heyden, A., Sparr, G., Nielsen, M., Johansen, P. (eds.) ECCV 2002. LNCS, vol. 2352, pp. 21–31. Springer, Heidelberg (2002)
5. Friedman, J.H., Bentley, J.L., Finkel, R.A.: An algorithm for finding best matches in logarithmic expected time. ACM Transactions on Mathematical Software (TOMS) 3(3), 209–226 (1977)
6. Haghani, P., Michel, S., Aberer, K.: Distributed similarity search in high dimensions using locality sensitive hashing. In: Proceedings of the ACM 12th International Conference on Extending Database Technology: Advances in Database Technology, pp. 744–755 (2009)

7. Hsu, C.H.: Improving the efficiency and accuracy of SIFT image matching. Master's thesis, National Taipei University (2009)
8. Lin, D.T.: Autonomous sub-image matching for two-dimensional electrophoresis gels using maxrst algorithm. Image and Vision Computing 28(8), 1267–1279 (2010)
9. Lowe, D.G.: Distinctive image features from scale-invariant keypoints. International Journal of Computer Vision 60(2), 91–110 (2004)
10. Maneewongvatana, S., Mount, D.M.: It's okay to be skinny, if your friends are fat. In: Proceedings of the 4th Annual Workshop on Computational Geometry, Center for Geometric Computing (1999)
11. Mikolajczyk, K., Schmid, C.: A performance evaluation of local descriptors. IEEE Transactions on Pattern Analysis and Machine Intelligence 27(10), 1615–1630 (2005)
12. Wei, Z., Weisheng, X., Youling, Y.: Area harmony dominating rectification method for SIFT image matching. In: Proceedings of the International Conference on Electronic Measurement and Instruments, pp. 2–935 (2007)

Morphological Characteristics of Cervical Cells for Cervical Cancer Diagnosis

Rahmadwati, Golshah Naghdy, Montse Ros, and Catherine Todd

Abstract. This paper investigates cervical cancer diagnosis based on the morphological characteristics of cervical cells. The developed algorithms cover several steps: pre-processing, image segmentation, nuclei and cytoplasm detection, feature calculation, and classification. The K-means clustering algorithm based on colour segmentation is used to segment cervical biopsy images into five regions: background, nuclei, red blood cell, stroma and cytoplasm. The morphological characteristics of cervical cells are used for feature extraction of cervical histopathology images. The cervical histopathology images are classified using four well known discriminatory features: 1) the ratio of nuclei to cytoplasm, 2) the diameter of nuclei, 3) the shape factor and 4) the compactness of nuclei. Finally, the images are analysed and classified into appropriate classes. This method is utilised to classify the cervical biopsy images into normal, pre-cancer (Cervical Intraepithelial Neoplasia (CIN)1, CIN2, CIN3) and malignant.

Keywords: cervical cancer, diagnosis, morphological characteristic.

Rahmadwati
Department of Electrical Engineering Brawijaya University, Malang, Indonesia
e-mail: rr516@uow.edu.au

Golshah Naghdy · Montserrat Ros
School of Electrical, Computer and Telecommunication Engineering,
University of Wollongong, New South Wales, Australia
e-mail: {golshah,montse}@uow.edu.au

Catherine Todd
Faculty of Computer Science and Engineering,University of Wollongong
in Dubai, Dubai, United Arab Emirates
e-mail: CatherineTodd@uowdubai.ac.ae

F.L. Gaol et al. (Eds.): Proc. of the 2011 2nd International Congress CACS, AISC 145, pp. 235–243.
springerlink.com   © Springer-Verlag Berlin Heidelberg 2012

1 Introduction

In 2009, there were 500,000 diagnosed cases. Approximately 250,000 women die annually from the condition [1]. Cervical cancer is caused by the Human Papillomavirus (HPV) type 16 and 18. Women are screened by medical doctors for detection of the virus, where early, accurate screening and detection through biopsy can reduce the likelihood of cervical cancer development, initiated from the virus. Effective screening and biopsy examination are impacted by the level of adequate laboratory facilities for tissue analysis, as well as the competency and experience of the health practitioner[2]. Variability of diagnosis and interpretation of the biopsy may result from pathologists subjectivity[3]. However, advances in imaging technology have offered great improvement in the analysis and understanding of medical image data and have contributed to earlier and more accurate diagnosis[4]. Through development of medical image processing techniques and computer-assisted examination, the impact of subjectivity in cervical examination could be reduced and provide the pathologist with assistance in the diagnosis process.

In nucleus segmentation and recognition of uterine cervical pap-smears, several researchers discuss the morphological feature characteristics for cervical cancer diagnosis in medical image decision support systems [5]. According to the mathematical morphological feature extraction of the digital image of cancerous pap smear, Thiran and Lassououi [6, 7] present the automatic method for cancerous tissue recognition based on morphological segmentation and cell shape analysis. Walker [8] conducted a study that describes the morphology of cervical cells based on the height and weight of subdivision of the squamous epithelium layer. Guillaud [9] presents 18 features for distinguishing the histopathology cervical images between normal and Cervical Intraepithelial Neoplasia (CIN) 3, koilocytosis and CIN 1, and all CIN 1, 2 and 3 using the Linear Discriminant Analysis (LDA). The result was good for differentiating between normal and CIN 3. In a 2003 study by G.J. Price (et al) [10] the CIN grades are determined, using the Decision Support System (DSS) that is designed based on 8 histological features. The DSS is constructed by a Bayesian belief network. The morphological characteristic of nuclei is reported by authors elsewhere[11]. The authors also reported on promising results using texture features obtained by Gabor filters for the classification of cervical cancer images [4]. Furthermore, this paper presents the classification of the cervical histology images based on the previous paper[11]. Image processing and analysis methods for data extraction from the histological images are provided to obtain prognostic features in the research which include nuclei cytoplasm ratio (NC ratio), nuclei diameter, shape factor of nuclei and compactness[11].

2 Data Collection

The cervical biopsy images were collected from the pathology anatomy laboratory of Saiful Anwar hospital in Indonesia and categorised by an expert pathologist based on the abnormal cell spread. The images were classified as normal, pre-cancer and malignant. The pre-cancer images were divided into: Cervical Intraepithelial Neoplasia (CIN) 1, CIN 2 and CIN 3. The cervical biopsies were scanned (resolution 4080 x

3072 pixels) at a magnification of 400x using a digital microscope. A total of 475 labelled images were used in this study, which consisted of 60 normal images, 70 CIN 1, 50 CIN 2, 50 CIN 3 and 245 malignant images. Figure 1 shows sample data for each of the normal, malignant and pre-cancer images (consisting of CIN 1, 2,3).

3   Segmentation Methodology

In this research, segmentation of regions of interest (nuclei) using a region based approach. The principle of this approach is based on the similarities between properties such as colour, intensity and texture that are classified in the same cluster. The aim of this algorithm is to segment the image into distinctive regions or patterns. Before the segmentation process, the pre-processing step is taken. The images are processed by a Median filter for improving the quality of image. The pre-processing step is described below:

3.1   Median Filter

A Median filter is used to reduce the noise of the images. The algorithm of the median filter involves sorting the value of each pixel of each neighbourhood in ascending order [12]. The median value of this ordered sequence is selected and becomes the middle value. Median filters are suitable filters for pre-processing requirements of this system because the filter preserves the edge of the image. The result of the median filter is shown in Figure 2, which presents an original image before and after filtering.

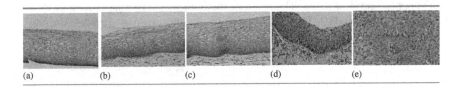

Fig. 1 Cervical images: benign (a) and malignant (b). The precancer is divided into images for CIN1(c),CIN2(d)andCIN3 These images were taken from Saiful Anwar Hospital, Indonesia

(a)   (b)

Fig. 2 (a) Original image (b) Filtered image

3.2   K-Means Clustering Algorithm

K-means clustering is one of the simplest methods for clustering. In this paper, the algorithm of K-means clustering is used to classify an image based on its features; the various classifications are represented by distinguished colours. The process of clustering is achieved by measuring the sum of distance between the data and the cluster centroid[12]. For the purpose of classification, the histopathology slides have five dominant colours: dark purple (nuclei), white (background), red (red blood cells), dark pink (cytoplasm) and light pink (stroma)[3,11]. The number of cluster is predefined, in this paper the number of class is five according to the number of dominant colour in histopathology images. According to the Euclidian distance metric of colour then each the regions are classified. The K-means clustering process is outlined below:

1. The filtered image is converted to CIE (L*a*b) colour space. The 'L' component describes the brightness of the colour. Component 'a' represents the colour placed along the red green axis and 'b' indicates the colour along the blue yellow axis.
2. There are five colours to be clustered. The number of clusters is determined to be five classes.
3. The initial cluster centroid locations are initialised and the colour distances are quantified by distances metric.
4. The segmented image is obtained.

The K-means result gives five separate images. Figure 3 shows the K-means clustering process. Figure 3(a), (b) and (c) show the original images of normal, pre-cancer and abnormal cervical tissue. The result of the K-means clustering process is shown in Figure 3 (d), (e) and (f). The segmentation results are shown as greyscale images.

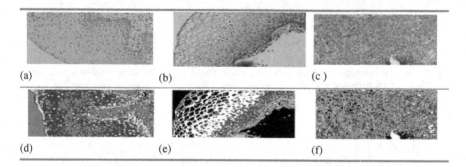

(a)   (b)   (c)

(d)   (e)   (f)

Fig. 3 (a),(b) and (c) are cervical cancer images normal, precancer and malignant respectively (d),(e) and (f) are normal, pre cancer and malignant images segmented by K-means clustering based on the colour.

3.3   Morphology Operation

The result of K-means clustering is the separation of the nuclei and cytoplasm into different regions. The aim of nuclei and cytoplasm separation is to determine the area of nuclei and cytoplasm. A morphological operation is used to improve the quality of the image; dilation and erosion operations are applied to the regions representing the nuclei and cytoplasm image data.

3.4   Matching Method for Correlation

The nuclei and cytoplasm have centroids that provide their positions. The correlation system is used to determine the nuclei and cytoplasm that correlate. The correlation matching method process is below:

1. Nuclei and cytoplasm are represented in separate images
2. The centroids of each nuclei and cytoplasm in the entire image are found
3. The nuclei and cytoplasm with the same location are identified.
4. The nuclei and cytoplasm that correlate are counted
5. The area, diameter and perimeter of the correlated nuclei are counted.

The areas of each corresponding nuclei and cytoplasm detected are calculated in order to calculate the feature referred to as nuclei to cytoplasm (N/C) ratio.

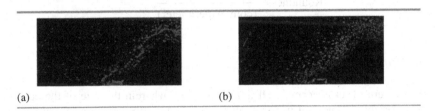

(a)   (b)

Fig. 4 (a) nuclei (b). cytoplasm

4   Morphological Features

Features are extracted from the segmentation results as discussed previously. Computer aided diagnosis of cancer depends on capturing variations of cell structure (cellular level) and changes in cells spreading in the tissue (tissue level) [13]. These changes are enumerated by the extraction of features. When the features are extracted, there are two types of information in the image: 1) the intensity values of the pixels and 2) the spatial arrangements of the pixels[3]. The following information describes the features that are extracted from the histopathology cervical cancer image based on the morphological characteristics [15].

4.1   Nucleus to Cytoplasm Ratio (N/C Ratio)

An important characteristic of malignancy is the value of nucleus to cytoplasm ratio
[6]. Commonly, abnormal cells have a high nuclei to cytoplasm ratio which may approach 1:1 instead of normal cell ratios which are much lower, between 1:4 and 1:6.
[5]. The first step for calculating the nuclei to cytoplasm ratio is to count the number
of nuclei and cytoplasm that are correlated. The next step is finding each area of nuclei
and cytoplasm, then calculating the N /C ratio using equation 3

$$N/C \text{ ratio} = \frac{\text{nuclei area}}{\text{cytoplasm area}} \qquad (3)$$

4.2   Shape

The second feature used to identify abnormal cells is shape. The abnormal cell
shape is irregular, wherein the cell shape is highly variable; for example it may assume shapes including that of: round, tadpole, bizarre, oval and caudate [15]. The
normal nuclei usually has a smooth and circular shape in comparison. The roundness factor is used to calculate the shape of the nuclei. The roundness factor of
normal nuclei is approximately 1 (circle). Equation 4 shows the calculation of
roundness of the nuclei.

$$\text{Roundness} = \frac{4\pi \text{ Area}}{(\text{convex perimeter})^2} \qquad (4)$$

4.3   Compactness

The structure of an abnormal cell is pleomorphic, wherein the size of the cell is
highly variable. Abnormal histopathology images tend to vary in terms of shape
and size of nuclei. The compactness is the ratio of the area of nuclei to the area of
the circle. In this paper, the variety of nuclei in shape and size are calculated by
the compactness factor.

$$\text{Compactness} = \frac{\text{perimeter}^2}{\text{Area}} \qquad \text{perimeter} = 2\pi r \qquad (5)$$

5   Result and Discussion

The number of samples that are used in this work is 75. In this work, all the
samples were segmented into nuclei and cytoplasm regions. The system then
measured the morphological features of each category. Table 1 represents the
morphological features of normal and abnormal cells.

Table 1 The result of morphological features

Type	C/N	Shape factor	Compactness	Diameter (pixel)
Normal	0.0265± 0.006	0.891	12.30	13.26±0.0045
Abnormal	0.2246±0.002	0.735	13.56	18.49±0.00593

It is interesting to note that the abnormal cells have the greatest values in the C/N ratio, compactness and diameter whilst the normal cells have the greatest roundness value. The system can distinguish the normal and abnormal cell based on the features above.

The morphological features of cervical cytology described previously were applied to 475 images. The classification process uses the threshold values that are found from table 1. All images are tested then the numbers of normal and abnormal cells in each image are obtained. Classification of cervical cancer is determined by the spread of abnormality in the image. The system classifies the images into normal, CIN 1, CIN 2, CIN 3 and malignant using the ratio of abnormal and normal cells. Classification is based on the following algorithm:

1. Benign: the ratio between normal and abnormal cells is less than 0.1
2. CIN 1: the ratio between abnormal and normal cells is not more than 1/3.
3. CIN 2: the ratio between abnormal and normal cells is more than 1/3 but not more than 2/3.
4. CIN 3: the ratio between abnormal and normal cells is more than 2/3 but not more than 1.
5. Malignant: the ratio between abnormal and normal cells is more than 1.

The performance of segmentation of cervical cancer nuclei based on the morphological features is determined by calculating the number of images correctly classified. This result is shown in Table 2.

Table 2 Confusion matrix of proposed method

Category	Normal	CIN 1	CIN 2	CIN 3	Malignant	Total
Normal	48	9	3			60
CIN 1		53	15	2		70
CIN 2		2	43	5		50
CIN 3		2	3	45		50
Malignant			8	18	219	245

The performance of the proposed methods is presented in Table 2 :for the normal case: 80%, CIN 1: 75%, CIN 2: 87%, CIN 3: 90% and malignant: 89%.

The proposed system does not give good result on CIN 1 because it is quite similar the appearance of the nuclei in normal and CIN 1. It is happens in CIN 1 and CIN 2, 15 images should be go to the CIN 1 but there is misclassification. The image classified in CIN 1 because it is quite difficult to distinguish the CIN 1 and CIN 2.

6 Conclusion

In this paper, the diagnosis of cervical cancer based on the morphological feature characteristics of cervical cells is investigated. The colour segmentation method and matching methods are used to obtain the nuclei and cytoplasm dimensions. Area, perimeter and diameter of nuclei were obtained to calculate the features of normal and abnormal cells of the histopathology cervical images. The results achieved using morphological feature characteristics is promising. The confusion between the normal and malignant samples is zero. While absolute false negative is close to zero, there is slight misclassification of malignant images into CIN3 or CIN2. While there is slightly higher false positive, the confusion is largely between normal and CIN. It is worth noting that all misclassifications are between normal versus pre-cancer or malignant versus pre-cancer and non between normal versus malignant.

Acknowledgments. The author gratefully acknowledge obtaining the histology images data in this work from pathology laboratory Saiful Anwar hospital in Indonesia specially to dr. Evianna Norahmawati SpPa and this work was supported by Indonesia Directorate General of Higher Education and Brawijaya University.

References

[1] W.H. Organisation, Strengthening Cervical Cancer Prevention and Control, Report of the GAVI-UNFPA-WHO meeting (June 27, 2010)
[2] Mukherjee, B.M.G., Bafna, U.D., Laskey, R.A., Coleman, N.: MCM immunocytochemistry as a first line cervical screening test indeveloping countries: a prospective cohort study in a regionalcancer centre in India. British Journal of Cancer 96, 1107–1111 (2007)
[3] Wang, Y.: Computer Assisted Diagnose of cervical intraepithelial neoplasia using histological virtual slide. Queen University of Belfast (2008)
[4] Rahmadwati, Naghdy, G., Ros, M., Todd, C., Norahmawati, E.: Cervical Cancer classification using Gabor filters. In: Proceeding in HISB Conference (2011)
[5] Kim, K.-B., Song, D.H., Woo, Y.W.: Nucleus segmentation and recognition of uterine cervical pap-smears. In: An, A., Stefanowski, J., Ramanna, S., Butz, C.J., Pedrycz, W., Wang, G. (eds.) RSFDGrC 2007. LNCS (LNAI), vol. 4482, pp. 153–160. Springer, Heidelberg (2007)
[6] Thiran, J.-P.: Morphological Feature Extraction for the Classification of Digital Images of Cancerous Tissues. IEEE (1996)
[7] Lassouaoui, L.H.N., Nouali, N.: Morphological Description of Cervical Cell Images for the Pathological Recognition. World Academy of Science, Engineering and Technology 5 (2005)
[8] Walker, D.C., Brown, B.H., Hose, D.R., Smallwood, R.H.: Modelling the electrical impedivity of normal and premalignant cervical tissue. Electronics Letters 36, 1603–1604 (2000)

[9] Guillaud, M., Cox, D., Malpica, A., Staerkel, G., Matisic, J., Niekirk, D.V., Adler-Storthz, K., Poulin, N., Follen, M., Macaulay, C.: Quantitative histopathological analysis of cervical intra-epithelial neoplasia sections: methodological issues. Cellular Oncology 26, 31–43 (2004)

[10] Price, G.J., Mccluggage, W.G., Morrison, M.L., Mcclean, G., Venkatraman, L., Diamond, J., Bharucha, H., Montironi, R., Bartels, P.H., Thompson, D., Hamilton, P.W.: Computerized diagnostic decision support system for the classification of preinvasive cervical squamous lesions. Human Pathology 34, 1193–1203 (2003)

[11] Rahmadwati, Naghdy, G., Ros, M., Todd, C., Norahmawati, E.: Classification Cervical Cancer using Histology Images. In: Proc. ICCEA, vol.1 (2010)

[12] Gonzales, R.C., Woods, R.E.: Digital Image Processing, 3rd edn. Prentice Hall (2007)

[13] Demir, C., Yener, B.: Automated cancer diagnosis based on histopathological images: a systematic survey. Technical Report, Rensselaer Polytechnic Institute, Department Of Computer Science

[14] Costa, S., et al.: Independent determinants of inaccuracy of colposcopically directed punch biopsy of the cervix. Gynecologic Oncology 90(1), 57–63 (2003)

[15] Koss, L.G.: Diagnostic Cytology and Its Histopathologic Bases, 5th edn., vol. 1. Lippincott Williams & Wilkin

A Study on a Method of Effective Memory Utilization on GPU Applied for Neighboring Filter on Image Processing

Yoshio Yanagihara and Yuki Minamiura

Abstract. In this paper, the methods of implementing neighboring filters on newly supplied Graphics Processing Unit (GPU) are described. In general, neighboring filters are always utilized in image processing. Mainly in consideration of memory accesses, four methods implementing neighboring filtering are proposed and compared. The experimental result shows that one of the proposed methods (called "4X-block") at the block size of 16 is the fastest among them, when loading and processing data in shared memory in GPU. It is also shown that this method is about 1.45X faster than the basic method implemented on GPU.

1 Introduction

Recently, Graphics Processing Unit (GPU) makes rapid progress to get higher performance. Most GPU has more than one hundred processing cores for calculation inside of one graphic chip. Its performance of calculation may become more 10X-100X than by Central Processing Unit (CPU). So, GPU can have high performance especially for applications of determinant calculation as array processing. And the parallel computing environment of Compute Unified Device Architecture (CUDA) is provided by NVIDIA Corporation to get high performance of such GPU [1]. In the field of manipulating huge data, CUDA is one of very useful techniques [2, 3, 4]. Also, as huge images must be maintained and calculated in image processing in medical applications [5, 6, 7, 8], such efficient environment must be useful.

Yoshio Yanagihara · Yuki Minamiura
3-3-138 Sugimoto Sumiyoshi Osaka Japan, Osaka City University
e-mail: yanagihara@info.eng.osaka-cu.ac.jp

F.L. Gaol et al. (Eds.): Proc. of the 2011 2nd International Congress CACS, AISC 145, pp. 245–251.
springerlink.com   © Springer-Verlag Berlin Heidelberg 2012

The field of Image processing contains various types of processing techniques, some of which is suitable for GPU calculation such as array processing, and others of which is difficult to be implemented on GPU. Examples of the former are filtering processing, mask process, neighboring filtering, and so on [9, 10, 11]. And high performance can be obtained on such processes. Examples of the latter are labeling processing [12], pixel discrimination processing and so on. But, also, good performance can be obtained in such processing using GPU. Although neighboring filter processing, which is one of the former processing, can be obtained high performance by implementation on GPU, the memory accessing of GPU must be considered under efficiency of the total processing of software.

Recently a new GPU architecture called Fermi is supplied by NVIDIA Corporation. Though this architecture has various functions added, some methods suitable for Fermi are not sufficiently studied yet, as contrast to its former architecture called GT200. In this paper, mainly in consideration of memory accesses, the four implementing methods for a basic process of neighboring filters are proposed and compared.

2 Mask Size and Memory on GPU

Process of neighboring filtering is one of basic processes of image processing and always utilized especially in the early stage of image processing. Process of neighboring filtering has several mask size of 3(width)x3(height), 5x5, and so on. As the size of mask becomes wider, the process becomes complicated, which means more generalized processing can be better to be applied. Actually, the most process of neighboring filtering having the wider size can be decomposited and represented by combination of plural processes of neighboring filtering having only the small mask size of, such as, 3x3. In this paper, the process of neighboring filtering of the mask size of 3x3 is the subject to study for the preliminary study. The equation of the process of neighboring filtering with the mask size of 3x3 is presented below, where m -1, 0, 1 and n = -1, 0, 1.

$$IMout[i][j] = \Sigma\Sigma \; IMin[i+m][j+n] \; mask[m+1][n+1] \qquad (1)$$

Recent Fermi GPU has various memories utilized for several calculation usages, which are shared memory (16/48KB), L1 cache memory, L2 cache memory, and global memory (except for register memory which is used for calculation in each core). By comparison of them, the access for the two of shared memory and L1 cache memory is fastest. The access for global memory is most slow. So, the effective usage of shared memory affects the performance of processing. In this study, the usage of shared memory is also discussed.

3 Method

In consideration of effective access of memory according to the mask size, the four methods are proposed and described in the following paragraphs. In (1), the count of the nine elements of 'mask' is few, so that all of the elements can be located in register memory for each core with fastest access in calculation.

3.1   Method-A (Line Access)

Method-A is basic implementation of the process of neighboring filtering. In this basic method, each core calculates the result of neighboring filtering for the corresponding pixel, f[i][j], on each line of image. That is, one core process one pixel. This method is like to an ordinary processing technique by CPU.

3.2   Method-B (Block Access)

In Method-B, each core calculates the result of neighboring filtering for the corresponding pixel, f[i][j], on each square block area of image. All pixel data in the square block is copied to the shared memory and processed on the shared memory. The size of the block is selected as 16x16 and 32x32 in consideration of the size of the shared memory and, also, effectiveness of memory transfer between global memory and shared memory.

3.3   Method-C (4X-Line Access)

In Method-C, each core calculates the result of neighboring filtering for the corresponding four pixels on each consecutive four lines of image. This means that each core processes the four points of interest of f[i][j], f[i][j+1], f[i][j+2], and f[i][j+3] simultaneously, where j is the number of the start line of the four consecutive lines and is presented as $j = 4k$ ($k = 0,1,2,...$). In this case, one core process for four pixels. The data of several common lines of image can be store in the shared memory, so that reduction of total size of memory to transfer between global memory and shared memory will be expected.

3.4   Method-D (4X-Block Access)

In this method, each core calculates the result of neighboring filtering for the corresponding four pixels on each square block area of image, such as described in Method-B. This means that the each core processes the four points of interest of f[i][j+0], f[i][j+1], f[i][j+2], and f[i][j+3], where all pixels are included in the square block area. In this case, one core process four pixels. The data of all common lines in a square block area of image can be store in the shared memory, so that reduction of the total size of memory to transfer between global memory and shared memory will be expected. The size of the block is selected as 16 or 32, similarly as described in Method-B.

By the theoretical restriction of neighboring processing, the first and the last line of images cannot be processed and also the pixels on both ends of image cannot. In all proposed methods, all of the processes are implemented under the satisfaction of the theoretical restriction. Also, on the cases of processing without loading and processing data in shared memory of GPU, selected pixels are transferred from global memory to shared memory under the satisfaction of the theoretical restriction.

4 Experiments and Result

The experiments are implemented and executed on GTX480 with Fermi architecture. The size of the processing image is 1024 x1024 pixels, where one pixel has 8 bit density. Each elements of the used mask has the number of double precision for presenting Gaussian distribution, so that whole calculation in filtering is implemented at double precision.

In the first experiment, the both of L1 and L2 caches of GPU are utilized. The block size of "block" and "4X-block" is selected at 32. In this experiment, shared memory is not utilized. Table 1 shows the processing times in micro seconds and the rating of fast in processing. The process times of the methods of "line" and "4X-line" are almost the same. The method of "4X-block" is the fastest among them, and its processing performance is almost 1.3X faster than the method of "line" or "4X-line".

Table 1 Processing time at each method

method	A (line)	B (clock) (32x32)	C (4X-line)	D (4X-block) (32x32)
time(usec)	349.5	303.4	343.3	265.1
Rate	1.0	1.15	1.01	1.31

In the second experiment, the effects of the block size of 16 or 32 in the methods of "block" and "4X-block" are compared. The table 2 shows the processing time in micro seconds and the rating of fast in processing. The both methods of "block" and "4X-block" at the block size of 32 are faster than at the block size of 16 (because of effectiveness of L1 and L2 caches).

Table 2 Processing time at "Block" and "4X-block" method

method	B (block) (16x16)	B (block) (32x32)	D (4X-block) (16x16)	D (4X-block) (32x32)
time(usec)	422.9	303.4	289.8	265.1
rate	1.0 : 1.39		1.0 : 1.09	

In the third experiment on using shared memory in GPU, the effects of the block size of 16 or 32 in the methods of "block" and "4X-block" are compared, where data of one block utilized in the "block" or "4X-block" is also loaded and processed in shared memory. The table 3 shows the processing time in micro seconds and the rating of fast in processing. On loading and processing data in shared memory of GPU, the both methods of "block" and "4X-block" at the block size of 16 are faster than at the block size of 32. This result is opposite to the contents in table 2.

Table 3 Processing time at "block" and "4X-block" method utilizing shared memory

method	B (block) (16x16)	B (block) (32x32)	D (4X-block) (16x16)	D (4X-block) (32x32)
time(usec)	287.9	350.0	217.1	249.0
rate	1.21 : 1.0		1.15 : 1.0	

In the fourth experiment on using shared memory in GPU, the results of the methods of "line" and "4X-line" are compared to the results of the methods of "block" or "4X-block". In this case, data of one line and the related lines utilized in the methods of "line" or "4X-line" is also loaded and processed in shared memory in GPU. Table 4 shows the processing time in micro seconds, the rating of fast in processing, and usage memory size in shared memory. The method of "4X-line" is faster than the method of "line". Furthermore, the method of "4X-block" is the fastest among them, and its process performance is about 1.45X faster than the method of "line". By comparison of usage memory sizes in shared memory, the method of "block" needs the smallest size for calculation and the method of "4X-block" is the next.

Table 4 Processing time at each method utilizing shared memory

method	A (line)	B (block) (16x16)	C (4X-line)	D (4X-block) (16x16)
time(usec)	314.5	287.9	249.7	217.1
rate	1.0	1.09	1.25	1.45
memory(bytes)	9216	768	18432	3072

5 Discussion

The recent architecture of GPU has various memories utilized for several calculation usages. Especially, shared memory has greater and the both of L1 and L2 cache memory are added, as contrast to the former architecture. For this reason, the authors considered that the processing suitable for the former architecture might not be suitable for the new architecture. So, the efficient processing method of neighboring filtering is studied.

In this paper, the four methods are proposed for processing basic neighboring filter of image processing implemented on GPU, which are called "line", "block", "4X-line", or "4X-block". Without loading and processing data in shared memory in GPU, the method of "4X-block" at the block size of 32 is the fastest among them. The experimental result shows that this method is about 1.31X faster than the basic method of "line".

On comparison of the block size of 16 or 32 in Table 2, the method of "4X-block" at the block size of 32 is faster than at the block size of 16. This result might be, the authors thought, caused by effectiveness of L1 and L2 caches according to the adequate size of memory transfer among memories included in GPU.

On loading and processing data in shared memory in GPU, the method of "4X-block" at the block size of 16 is the fastest. This method is about 1.45X faster than the basic method of "line". The method of "4X-block" at the block size of 32 is slow and has almost the same performance to the method of "4X-line", shown in Table 3 and Table 4.

The Table 3 on utilizing shared memory shows that the method of "4X-block" at the block size of 32 becomes slower than at the block size of 16. Visual profiler also supplied by NVIDIA Corporation shows the occupancy in the method of "4X-block" at the size of 32 is 66%, and only one thread block is assigned to each streaming multi-processor in GTX480. On the other side, the occupancy at the size of 16 is 100%, and six thread blocks can be assigned to each streaming multi-processor in GTX480. This means that GPU is effectively utilized on using the method of "4X-block" at the size of 16, the author think, in consideration of mechanism of GPU that, on the case of a thread waiting, the next thread replaced and could run.

6   Conclusion

In this paper, the four methods of implementing processes of neighboring filtering of image processing on GPU are proposed and compared. The experimental result shows that the method (called "4X-block") processing on each square block area of image, on which each core processes four points on consecutive lines simultaneously, is fastest among them. It is shown that the method of "4X-block" at the block size of 16x16, when loading and processing data in shared memory in GPU, is about 1.45X faster than the basic method.

References

1. Sanders, J., Kandrot, E.: Cuda By Example. Addison-Wesley (2010)
2. Munekawa, Y., Ino, F., Hagihara, K.: Acceleration Smith-Waterman Algorithm for Biological Database Search on CUDA-Compatible GPUs. IEICE Trans. Inf. & Syst., E93-D 6 (2010)
3. Chen, G., Li, G., Pei, S., Wu, B.: Gpgpu supported cooperative acceleration in molecular dynamics. In: Proc. Conf. of Computer Supported Cooperative Work in Design, pp. 113–118 (2009)
4. Nukada, S.M.A.: Auto-tuning 3-d fft library for cuda gpu. In: Proc. High Performance Computing Networking (2009)

5. Fontes, F.P.X., Barroso, G.A., Coupe, P., Hellier, P.: Real time ultrasound image de-noising. J. Real-Time Image Proc. (2010)
6. Yang, Y., Zhong, Z., Wang, J., Sorberg, T.: Real-Time GPU-Aided Ling Tumor Tracking. In: Fourth Symp. on Image and Video Technology (2010)
7. Yanagihara, Y.: A Study of Region Extraction and System Model on an Observation System of Time-Sequenced 3-D CT Images. In: Proc. ISITA, M-TA-4 (2008)
8. Yanagihara, Y.: A study about software architecture for realtime processing and smoothed presentation on an observation system of time-sequenced 3-D CT images. CARS 5,1, S341 (2010)
9. Fialka, O., Cadik, M.: FFT and Convolution Performance in Image Filtering on GPU. In: Proc. of ICIV, pp. 609–614 (2006)
10. Ogawa, K., Ito, Y., Nakano, K.: Efficient Canny Edge Detection Using a GPU. In: Proc. of ICNC, pp. 279–280 (2010)
11. Zhang, N., Chen, Y., Wang, J.: Image parallel processing based on GPU. In: Proc. of ICACC, pp. 367–370 (2010)
12. Kalentiv, O., Rai, A., Kemniz, S., Achneider, R.: Connected component labelling on a 2D grid using CUDA. J. Parallel Distrib. Compt. 71, 615–620 (2011)

A New Design for Digital Audio Effect of Flange Based on Subband Decomposition

Jianping Chen, Xiaodong Ji, JianliSn Qiu, and Tianyi Chen

Abstract. Traditional flange processing is carried out only from the time domain. The frequency characteristics of audio signals is not considered. This paper presents a new time-frequency processing scheme for flange effect, which is based on subband decomposition. An audio signal is broken down into several bands of low, middle, and high frequencies to be flanged separately. In each frequency band, the flanging parameters are selected according to the characteristics of the signal in that band. Theoretical analysis and experimental results show that the new flange processing system fits the human hearing sense and improves the effect of flange processing effectively.

1 Introduction

Digital audio effect is a digital signal processing technique used to modify an audio signal to produce special sound effects of virtual spaces and hallucinogenic coloring, which includes echo, reverberation, chorus and flange [6]. As one of the most often used special audio effects, flange can produce a sound effect of periodic wobbling and drifting, and is widely used in electric guitar and rock music.

Flange processing is actually a comb filtering effect in which an audio signal is dynamically delayed and then superimposed onto the original signal [4]. The technique of modulation is commonly used for flange processing. The delay time

Jianping Chen · Xiaodong Ji · Jianlin Qiu
School of Computer Science & Technology, Nantong University, Nantong, Jiangsu,
226019, China
e-mail: {chen.jp,ji.xd,qiu.jl}@ntu.edu.cn

Tianyi Chen
School of Electronic and Optical Engineering, Nanjing University of Science
and Technology, Nanjing, Jiangsu, 210094, China
e-mail: chenty-nt@qq.com

F.L. Gaol et al. (Eds.): Proc. of the 2011 2nd International Congress CACS, AISC 145, pp. 253–259.
springerlink.com   © Springer-Verlag Berlin Heidelberg 2012

of an audio input is modulated by a signal of very low frequency and varies from short to long periodically [1]. In the traditional method, the flange processing is conducted purely in time domain. The factor of frequency domain is not considered. However, the perception of the human ear to different frequencies of sound is different. In addition, it can be proved that the flanging effect of different frequency ranges of sound is not only related to the parameters of the flange processing, but also related to the frequency of the sound signal being processed. Therefore, it is necessary to consider a processing in frequency domain. We propose a new time-frequency domain flange processing system, in which an audio signal is broken down into several frequency bands to be flanged separately. To each band of the sound signal, proper flanging parameters are selected according to the characteristic of the signal in that band. Thus, the frequency characteristic and the human hearing sense of the audio signal are well considered, and the effect of the flange processing is improved.

2 Analysis of Traditional Flange Processing

The traditional model for digital flange processing is shown in Figure 1 [5].

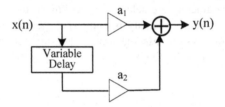

Fig. 1 Traditional model for flange processing

The input audio signal $x(nT)$ is passed through a variable delay unit to produce a dynamically delayed signal $x(nT-d(nT))$, where T is the sampling period and $d(nT)$ is a modulation signal with a very low frequency. Controlled by an attenuation coefficient a_2 ($0< a_2<1$), the delayed signal is superimposed onto the original signal $a_1x(n)$ (a_1 is an attenuation coefficient of x(n) with $0<a_1<1$). The output signal is as follows

$$y(nT) = a_1x(nT) + a_2x(nT - d(nT)).$$ (1)

The modulation signal could be a sinusoidal, triangle, logarithmic, or exponential wave. For convenience of analysis, we assume that a sinusoidal wave is used as the modulation signal:

$$d(nT) = \frac{D}{2}(1 - \cos(2\pi F_d nT)).$$ (2)

In equation (2), F_d is the modulation frequency and D is the maximum delay time. Without loss of generality, take a single-frequency sinusoidal signal as the input:

$$x(nT) = \cos(2\pi f_0 nT) \cdot \tag{3}$$

The second tern in (1) can be now expressed as

$$a_2 x(nT - d(nT)) = a_2 \cos\left[2\pi f_0 (nT - d(nT))\right]$$

$$= a_2 \cos(2\pi f_0 nT)\sin[2\pi f_0 d(nT)] + a_2 \sin(2\pi f_0 nT)\cos[2\pi f_0 d(nT)] \tag{4}$$

Let

$$\theta_1(nT) = a_2 \sin[2\pi f_0 d(nT)] \tag{5}$$

$$\theta_2(nT) = a_2 \cos[2\pi f_0 d(nT)] \tag{6}$$

Equation (1) becomes

$$y(nT) = a_1 x(nT) + \theta_1(nT)x(nT) + \theta_2(nT)x'(nT) \tag{7}$$

where $x'(nT) = \sin(2\pi f_0 nT)$ is the orthogonal signal of the input $x(nT)$.

As seen from (7), the output of the flanger is controlled by the modulation parameters $\theta_1(nT)$ and $\theta_1(nT)$. According to (5) and (6), the amplitude of the modulation is determined by the gain factor a_2 which is constant, and the variation rate of the modulation is determined by the modulation signal $d(nT)$ and the frequency f_0 of the input signal. The modulation signal $d(nT)$ is set at the beginning. The frequency f_0 is related to the input signal. Since in practice the audio signal to be processed is generally composed of many frequency components, it is implied that for different frequency components of the input signal, the modulation variation rates are different while the modulation amplitudes are the same. Such kind of likeness and difference of the modulation parameters between different frequency components is determined inherently by the flange processor, and the characteristic of the processed signal itself is not considered. Studies have shown that the sensitivity of the human ears is not the same for different frequency ranges of the audible sound from 20Hz to 20kHz. In order to achieve a good flanging effect, the modulation parameters should be selected separately for different frequency ranges of an audio signal. Therefore, it is necessary to consider the flange processing from the frequency domain.

3   New Flange Processing Based on Subband Decomposition

Studies have shown that the human ear is more sensitive to the sound of middle frequency range of 600Hz-4kHz, and less sensitive to the lower and higher frequency ranges [7]. It is also found that when the delay time of the superimposed signal is less than 1ms, the flanging effect is concentrated on the high frequency range; when the delay time is 3-17ms, the flanging effect is concentrated on the middle frequency range; and when the delay time is 17-35ms, the flanging effect is concentrated on the low frequency range [3]. Applying the method of subband decomposition, we use quadrature mirror filter banks to break down the input audio signal into three bands of low, middle, and high frequencies. Each band of the signal is applied flange processing separately. The modulation parameters are selected carefully to achieve a better flanging effect.

The block diagram of the subband based flange processing system is shown in Figure 2. The system is made of two dual-channel quadrature mirror filter banks [2]. $H_1(z)$ and $G_1(z)$ are the analysis filters of the first dual-channel quadrature mirror filter bank, and $B_1(z)$ and $K_1(z)$ are the corresponding synthesis filters. $H_2(z)$ and $G_2(z)$ are the analysis filters of the second dual-channel quadrature mirror filter bank, and $B_2(z)$ and $K_2(z)$ are the corresponding synthesis filters. The outputs of $H_2(z)$, $G_2(z)$ and $G_1(z)$ correspond to the low, the middle, and the high frequency bands of the input audio signal, respectively. They are then applied flange processing separately as shown with blocks FL1-3 in Figure 2. For the low frequency and the high frequency bands, to which the human hearing is not sensitive, we can increase the gain factor a_2 to increase the modulation amplitude. As stated previously, the flanging effect is concentrated on different frequency ranges for different delay time of the superimposed signal. We can modify the modulation signal $d(nT)$ appropriately to improve the flanging effect of the corresponding frequency range. Take the middle frequency range as an example. The traditional method is to set the maximum delay time at 17ms, such that $D=17$ms in (2). Thus, the delay time varies within the time limits of 0 to 17ms repeatedly at the frequency F_d. However, the flanging effect is concentrated on the middle frequency range only when the delay time is 3-17ms. The effect of the time delay between 0-3ms is not on the middle frequency range. So the superposition of the delayed signal between 0-3ms is not necessary for the middle frequency range. In order to set the delay time accurately for each frequency range, we modify (2) as follows

$$d(nT) = \frac{Dh - Dl}{2}(1 - \cos(2\pi f_d nT)) + Dl \tag{8}$$

where Dl and Dh correspond the lower limit and the upper limit of the delay time, respectively. For the middle frequency band, we may set $Dl=3$ms and $Dh=17$ms. Appropriate values of Dl and Dh for the low frequency and the high frequency bands can be set similarly. Thus, using the modified equation in (8) we can set the delay time flexibly and effectively for different frequency ranges.

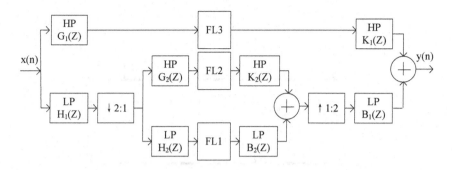

Fig. 2 New flange processing system based on subband decomposition

4 Experimental Results

The presented new flange processing system is simulated using Matlab. The sampling rate is set at 22.05kHz and the input signal is broke down into three frequency bands of 0-2.756kHz, 2.756-5.513kHz, and 5.513-11.025kHz. A mixed signal of three single-frequency sinusoidal signals of 300Hz, 3kHz and 10kHz is used as the input signal.

Figure 3 shows the frequency spectrum of the mixed-signal after the flange processing. The sub-figure (a) is the output of the traditional flange processing with $a_1=a_2=0.5$, $D=35$ms and $F_d=0.2$Hz. The sub-figure (b) and (c) are the outputs of the new flange processing system. The gain factor for both the low frequency and the middle frequency bands are $a_1=a_2=0.5$. The gain factor for the high frequency band is $a_1=0.8$ and $a_2=0.2$ in figure (b) and $a_1=0.2$ and $a_2=0.8$ in figure (c). The first three spectrum lines in Figure 3 represent the main components of the output signal at 300Hz, 3kHz and 10kHz (the other three lines are their mirrors). Around each main component, there are many side components distributed which are caused by the flange processing. Comparing figure (b) with figure (c), it can be seen that increasing or decreasing the gain factor a_1 of the high frequency band can independently increase or decrease the main component in the high frequency band, and increasing or decreasing the gain factor a_2 can independently increase or decrease the side components in the high frequency band, while the outputs of the low frequency and the middle frequency bands are not affected.

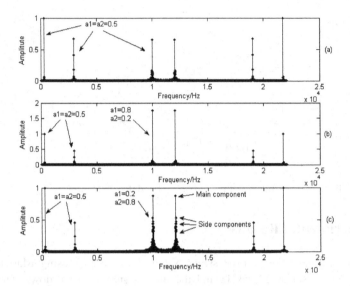

Fig. 3 Comparison of outputs of two flange processing systems

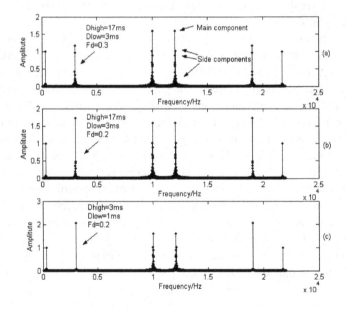

Fig. 4 Comparison of outputs of new flange processing system under different parameters

The flange processing for the different frequency bands can also be independently controlled by setting the frequency F_d and the time limit Dl and Dh of the modulation wave. Figure 4 shows the case that the middle frequency band is adjusted by means of selecting the modulation parameters. Comparing figure (a) with figure (b), it is seen that the distribution of the side components can be changed by adjusting the modulation frequency F_d. Comparing figure (b) with figure (c), it is seen that the amplitudes of the side components can be changed by adjusting the time limit Dh and Dl. In fact, the side components around the second spectrum line at 3kHz that is in the middle frequency band are almost disappeared in figure (c). This is because that the time limits in figure (c) is 1-3 ms. As indicated previously, the effect of the superposition of the delayed signal between 0-3ms is not acted on the middle frequency band. This proves the correctness that we modify the equation (2) into (9).

5   Conclusion

Theoretical analysis indicates that the traditional flange processing is conducted purely in time domain, and the characteristic of audio signals in frequency domain is not considered. We propose a new time-frequency domain processing system for flanging effect generation, which is based on the method of subband decomposition. In accordance with the human hearing sense, the audio signal is broken down into three bands of low, middle, and high frequencies to do flange processing separately. In each frequency band, the flanging parameters are selected according to the characteristic of the sound signal in that band. Theoretical analysis and experimental results show that with this subband decomposition based flange processing scheme, the flanging parameters for different frequency bands can be adjusted flexibly and effectively to enhance and improve the effect of flange processing.

Acknowledgments. Financial support from the Natural Science Foundation of Jiangsu Province (No. BK2010277) and the Application Research Program of Nantong City (No. K2007007) in China is acknowledged.

References

1. Caputa, M.: Developing Real-Time Digital Audio Effects for Electric Guitar in an Introductory Digital Signal Processing Class. IEEE Transactions on Education 41(4), 341–346 (1998)
2. Hoang, D., Le, H., Hoang, T., et al.: Multicriterion optimized QMF Bank design. IEEE Transactions on Signal Processing 51(10), 2582–2591 (2003)
3. Huang, C.: Processing Mode Analysis of Flanging Effect. Audio Technology (2), 58–61 (2007)
4. Sophocles, J.: Introduction to Signal Processing. Tsinghua University Press and Prentice Hall, Beijing (1998)
5. Udo, Z.: DAFX Digital Audio Effect. John Wiley & Sons, Sussex (2002)
6. Verfaille, V., Zolzer, U., Arfib, D.: Adaptive Digital Audio Effects (A-DAFx): A New Class of Sound Transformations. IEEE Transactions on Audio, Speech and Language Processing 14(5), 1817–1831 (2006)
7. Zhao, L.: Speech Signal Processing. Machinery Industry Press, Beijing (2003)

A Robust Algorithm for Arabic Video Text Detection

Ashraf M.A. Ahmad, Ahlam Alqutami, and Jalal Atoum

Abstract. In this paper, we propose an efficient Arabic text detection method based on the Laplacian operator in the frequency domain. The zero crossing value is computed for each pixel in the Laplacian-filtered image to found edges in four directions. K-means is then used to classify all the pixels of the filtered image into two clusters: text and non-text. For each candidate text region, the corresponding region in the canny edge map of the input image undergoes projection profile analysis to determine the boundary of the text blocks. Finally, we employ empirical rules to eliminate false positives based on geometrical properties. Experimental results show that the proposed algorithm is able to detect texts of different fonts, contrasts and backgrounds. Moreover, it outperforms four existing algorithms in terms of detection and false positive rates.

1 Introduction

Digital Arabic videos play an important role in entertainment, education and other multimedia applications. Hence, there is an urgent demand for tools that allow efficient browsing, indexing and retrieving of video data. In response to such needs, various video analysis techniques using single or multiple information sources like images, audio, and text in videos have been proposed to parse, index and abstract massive amount of data [12]. Among these information sources, texts in video frames play an important role in content understanding and in automatic indexing. In summary, texts in video frames provide highly condensed information about the content of a video and can be used for video skimming, browsing, and retrieval in large video databases.

Ashraf M.A. Ahmad · Ahlam Alqutami · Jalal Atoum
Princess Sumaya University for Technology, Amman, Jordan
e-mail: {a.ahmad,atoum}@psut.edu.jo

F.L. Gaol et al. (Eds.): Proc. of the 2011 2nd International Congress CACS, AISC 145, pp. 261–266.
springerlink.com   © Springer-Verlag Berlin Heidelberg 2012

There exist mainly two types of texts occurrences in videos, namely artificial and scene texts. Artificial text is artificially added in order to describe the content of the video or to give additional information related to it. This makes it highly useful for building keyword indexes. Scene text is textual content that was captured by the camera as part of the scene. Although embedded texts provide important information about the image, it is not an easy problem to reliably detect and localize the text embedded in images dues to the following reasons: in a single frame, the size of the characters can vary from very small to very big, the font can be different, the text presented in the same image can have multiple colors, and a text can occur in a much cluttered background.

For video sequences, the text can be either still or moving in an arbitrary direction. The same text may vary in its size from frame to frame, due to some special effects. The background can also be moving or changing, independent of the text. Hence, these problems are very challenging and important for several applications. Moreover, until now no method had been proposed to handle the Arabic texts embedded in Arabic video frames. From the literature review for other languages such as English, Chinese etc..., we realized that connected component based methods assume that the text pixels belonging to the same connected region share some common features such as color or grey intensity [1, 3]. On the other hand texture based methods may be unsuitable for small fonts and poor contrast text [6, 12] while edge based methods return more false alarms and are not robust for complex background images.

Wong and Chen had proposed a new robust algorithm for video text extraction [4]. They have considered too many thresholds to detect the text pixels. Hence, their method is too sensitive to threshold setting. Liu et al had proposed text detection in images algorithm that rely on unsupervised classification of edge based features [2]. Their method had considered both texture and edge features for detecting text in video images. However, this method requires more computational time because of large feature set, also it fails to detect small fonts and low resolution texts as the Sobel edge method which detects only edges that have high contrast [7]. Mariano and Kasturi had presented a method for locating uniform colored text in video frames [5]. Their method expects uniform color for text lines. In addition, this method is based on the transition between the characters and words. However, this method may be unsuitable for scene text due to non-uniform spacing between the characters and the words. A generalized method for text detection had proposed by Antani et al [8]. This method extracts almost all kinds of texts. However, this method assumes that the text and the background in a localized region have consistent grey levels in which all characters are either lighter than or darker than the background. In addition, connected component analysis is used in this work may eliminate single characters like digits in sports video frames. Thus the detection rate may get reduced.

Therefore, there is a need for more research in issues related to detecting texts in video images. Some of these issues are listed below:

1 Better recall and precision rate for both graphic and scene text detection.
2 Simple and efficient methods for detecting different fonts, styles, sizes, scripts and small fonts with poor contrast.

In this paper, we consider four existing methods [1, 4, 9, and 11] for comparative study. Liu et al. [4] extract edge features by using the Sobel operator. This method is able to determine the accurate boundary of each text block. However, it is sensitive to the threshold values for edge detection. Wong et al. [11] compute the maximum gradient difference values to identify candidate text regions. This method has a low false positive rate but uses many threshold values and heuristic rules. Therefore, it may only work well for specific datasets. Finally, Mariano et al. [1] perform clustering in the L*a*b* color space to locate uniform colored text. Although it is good at detecting low contrast text and scene text, this method is extremely slow and produces many false positives. In [9] proposed an efficient method for text detection based on the Laplacian operator. The K-cluster algorithm used to identify the text boundary box and the edge information serves to determine the accurate boundary of each text block.

In this paper, we propose an algorithm that is based on two algorithms presented in [9, 10]. These two algorithms are suitable for detecting Arabic letter characteristics in which regions typically are assumed to have large number of discontinuities. For instance, transitions between corresponding letters and their background in [10] handle the low contrast of video Arabic text against complex local backgrounds. These two algorithms will be combined in hybrid manner to build our proposed robust algorithm for detecting an Arabic text embedded in a video scene.

This paper is organized as follow: section 2 presents the proposed algorithm, experimental results are presented in section 3 and finally the conclusion is presented in section 4.

2 Proposed Methodology

We propose a text detection algorithm that consists of three steps as shown in Fig.1: text detection, boundary refinement and false positive elimination. In the first step, candidate text regions are identified using the Laplacian operator in frequency domain. In the second step, the accurate boundaries of each text block are determined using the projection profile analysis. Finally, in the third step, false positives are removed based on the geometrical properties. Each of these three steps will be explained in the following subsections.

2.1 Text Detection

Text regions typically have a large number of Discontinuities, and because video text can have a very low contrast against complex local backgrounds, it is important to preprocess the input image to highlight the differences between text and non-text regions. Fourier-Laplacian is used initially for filtering in order to smoothes the noise in the image and then to detect candidate text regions.

Fig. 1 The proposed Arabic Text Detection algorithm.

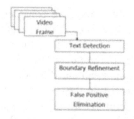

Ideal lowpass filter in the frequency domain is used to smooth the noise; the rationale is that the higher the frequency components of the Fourier transform the higher the information obtained about the noise and thus it should be cut off. The input image then should be converted again into grayscale and filtered by a 3 × 3 Laplacian mask to detect the discontinuities in the four directions: horizontal, vertical, up-left and up-right. To combine this step with the previous step, we use the Laplacian operator in the frequency domain (instead of the spatial domain).

As shown in Fig.2, the Laplacian masking process produces two values for every edge. The transitions between these values (the zero crossings) correspond to the transitions between the text and the background. In order to capture the relationship between positive and negative values, our proposed algorithm, unlike [9,10], does not rely on the maximum gradient value in which its accuracy depends on the sliding windows size. The proposed algorithm finds the zero crossing in four directions by multiplying the image by four matrices shown in Fig.2. The matrix in Fig.2.a is used to find the horizontal edge. The Matrix in Fig.2.b is used to find the vertical edge. In order to find all edges in the four directions; the matrices in Fig.2.c and 2.d are used for detecting edges with 45 and 135 degrees (masks), respectively. Consequently, the K-mean cluster is used to classify text and non text regions. The k-mean cluster is used in our proposed algorithm due to its simplicity and its independence of a threshold value.

Fig. 2 a) Orignale frame image. **b)** grayscale image.
c) Laplacian-filtered image. **d)** final result.

3   Experimental Results

As there is no standard benchmarking dataset available, we have selected a variety of video images, extracted from news programs, sports videos and movie clips. These video images dataset have both graphic texts and scene texts. The image sizes range from 320X240 to 816X48. The parameter values are empirically determined as: T1 = 0.5 and T2 = 0.1.

3.1 Performance Measures

The performance at the block level is evaluated rather than at the word or character levels because. The following categories are defined for each detected block by the proposed text detection algorithm.

1. Truly Detected Block (TDB): A detected block that contains at least one true character. Thus, a TDB may or may not fully enclose a text line.
2. Falsely Detected Block (FDB): A detected block that does not contain text.
3. Text Block with Missing Data (MDB): A detected block that misses more than 20% of the characters of a text line (MDB is a subset of TDB). The percentage is chosen according to [2], in which a text block is considered correctly detected if it overlaps at least 80% of the ground-truth block.

For each image in the dataset, we also manually count the number of Actual Text Blocks (ATB), i.e. the true text blocks.

The performance measures used in the experiments are defined as follows.

$$\text{Recall (R)} = \text{TDB} / \text{ATB}$$

$$\text{Precision (P)} = \text{TDB} / (\text{TDB} + \text{FDB})$$

3.2 Sample Results

Table 1 shows that our method has a high Recall with minimum Precision; Therefore, the proposed method has achieved better detection results than the four existing methods on the dataset.

Table 1 Performance on the Dataset (Precision and Recall)

Method	Recall	Precision
Edge-based [4]	0.8	0.82
Gradient-based [11]	0.71	0.88
Uniform-colored [1]	0.51	0.73
[9]	0.93	0.92
Proposed	0.96	0.95

4 Conclusion

We have proposed an efficient a simple and novel method for detecting Arabic text in a video frames based on the Fourier- Laplacian operation, the gradient information obtained from zero crossing operation helps to identify the candidate text regions and the edge information obtained serves to determine the accurate boundary of each text block.

Experimental results show that the proposed algorithm outperforms the four existing methods in terms of detection and false positive rates.

References

1. Jain, A.K., Yu, B.: Automatic Text Location in Images and Video Frames. Pattern Recotion 31(12), 2055–2076 (1998)
2. Liu, C., Wang, C., Dai, R.: Text Detection in Images Based on Unsupervised Classification of Edge-based Features. In: IEEE ICDAR, pp. 610–661 (2005)
3. Lee, C.W., Jung, K., Kim, H.J.: Automatic text detection and removal in video sequences. Pattern Recognition Letters 24, 2607–2623 (2003)
4. Jung, K., Kim, K.I., Jain, A.K.: Text information extraction in images and video: a survey. Pattern Recognition 37, 977–997 (2004)
5. Mariano, V.Y., Kasturi, R.: Locating Uniform-Colored Text in Video Frames. In: IEEE 15th ICPR, vol. 4, pp. 539–542 (2000)
6. Ye, Q., Huang, Q., Gao, W., Zhao, D.: Fast and robust text detection in images and video frames. Image and Vision Computing 23, 565–576 (2005)
7. Ye, Q., Gao, W., Wang, W., Zeng, W.: A Robust Text Detection Algorithm in Images and Video Frames. In: IEEE ICICSPCM, pp. 802–806 (2003)
8. Antani, S., Crandall, D., Kasturi, R.: Robust Extraction of Text in Video. In: IEEE 15th ICPR, vol. 1, pp. 831–834 (2000)
9. Phan, T.Q., Shivakumara, P., Tan, C.L.: A Laplacian Method for Video Text Detection. In: 2009 10th International Conference on Document Analysis and Recognition (2009)
10. Phan, T.Q., Shivakumara, P., Tan, C.L.: A Laplacian Approach to Multi-Oriented Text Detection in Video. In: IEEE Transactions on Pattern Analysis and Machine Intelligence (2010)
11. Mariano, V.Y., Kasturi, R.: Locating Uniform- Colored Text in Video Frames. In: 15th ICPR, vol. 4, pp. 539–542 (2000)
12. Zhong, Y., Zhang, H., Jain, A.K.: Automatic Caption Localization in Compressed Video. IEEE Trans. Pattern Analysis and Machine Intelligence 22(4), 385–392 (2000)

Detection of Lip Area and Smile Analysis for Self Smile Training

Won-Chang Song, Sun-Kyung Kang, Jin-Keun Dong, and Sung-Tae Jung

Abstract. In this study, we propose an automated smile analysis system for self smile training. The proposed system detects face area from the input image with the AdaBoost algorithm. It then identifies facial features based on the face shape model generated by using ASM(Active Shape Model). Once facial features of the face are identified, the lip line and teeth area necessary for smile analysis are detected. An analysis of the lip line and individual teeth areas allows for an auto-mated analysis of smiling degree of users, enabling users to check their smiling degree on a real time basis. The developed system exhibited an average error of 0.7 compared to previous smile analysis results released by dental clinics for smile training, and it is expected to be used directly by users for smile training.

1 Introduction

Smile is one of the important factors determining the impression of a person. On this account, there is a growing need for smile training which helps people smile better. Another advantage of smile training is the adjustment of tension and facial muscles. Sufficient smile training improve the overall look, making the face full of vitality[1]. This is why many people make efforts to find a way train good smile.

For training to practice good smile, some methods have been proposed; users can use a tool to maintain smile line and practice repeatedly as in [2]. They also can practice smile looking at the smile line drawn in the mirror as in [3], but in this method, it is impossible to check whether the smile is good or not because us-ers simply copy smile in the mirror. Another training method is to go to a smile

Won-Chang Song · Sun-Kyung Kang · Sung-Tae Jung
Dept. of Computer Eng., Wonkwang Univ. Iksan, Korea
e-mail: {colorart,doctor10,stjung}@wku.ac.kr

Jin-Keun Dong
Dept. of Prosthodontics, Wonkwang Univ., Iksan, Korea
e-mail: dong@wku.ac.kr

F.L. Gaol et al. (Eds.): Proc. of the 2011 2nd International Congress CACS, AISC 145, pp. 267–272.
springerlink.com   © Springer-Verlag Berlin Heidelberg 2012

clinic offered by a dentist office for diagnosis and regular treatment. This method helps prevent mistakes to practice misled smiles because the doctor informs the patient which is a good smile.

In this paper, an automated smile analysis system is proposed to solve the problem of the difficulty for users to judge good smile by themselves and save cost arising from diagnosis process at a clinic.

2 Detection of Face Features

In this study, we employed AdaBoost algorithm to detect face area from image input[5]. Following the detection of the face area, it is necessary to detect facial features from the detected face area. In this study, we detected facial features with the ASM[6]. In order to express facial features of this model, we set 86 landmarks along the boundary of the facial features for training purposes.

Once face area is detected from the images, the average face shape obtained from the ASM training module is placed over the detected face image. As the average face shape may be different from the detected face image in terms of size or direction, scaling and rotation are applied. From the face shape placed like this, the system begins to search the face shape fit to facial features in the image, changing the shape. The searching process stops when the change of the detected face shape remains lower than a certain level. In order to search for a new shape changed from a certain shape, the system find a profile vector g, moving landmarks along the line vertical to the boundary of each landmark and then finds the most similar location with the trained profile model. To know at which point profile vector is similar to profile model, the Mahalnanobis distance is calculated with the average profile vector $\bar{g}$ of the profile model and covariance matrix S_g as in Equation(1):

$$Mahalnanobis\ distance = (g - \bar{g})^T S_g^{-1}(g - \bar{g}) \qquad (1)$$

By moving individual landmarks using the profile model, the system creates a new shape. The new shape, however, is created by moving the landmarks individually, and therefore, some of the landmarks may be situated at an unstable location. Accordingly, it is necessary to adjust the new shape so that is can be placed within the range of the face shape model.

The new shape is called X, and an approximate shape to X can be calculated from the training shaped model. Using an iterative method, T and ω which minimize the value of Equation (2) are calculated to generate, $\hat{X} = T(\bar{X} + P\omega)$ which approximates X is a conversion that maps the shape model into shape space of the image. Now, the $\hat{X}$ shape is used to repeat the above process.

$$distance\ (X, T(\bar{X} + P\omega)) \qquad (2)$$

3   Detection of Lip Line and Teeth

The detection of lip line and teeth is an important part of smile training and smile measurement. In this paper, we used the following method to detect teeth. First of all, we detected teeth area surrounded by 73-84 landmarks.

Next, six teeth of the upper jaw in the middle were detected as they are easy to detect. As a preprocessing, the 2nd differentiation was conducted in the detected teeth area using a 3x3 size Sobel mask for edge detection to obtain the edge information image both in the vertical and horizontal directions as shown in Fig. 1. From the image of the differentiation in Fig. 1, we see that the pixel of the vertical and horizontal boundaries of teeth looks lighter than inside of teeth.

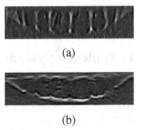

(a)

(b)

Fig. 1 Differentiation image (a)Horizontal differentiation (b) Vertical differentiation

Calculating the average pixel values of each columns result in the histogram projection as in Fig. 2 (a). Removing columns with smaller values than surrounding columns leads to the results in Fig. 2 (b). Among them, column closest from both left and right to landmark 81 is determined as the centermost boundary of teeth. Ones the centermost boundary of the teeth is determined, the outer boundary of two teeth in the middle is to be found. As shown in Fig. 2 (c), rows other than seven lightest rows exceeding average brightness are removed from the rows within a certain area on both sides. Since size of teeth is bilaterally symmetric, in general, two rows which are at a similar distance from the centermost boundary of the teeth and of which average brightness ranks high are selected as the boundary of teeth, among remaining rows. After finding the vertical boundary of two teeth in the middle, the horizontal boundary of the two teeth is to be found.

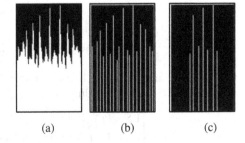

Fig. 2 Detection of vertical boundary (a) histogram of the distribution of average pixels of columns (b) removal of noises (c) boundary detection

(a)   (b)   (c)

Pixel averages of rows for the purpose of finding the horizontal boundary are shown in the histogram projection results in Fig. 3 (a). Removing noisy columns except for those with peak values in Fig. 3 (a) results in the horizontal boundary of teeth, as in Fig. 3 (b).

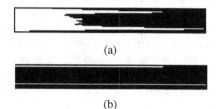

(a)

Fig. 3 Detection of horizontal boundary (a) histogram of the distribution of average pixels of rows (b) boundary detection

(b)

4 Smile Analysis and Smile Assessment

Smile level is assessed on a 9-point scale, considering the following six conditions[4]. Fig. 4 shows the locations of main landmarks CLow(81), LCh(84), Rch(78), and CLab(75) which are used for smile assessment.

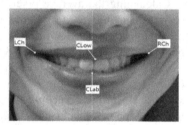

Fig. 4 Main landmarks for smile assessment.

- 1. Location of upper lip(A): The location of upper lip is assessed using Equation (3), where T_y refers to y coordinate of the upper boundary of upper central teeth, $CLow_y$ to y coordinate of Landmark 81, and T_w and T_h to horizontal and vertical width of upper central teeth, respectively.

$$A = \begin{cases} 1.5 & CLow_y - T_y \leq 3 \quad \text{and} \quad T_w < T_h \\ 1.0 & CLow_y - T_y > 3 \\ 0.5 & CLow_y - T_y \leq 3 \quad \text{and} \quad T_w \geq T_h \end{cases} \tag{3}$$

- 2. Bend of upper lip(B): Bend of upper lip is calculated using Equation (4). RCh_y and LCh_y are y coordinates of both corners of the mouth.

$$B = \begin{cases} 1.5 & RCh_y > CLow_y \quad \text{and} \quad LCh_y > CLow_y \\ 1.0 & RCh_y < CLow_y \quad \text{or} \quad LCh_y < CLow_y \\ 0.5 & RCh_y < CLow_y \quad \text{and} \quad RCh_y < CLow_y \end{cases} \tag{4}$$

- 3. Relationship between upper teeth and lower lip(C): If the angle between lower incisor tip of the six upper teeth and horizontal axis is 10 degree or over, and the lower incisor tip of the teeth bends upward and placed evenly with superior border of lower lip, then the aesthetic value is higher, so 2.0 is given. Meanwhile, if the angle is below 10, the lower incisor tip of the upper teeth becomes close to a straight line, lowering the aesthetic value, so 1.0 is given. . 12 depicts examples by the relationship between upper teeth and lower lip.
- 4. Contact between upper teeth and lower lip(D): Contact between lower lip and upper teeth is calculated with Equation (5), where T_{1y} refers to y coordinate of the lower boundary of two teeth in the middle and $CLab_y$ to y coordinate of landmark 75.

$$
D = \begin{cases} 1.0 & |T_{1y}-CLab_y| \leq 1 \ and \ T_w < T_h \ \ or \\ & |T_{1y}-CLab_y| > 1 \ and \ |T_{1y}-CLab_y| < 6 \\ \\ 0.5 & |T_{1y}-CLab_y| \leq 1 \ and \ T_w \geq T_h \ \ or \\ & |T_{1y}-CLab_y| > 6 \end{cases}
\tag{5}
$$

- 5. Teeth exposed when smiling(E): The assessment of the teeth exposed when smiling is calculated using the relationship between the section of the six teeth in the upper jaw and the section of both corners of the mouth as in Equation (6), where T_{Lx}, T_{Rx}, RCh_x, and LCh_x mean x coordinates of the relevant points in Fig. 5, respectively.

$$
E = \begin{cases} 2.0 & \dfrac{RCh_x - LCh_x}{T_{Rx} - T_{Lx}} < 0.7 \\ \\ 1.5 & 0.7 \leq \dfrac{RCh_x - LCh_x}{T_{Rx} - T_{Lx}} < 0.8 \\ \\ 1.0 & 0.8 \leq \dfrac{RChx - LChx}{T_{Rx} - T_{Lx}} < 0.9 \\ \\ 0.0 & 0.9 \leq \dfrac{RChx - LChx}{T_{Rx} - T_{Lx}} \leq 1.0 \end{cases}
\tag{6}
$$

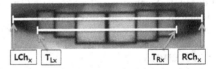

Fig. 5. Teeth exposed when smiling.

- 6. Symmetry of smile(F): The left-right symmetry degree of smile is calculated as in Equation (7). If the value is between 0.9 and 1.1, which is considered symmetric, then 1.0 point is given, for other values, which are considered asymmetric, the 0.5 point is given.

$$
Symmetry = \frac{\text{Distance between RCh and CLow} + \text{Distance between RCh and CLab}}{\text{Distance between LCh and CLow} + \text{Distance between LCh and CLab}}
\tag{7}
$$

5 Results

In this study, we used 500 training images taken with a digital camera, which consist of 10 various face images of 50 people. Each image is marked with 86 landmarks. Separate ten images taken by a dental clinic were used to test the accuracy of smile assessment. For the test images, facial features were correctly detected. Individual teeth areas for smile analysis were also detected correctly.

In an attempt to test the accuracy of smile assessment, we compared scores given manually to the ten images by dental clinic professional with the assessment results of the proposed system and found the average error of 0.7 between the two assessment results, which indicates that there was no big difference between them, as in Table 1. From these results, we conclude that the proposed system can help users analyze and assess their own smile successfully.

Table 1 Comparison of resluts of smile analysis by expert and by the developed system

	LT1	LT2	LT3	LT4	LT5	LT6	LT7	LT8	LT9	LT10
Dental expert	7.0	8.5	7.0	8.5	6.5	8.0	7.0	6.0	7.0	7.0
The developed system	7.5	7.5	9.0	8.0	6.5	7.5	7.0	7.0	7.5	8.0
Error	0.5	1.0	2.0	0.5	0.0	0.5	0.0	1.0	0.5	1

6 Conclusion

In this paper, we developed a system that helps detect lip area and analyze smile for self smile training. Face area was detected from the images, and facial features were detected from the face area, which were followed by the detection of lip line and teeth area. Individual teeth were separated from the detected teeth area for smile assessment. The proposed system is expected to be used for self smile training.

Acknowledgments. "This work was supported by the Grant of the Korean Ministry of Education, Science and Technology" (The Regional Core Research Program/Center for Healthcare Technology Development).

References

1. Gibson, R.M.: Smiling and Facial Exercise. Dent. Clin. North Am. 33(2), 157–164 (1989)
2. Kim, Y. H.: Beauty smile trainer. Samkuk Trading Co. Ltd., 20-2004-0026216
3. Kim, H.-J.: Smile line mirror, 20-2006-0009758
4. Yoon, M.E., Jin, T.H., Dong, J.K.: A Study on The smile in korean youth. The Korean Academy of Prosthodontics 30(2), 259–271 (1992)
5. Viola, P., Jones, M.: Rapid object detection using a boosted cascade of simple features. Computer Vision and Pattern Recognition 1, I-511– I-518 (2001)
6. Cootes, F., Taylor, C.J., Cooper, D.H., Graham, J.: Active Shape Models - Their training and application. Comput. Vis. Image Understating 61(1), 38–59 (1995)

Algorithm for the Vertex-Distinguishing Total Coloring of Complete Graph

Li Jingwen, Xu Xiaoqing, and Yan Guanghui

Abstract. A new algorithm whose name is algorithm of classified order coloring is proposed on the base of the characteristics of the vertex-distinguishing total coloring of complete graph in this paper. All of its elements are classified according to some rules and then are colored in proper sequence. Moreover, a relate-lock-table is presented to judge whether the results are correct. The experimental results show that the algorithm can effectively solve the vertex-distinguishing total coloring of complete graph.

1 Introduction

Graph coloring is one of the important topics in graph theory all the time. In the real world, the problem which is about classifying sets of some objects according to some rules can turn into graph coloring. The paper is about the study of classified coloring for complete graph.

Definition[1]. A proper k-total-coloring of graph G is a mapping f from $V(G) \cup E(G)$ into $\{1,2,...,k\}$ such that every two adjacent or incident elements of $V(G) \cup E(G)$ are assigned different colors. Let $C_f(u) = f(u) \cup \{f(uv) \mid uv \in E(G)\} \setminus$ be the neighbor color-set of u, if $C_f(u) \neq C_f(v)$ for any two vertices u and v of $V(G)$, we say f a vertex-distinguishing proper k-total-coloring of G, or a $k - VDTC$ of for short. The minimal number of all over $k - VDTC$ of G is denoted by $\lambda_{vt}(G)$, and it is called the $VDTC$ chromatic number of G.

Li Jingwen · Xu Xiaoqing · Yan Guanghui
College of Information and Electronic Engineering, Lanzhou Jiaotong University,
Lanzhou, China
e-mail: leejwcn@yahoo.com.cn, xuxiaoqing116@163.com

F.L. Gaol et al. (Eds.): Proc. of the 2011 2nd International Congress CACS, AISC 145, pp. 273–278.
springerlink.com   © Springer-Verlag Berlin Heidelberg 2012

Lemma[1]. For complete graph K_n, we have $\chi_{vt}(K_n) \geq n+1$.

Proof. Let u and v be two vertices of K_n, to satisfy total coloring, there must be n elements in $C(u)$ and $C(v)$, and to satisfy $C(u) \neq C(v)$, thus there exists $\chi_{vt}(K_n) \geq n+1$.

Conjecture[1]. Let G be a connected graph with $|V(G)| \geq 3$, then $\mu_t(G) \leq \chi_{vt} \leq \mu_t(G)+1$, where $\mu_t(G) = \max\left\{\min\left\{m \middle| C_m^{i+1} \geq n_i, \delta \leq i \leq \Delta\right\}\right\}$ is called the combination of all degrees, n_i whose degree is i is the number of vertices, C_m^{i+1} is called combination number and δ, Δ is the minimum degree and the maximum degree respectively.

For complete graph, $\mu_t(K_n) = n+1$, then the Conjecture turns into $n+1 \leq \chi_{vt}(K_n) \leq n+2$.

2 Algorithm of Vertex-Distinguishing Total Coloring of Complete Graph

2.1 Data Structure Definition and Introduction

$graph[][]$: Indicate Adjacency Matrix. In $graph[i][j]$, i indicates vertex V_{i+1} and j indicates V_{j+1}, that is, vertex number starts with 1.

$\deg ree[][]$: Indicate the set of the degrees of every vertex. In $\deg ree[p][q]$ ($p = 0,1; q = 0,1,...,vex\_num-1$), $\deg ree[0][q]$ stores the vertex number $q+1$ of V_{q+1} and $\deg ree[1][q]$ stores the degrees of V_{q+1}. Traversing $graph[][]$ can obtain $\deg ree[][]$.

$adj\_vex[][]$: Indicate the set of every vertex and its adjacent vertices. In $adj\_vex[i][j]$, i indicates the vertex number V_{i+1} and $j = 0$ indicates the number of vertex which is adjacent with V_{i+1}.

$E[]$: Indicate the sets of color number. When we color an element, all of available color numbers is stored in $E[]$. Once coloring is correct, the Array is emptied.

$A[][]$: Indicate the sets of being adjacent with the vertex but not being repeated. In $A[i][j]$, i indicates V_{i+1}, $j = 0$ indicates the numbers of vertex being adjacent with V_{i+1}, $j = 1$ indicates whether the vertex is colored, initiate value is 1. After colored, the value is 2. $j = 2, 3, \cdots, vex\_num$ indicate the vertex number. Moreover, make sure that the adjacent vertex having appeared does not appear again.

$B[][][]$: Indicate the sets of all elements of being colored. In $B[i][j][k]$ $(i = 0, 1, ..., color\_num - 1; j = 0, 1)$, i indicates color number. $B[i][0][k]$ with $B[i][1][k]$ stores the vertex $V_{B[i][0][k]}$ or the edge $V_{B[i][0][k]} V_{B[i][1][k]}$ which is colored the ($i+1$)- color . $B[i][0][0]$ stores the number of the element being colored the($i+1$)-color.

$lock[][][]$: A relate-lock-table, $lock[x][y][z](y = 0, 1)$ being corresponding with $B[i][j][k]$ indicates a relate lock of every color. $lock[x][0][z]$ indicates the relate-lock of vertex of being colored the $(x+1)$ -color and $lock[x][1][z]$ indicates the relate-lock of edge of being colored the $(x+1)$ -color. $lock[x][0][0]$ and $lock[x][1][0]$ are respectively the counter of the relate-lock of vertex and the relate-lock of edge.

$C[][]$: In $C[i][j]$, i indicates vertex V_{i+1}. $C[i][0]$ stores the numbers of the element and $C[i][1, 2, ..., vex\_num]$ store the color numbers.

2.2 Algorithm Steps

Step1: Generate deg $ree[][]$, $adj\_vex[][]$ and calculate the degrees of vertices according to $graph[][]$.

Step2: Decide if a given graph G is a complete graph or not and compute the chromatic number $color\_num$ basing on the quantitative relations of d_num and vex_num .

Step3: Put the elements of needing to be colored into $A[][]$.

Step4: Remove the elements of needing to be colored, starting with the first vertex V_1 and color them, go to Step5; when finished, go to Step7.

Step5: Label color number of the previous vertex with w and search available color number from w to $color\_num$ and then from 1 to $w-1$, then put them in $E[]$.

Step6: Decide if the element of needing to be colored is the last element of the set of colors: if the element of needing to be colored is the last element of the set of colors first, search the color which can color the element from $E[]$, then color it, that is, put the element number in $B[E[r]-1][j][k]$ and $lock[][][]$, moreover, put $E[r]$ in $C[][]$. if not, directly, put the element number in $B[E[0]-1][j][k]$ and $lock[][][]$, moreover, put $E[0]$ in $C[][]$.

Step7: Output the related values and the Arrays and the coloring finishes.

3 Experimental Results

Take vertex-distinguishing total coloring (see [2][3][4] and [5]) of complete graph which has odd number or even number example to test experimental results as follows:

Example 1: $vex\_num = 7$

Table 1 $B[][][]$

Color	Counter	Colored Element			
B_0	3	1	4	5	
		0	7	6	
B_1	3	1	5	6	
		2	7	0	
B_2	3	1	2	6	
		3	0	7	
B_3	3	1	2	7	
		4	3	0	
B_4	3	1	2	3	
		5	4	0	
B_5	3	1	2	3	
		6	5	4	
B_6	4	1	2	3	4
		7	6	5	0
B_7	3	2	3	4	
		7	6	5	
B_8	3	3	4	5	
		7	6	0	

Table 2 $lock[][][]$

lock		Adjacent Vertex Number							
	7	1	2	3	4	5	6	7	
$lock_0$	5	1	4	7	5	6			
$lock_1$	7	1	2	5	7	6	3	4	
	5	1	2	5	7	6			
$lock_2$	7	1	3	2	4	5	6	7	
	5	1	3	2	6	7			
$lock_3$	7	1	4	2	3	7	5	6	
	5	1	4	2	3	7			
$lock_4$	7	1	5	2	4	3	6	7	
	5	1	5	2	4	3			
$lock_5$	6	1	6	2	5	3	4		
	6	1	6	2	5	3	4		
$lock_6$	7	1	7	2	6	3	5	4	
	7	1	7	2	6	3	5	4	
$lock_7$	6	2	7	3	6	4	5		
	6	2	7	3	6	4	5		
$lock_8$	7	3	7	4	6	5	1	2	
	5	3	7	4	6	5			

Table 3 $C[][]$

Vertex	Counter	Color Number						
V_1	7	1	2	3	4	5	6	7
V_2	7	2	3	4	5	6	7	8
V_3	7	3	4	5	6	7	8	9
V_4	7	4	5	6	7	8	9	1
V_5	7	5	6	7	8	9	1	2
V_6	7	6	7	8	9	1	2	3
V_7	7	7	8	9	1	2	3	4

The experimental results show that the algorithm effectively solves the problem of vertex-distinguishing total coloring ([6] and [7]) of complete graph, and we have the results that

$$\chi_{vt}(K_n) = \begin{cases} n+1 & n \equiv 0 (\mathrm{mod}\ 2) \\ n+2 & n \equiv 1 (\mathrm{mod}\ 2) \end{cases}$$

4   Conclusions

In this paper, we have proposed the algorithm of classified order coloring based on the characteristics of the vertex-distinguishing total coloring of complete graph, have used relate-lock-table and sums of twice power to control the algorithm convergence and have made the algorithm compute out vertex-distinguishing total chromatic number ([8]) of complete graph which has any vertices.

Acknowledgements. This work was supported by the National Natural Science Foundation of China (NO.61163010) and the program for Longyuan Youth innovative talents under grant number252003.

References

[1] Zhang, Z.-F., Qiu, P.-X., Xu, B.-G., et al.: Vertex-distinguishing toal coloring of graphs. Ars. Com. 87, 33–45 (2008)
[2] Burris, A.C., Schelp, R.H.: Vertex-distinguishing proper edge-colorings. J. of Graph Theory 26(2), 73–82 (1997)
[3] Balister, P.N., Gyori, E., Lehel, J., et al.: Adjacent vertex distinguish edge-colorings. Journal on Discrete Mathematics 21(1), 237–250 (2007)
[4] Hatami, H.: Δ+300 is a bound on the adjacent vertex distinguishing edge chromatic number. Journal of Combinatorial Theory(Series B) 95(2), 246–256 (2005)

[5] Zhong, Z.-F., Liu, L.-Z., Wang, J.-F.: Adjacent strong edge coloring of graphs. Appl. Math. Lett. 15(5), 623–626 (2002)
[6] Zhang, Z.-F., Li, J.-W., Chen, X.-E., et al.: D(β)-vertex-distinguishing proper edge coloring of graphs. Acta Mathematica Sinica 49(3), 703–708 (2006)
[7] Zhang, Z.-F., Li, J.-W., Chen, X.-E., et al.: D(β)-vertex-distinguishing proper total coloring of graphs. Sci. China 49(10), 1430–1440 (2006)
[8] Bondy, J.A., Marty, U.S.R.: Graph theory with application. The Macmillan Press Ltd., New York (1976)

A Comparison of Local Linear Feature Extraction

Hou Guo-qiang, Fu Xiao-ning, and He Tian-xiang

Abstract. A number of studies have been carried out on the passive ranging method based on the monocular imaging features to non-cooperative target. This paper deals with target ranging estimation system focus on image linear feature. The ranging system implements target ranging by means of adjacent image matching method to extract target feature points, then obtain target rotational invariant linear feature, and combined target azimuth and pitching relative to camera when image is taken with camera space coordinate, the target distance to image pickup system is gained by solving the certain target ranging equation. As for target linear feature extraction, the paper applies three algorithms of the sub-pixel Harris corners method, the Simplified Scale Invariant Feature Transform (SSIFT) method and Speeded Up Robust features (SURF) method to extract linear feature and makes an analysis to ranging performance. It implied by our experiment that the SURF algorithm is the best one in the three methods. Its computational error is relatively small, and the time consumed is shorter compared with other two algorithms. The error of the sub-pixel Harris algorithm is a bit of larger than SSIFT algorithm while the real time realization performance is better than SSIFT algorithm.

1   Introduction

That estimating an imaged target distance from the camera based on the target's image feature and the imaging parameter is called passive ranging, it is a typical application of machine vision. For the target as aircraft, the linear feature using in passive ranging used to be the length of aircraft or the wing span [2,3]. With these linear features has previously known in the aircraft, the target distance estimation can be finished either based on geometrical optics [2] or combined with Pattern Recognition [3]. These linear features belong to the geometric size of the target itself, is valid only to the known type of aircraft. In the imaging tracking, these linear features have poor rotation invariance. Besides the occurrence of block, these

Hou Guo-qiang · Fu Xiao-ning · He Tian-xiang
Xi'dian University, Xi'an China, 710071
e-mail: {Coldwindsoftrain12,xning_fu,xning_fu}@163.com

F.L. Gaol et al. (Eds.): Proc. of the 2011 2nd International Congress CACS, AISC 145, pp. 279–285.
springerlink.com   © Springer-Verlag Berlin Heidelberg 2012

linear features are hard to obtain. Therefore, the target distance estimation based on local linear feature extraction produces a more practical value.

WANG [4] have developed a linear feature which has been successfully applied to passive ranging. The linear feature can be get from the circumcircle determined by matched points conforms to certain conditions. Supported by a grant from the National Natural Science Foundation of China (No. 60872136) and Natural Science Basic Research Plan in Shaanxi Province of China (Program No. 2011JM8002), the author perform a further depth research on the linear feature extraction for DSP realization in real-time. This paper will give a comparison of three linear feature extraction algorithms.

2   Ranging Local Linear Feature Introduction

It is supposed that the points A, B, C and A', B', C' are the matched points respectively in adjacent frames. We can determine the linear feature from these three points A, B and C for example, as shown in Fig 1. A circumcircle of points A, B and C with the center named O' could be determined, while O is the center of gravity for triangle ABC. If the direction $\overrightarrow{BM}$ is main SIFT direction of key point A, B and C. Parallel to the direction $\overrightarrow{BM}$, circumference string DE traversed point O can be determined. In Fig. 1, FG is another circumference string traversed point O but it is vertical to $\overrightarrow{BM}$.

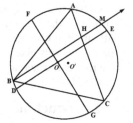

Fig. 1 Feature points A,B,C and the linear feature selecting

It has been proved by Wang [4] that chord FG in the circumcircle O' could be used as the linear feature.

Counting on the depth cues the linear feature has, combined observer's coordinate in space with direction cosine function of the target to observer in each sampling instant, a novel passive ranging algorithm was built in literature [5]. According to this algorithm, the target distance can always be made out if the observer has nonzero displacement within two adjacent sampling instant.

After three SIFT key points were matched, the circumference would be determined, and then, the very linear feature will be correctly obtained [4]. As we known that the SIFT matching algorithm is rather time-consuming. So, it is necessary to explore other algorithms which could perform the line feature extraction with the same precision, and easy for real-time realization.

3   Three Alagorithms

3.1.   Sub-pixel Harris Algorithm

The Harris corners algorithm can only check the pixel-level fusion coordinates while it has high performance such as rotation-invariant, scaling-invariant and so on. For this reason, this paper proposes a sub-pixel detection algorithm. Allow for the program's run time, the length of procedure code and the accuracy of sub-pixel, this paper adopts the interpolation algorithm [6] described as follows:

With the quadratic polynomial to approach the response function R(x, y), the sub-pixel precise location of R can be gained.

$$ax^2 + by^2 + cxy + dx + ey + f = R(x, y) \qquad (1)$$

Using nine pixels around the corner (x, y) that has been detected to build a super-lattice dislocation which contains six unknown quantity a ~ f. It can solve the su-perlattice dislocation by the least square method. Sub-pixel corner (x, y) corres-ponds to the local maximum point of quadratic polynomial. To find out the coordinate point , we can get the sub-pixel coordinate at first hand while deriva-tive the superlattice dislocation.

$$\begin{cases} \dfrac{\partial R}{\partial x} = 2ax + cy + d = 0 \\ \dfrac{\partial R}{\partial y} = 2by + cx + e = 0 \end{cases} \qquad (2)$$

The procedure of solving sub-pixel corner is as follows [7]:

(1) Smoothing every pixel in the image by using horizontal and vertical differ-ence to figure out the gray gradient I_x and I_y along the direction of X and Y.
(2) Smoothing I_x^2, I_y^2 and I_{xy} with Gaussian filter, while $I_x^2 = I_x \times I_x$, $I_y^2 = I_y \times I_y$, $I_{xy} = I_x \times I_y$.
(3) Figure out c(i, j) with the formula

$$c(i, j) = \frac{I_x^2(i, j) \times I_y^2(i, j) - I_{xy}(i, j) \times I_{xy}(i, j)}{I_x^2(i, j) + I_y^2(i, j) + 0.000001} \qquad (3)$$

(4) The point is a corner if it satisfying both c(i, j) greater than fixed threshold and $c(i, j)$ is the local maximum in its neighborhood.

(5) Take the value of nine pixels that around the corner into equation (1) to figure out a ~ f, and then, take them to equation (2) the sub-pixel coordinate of the corner can be solved.

3.2   Simplified SIFT Algorithm(SSIFT)

Based on literature [8, 9], after a good many experiment examination to the SIFT algorithm, the establishment of the scale space by the convolution operation of the image and the Gauss nucleus and the generation of feature descriptors in the algorithm takes the massive time. So it is necessary to simplify the SIFT algorithm for real-time application.

3.2.1   A Simplification to Scale Space

As for establishment of scale space, with the enlargement of Gauss template, the time consumption will also increase. Through a large number of experiments, among the 6 layers Gauss Pyramid the 80% matching points focus on the second DOG, while the time consumed to establish 4 layers Gauss Pyramid account for only one third of total time to establish 6 layers Gauss Pyramid. The experiment also shows that the establishment of 4-layer Gauss Pyramid nearly has no effect on ranging algorithm. For this reason, the paper simplified the SIFT algorithm in the way only to establish 4-layer Gauss Pyramid and then subtract two adjacent scale space to generate 3 layers DOG Pyramid[9], thus only need to examine the extreme points on the middle DOG layer.

3.2.2   A Simplification to Feature Descriptor

The literature [10] introduced in detail the construction of the SIFT feature descriptor within the circle area around the key-point as shown in Fig. 2. This paper made a further simplification to the SITF feature descriptor. According to the circle rotary invariability and combined with the image sequence characteristic to generate 13-dimension feature   vector quickly, then implement image matching.

First, produces 1×37 feature vector: take the feature point picture element value as the first element, after the 8 picture element value sorting of the first ring, take them as the 2nd to the 9th element; After the 12 picture element value sorting of the second ring, take them as the 10th to the 21st element; Similarly, after 16 picture element value sorting of the third ring, take them as the 22nd to the 37th element. Under the circumstances the circle rotates, the 37-dimension vector maintains invariable basically. Based on this, to reduce the feature descriptor's dimension further, selects 13 elements from this 37-dimension vector: take the key point picture element value as the first element, then takes 4 elements in every ring by bilateral symmetry around the key point. Thus constitutes a 13-dimension vector with rotational invariance. Finally, after making normalized processing to the 13-dimension vector, the rotary invariable feature descriptor generated.

Fig. 2 Descriptor

			89	90	91	92	93			
		88	61	62	63	64	65	94		
	87	59	60	33	34	35	66	67	95	
86	57	58	32	18	19	20	36	37	68	96
85	56	31	17	8	7	8	9	21	38	69
84	55	30	16		**0**	1	10	22	39	70
83	54	29	15	4		2	11	23	40	71
82	53	52	28	14	13	12	24	42	41	72
	81	51	50	27	26	25	44	43	73	
		80	49	48	47	46	45	74		
			79	78	77	76	75			

3.3 *SURF*

The SURF algorithm in this paper adopted from literature [11].

4 Experiment Result

Experiments with a reduced model indoors have been developed to test and verify the passive ranging estimation algorithm.

A set of integrated data from one of our experiments is given in Table 1, and diagram shown as Fig. 4 and Fig. 5. For saving space, only even number of the 33 frames was shown in Fig. 3. The trendline of linear feature ratio error and ranging error shown in Fig. 4 and Fig. 5.

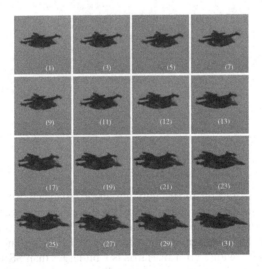

Fig. 3 Descriptor Target sequence one for experiment, (1), (3), et al, is frame numbers

Table 1 Experiment Data and Ranging Result.

Algorithm	Successful matches (class number)	File vo-lume /KB	Average time/s	The biggest ranging error /% [feature length]	The biggest ranging error /% [distance estimation]
Harris	24	75	0.165	4.13%	9.89%
SSIFT	23	77	1.228	4.07%	9.49%
SURF	29	98	0.087	2.82%	7.57%

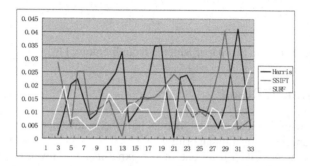

Fig. 4 Linear feature ratio error

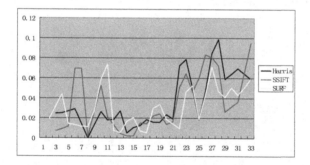

Fig. 5 Ranging error

In linear feature extraction, three algorithms applied to match two 256×256 images.

From the first to the end column, there are algorithm, successful matches, file volume, average time, and the biggest ranging error in Tab. 1. According to Fig. 4, it can be see that the linear feature ratio error of SURF algorithm method is minimum, and the linear feature error of Harris is a bit larger than SSIFT. From Fig. 5, the ranging error of sub-pixel Harris, SSIFT and SURF are 9.89%, 9.49% and 7.57%, and the average time are 0.165s, 1.228s and 0.087s respectively. The tread of ranging error is similar with linear feature ratio error.

Because of parameters measurement error, image matching error, in some conditions, the ranging equation has no root. But this occurs in very small chance.

5   Conclusion

This paper used three algorithms to extract linear feature and combined target azimuth, pitching relative to camera when image is taken with camera space coordinate the target distance to image pickup system can be obtained by solving the certain target ranging equation. The experiment shows that with SIFT key point, the sub-pixel Harris corner or the SURF matching, the ranging result can be achieved perfectly. The relative experiments suggested that a relative ranging error less than ± 6% could achieved in most cases, the biggest error is 9.89%. It implied by experiment that the SURF algorithm is the best one except for relatively larger file volume in the three methods. Its ranging error is small and the run time is shorter. The ranging error of the sub-pixel Harris algorithm is a bit of larger than SSIFT while the real time realization performance is better than SSIFT.

Acknowledgment. This work was supported both by the National Natural Science Foundation of China under Grant No. 60872136 and by Natural Science Basic Research Plan in Shaanxi Province of China (Program No. 2011JM8002).

References

1. Fu, X.-N.: Research on Infrared Passive Location Technology from Mono-station. Doctor dissertation. Xidian University, Xi'an (2005)
2. Huang, S.-K., Xia, T., Zhang, T.-X.: Passive ranging method based on infrared images. Infrared and Laser Engineering 126(1), 109–112+126 (2007)
3. Guo, L., Xu, Y.-C., Li, K.-Q., Lian, X.-M.: Study on Real-time Distance Detection Based on Monocular Vision Technique. Journal of Image and Graphics 11(1), 74–81 (2006)
4. Wang, D., Fu, X.-N.: A passive ranging system based on image sequence from single lens and the imaging direction - Introduction and performance. In: ICNNT, Taiyuan, China (2011)
5. A method for distance estimation based on the imaging system. Chinese Patent (February 2010)
6. Liang, Z.-M., Gao, H.-M., Wang, Z.-J., Wu, L.: Sub-pixels corner detection for camera calibration. Transactions of the China Welding Institution 27(2), 102–104 (2006)
7. Zhao, W.-B., Zhang, Y.-N.: Survey on Corner Detecton. J. of Application Research of Computers 38(10), 17–19 (2006)
8. Zhang, S.-Z., Song, H.-L., Xiang, X.-Y., Zhao, Y.-N.: Fast SIFT Algorithm for Object Recognition. J. of Computer Systems & Applications 19(6), 82–85+186 (2010)
9. Liu, L., Peng, F.-Y., Zhao, K., Wan, Y.-P.: Simplified SIFT algorithm for fast image matching. J. of Infrared and Laser Engineering 37(1), 181–184 (2008)
10. Tang, Y.-H., Lu, H.-Z., Hou, W.-J.: Serial Images Matching Algorithm Based on DOG Feature Points. J. of Modern Electronics Technique 136(4), 128–130+136 (2008)
11. Lu, X.-M., Sun, Z.-J., Wu, J., Wang, J.-B., Yu, T.-X., Zhao, L., Ding, X.-H.: An Improved Algorithm of Image Registration based on SURF. Dunhuang Research (6), 88–92 (2010)

Reducing X-Ray Exposure of 3D Hepatic MDCT Images by Applying an Optimized Feature Preserving Strategy[*]

Jiehang Deng, Min Qian, Guoqing Qiao, Yuanlie He, and Zheng Li

Abstract. The X-ray exposure of hepatic MDCT (Multidetector-Row Computed Tomography) images threatens human health. In this paper, the 3D adaptive median filter with local averaging is optimized to improve the image quality of low dose hepatic MDCT images. The modified strategy is designed according to the subtle texture of a hepatic tumor. Filtering experiments are based on phantom images with different dose levels and clinical hepatic ones with a certain dose level. Radiologists' visual evaluation and voxel value profiles show that the x-ray exposure can be reduced to 60% at least without compromising the quality of diagnostic images.

1 Introduction

The aim of this paper is to optimize a novel feature preserving strategy based on the subtle texture of a hepatic tumor and to investigate how much radiation dose of hepatic 3D multidetector-row computed tomography (MDCT) images can be reduced without the expense of feature preservation of tumors.

In 1996, Gray reported the hazard that 12.5 per 10,000 patients undergoing each single-phase CT scan of the abdomen died of cancer [1]. An efficient method

Jiehang Deng
Computer Faculty Guangdong University of Technology Guangzhou China
e-mail: dengjiehang@gdut.edu.cn

Min Qian
X-ray Department General Hospital of Guangzhou Military Command Guangzhou China
e-mail: qmgzfsk@163.com

[*] This work is supported by Natural Science Foundation of Guangdong Province, China (9451009001002711 and S2011040004295).

F.L. Gaol et al. (Eds.): Proc. of the 2011 2nd International Congress CACS, AISC 145, pp. 287–293.
springerlink.com © Springer-Verlag Berlin Heidelberg 2012

of mitigating the hazard is to lower the X-ray exposure level. However, low level of the X-ray dose leads to an increase in quantum noise and quality degradation in the obtained image, which compromises the quality of diagnostic images.

Since the contrast detectability is relatively high in lung and nasal cavity, low dose imaging technique is usually applied in thorax [2-5], head [6] regions. In the abdomen, the tissue structure is complex and the contrast detectability is low, the low dose imaging technique still remains an unsolved difficult problem.

Funama et al. [7-8], Martinsen et al. [9], denoised the hepatic low-dose CT images with different filters. However, only the general filtering procedures were shown in their studies, without a description of the detailed structure or the basis of the parameters. Furthermore, these specific methods only aimed at achieving effective trade-offs between smoothing efficiency, feature preservation over the whole image or the entire 3D dataset. However, it is paramount to discern tumors of less than 5 mm, preferably down to 2 mm in diameter for radiologists. Therefore, how to design filtering methods for reducing the quantum noise according to the size and the distribution of voxel value of small regions being suspected of tumors is a serious issue.

We previously proposed a feature preserving strategy -- 3D adaptive median filter with local averaging (3D AMLA) [10]. In that study, we used computer-simulated clinical hepatic datasets to achieve smoothness and sharpness. We found that the radiation dose could be reduced to a large degree without the expense of the feature preservation. However, radiologists pay great attention to the subtle texture of a tumor for routine clinical practice. Therefore, optimizing AMLA filtering scheme according to the subtle texture of tumors is required.

The AMLA strategy is optimized according to the subtle texture of tumors in this paper. 3D phantom datasets with five levels of X-ray exposure from 20% to 100% are examined to investigate how much radiation dose can be reduced without the expense of feature preservation of tumors. A certain low dose level validated by phantom images is selected to create low dose clinical hepatic MDCT images. The image quality of the low dose hepatic datasets is improved by optimized AMLA strategy. Radiologists' visual evaluation and voxel value profiles are used to evaluate the filtering performance.

2   The Optimized Adaptive Median Filter with Local Averaging

The AMLA works not so well to remove the quantum noise while preserving the subtle texture of a clinical hepatic tumor according to our experimental results. AMLA is optimized to enhance the low dose MDCT images in this study which is called the optimized adaptive median filter with local averaging (OAMLA). The local filtering region of OAMLA is defined as a 2D/3D filtering window R^n_{xyz} :

$$R^n_{xyz} = \{(x', y', z') \mid (x'-x)^2 + (y'-y)^2 + (z'-z)^2 \leq r_n^2\} \tag{1}$$

where r_n is the radius of the nth neighbors whose origin is at the point (x, y, z), e.g., $r_1^2 = 1$, $r_2^2 = 2$, and so on.

Let us denote that n_{max} is the maximum order of neighbors, and v_{xyz} is the current voxel value at (x, y, z). The OAMLA identifies and replaces noise candidates with the median value in R_{xyz}^n .

The OAMLA is carried out as follows:

(1)   Initialize the order of neighbors $n = 1$.

(2)   Compute S_{xyz}^{min,R^n}, S_{xyz}^{med,R^n} and S_{xyz}^{max,R^n} , which are the minimum, median, and maximum values with adaptive local averaging in R_{xyz}^n , respectively.

(3)   If $S_{xyz}^{min,R^n} < S_{xyz}^{med,R^n} < S_{xyz}^{max,R^n}$, go to step 5; otherwise, the values of current voxel neighborhood are almost the same, e.g., the current voxel is in air region or intro regions of a tissue. Then the size of this neighborhood needs to be enlarged to identify noise candidates, further. Therefore, set $n = n + 1$.

(4)   If $n < n_{max}$, go to step 2; otherwise, replace v_{xyz} with S_{xyz}^{med,R^n} .

(5)   If $S_{xyz}^{min,R^n} < v_{xyz} < S_{xyz}^{max,R^n}$, then v_{xyz} is kept unchanged; otherwise, v_{xyz} is replaced by S_{xyz}^{med,R^n} .

To calculate S_{xyz}^{min,R^n}, S_{xyz}^{med,R^n}, and S_{xyz}^{max,R^n} , the voxel values in the local region R_{xyz}^n are sorted in an increasing order, saved in an array with one dimension, and denoted as $A = \{v_i \mid i=1, 2, ..., M\}$, where v_i is the ith voxel value in the array and M is the size of R_{xyz}^n . Let us denote $A(v_{min})$, $A(v_{med})$ and $A(v_{max})$ as the minimum, median and maximum in A, respectively. We prepare three sets, S^{min}, S^{med} and S^{max} which are the first, median and last N elements of A, respectively.

The local distribution of voxel values in $\{A(v_1), A(v_2),..., A(v_M)\}$ can be divided into three types according to the voxel value distribution in local region. S_{xyz}^{min,R^n} and S_{xyz}^{max,R^n} are decided according to these three types [10]. S_{xyz}^{med,R^n} is set at the median value of S^{med} in this paper.

3   Experiments

By using a 64-row DSCT system (Somatom Definition 2008G; SIEMENS), we perform normal phantom and clinical helical scanning under the following scan conditions: tube voltage of 120 kV, tube current of 420mA, image slice thickness of 1, 5 and 8 mm, and field of view of 374 mm with a reconstruction function of B30f. The Ethics Board of the hospital approved the study, and all participants join this study voluntarily.

For phantom images, the X-ray exposure received in the normal abdominal routine examination is 350mAs, and assigned to 100% dose level. The exposure is decreased to 80%, 60%, 40% and 20%, i.e. 280, 210, 140 and 70mAs. For clinical datasets, 100% and a certain low dose level datasets are created after validating which low dose level is preferable by processing phantom datasets with different low dose levels.

In Fig. 1, phantom images created in different dose levels are shown. The diameter of five circles varies from 30mm to 60mm. The content inside each circle corresponds to the air, bone, calcified material, parenchyma and fat. In comparison with the normal dose images, a large amount of quantum noise distributes randomly on every slice of low-dose images. The noise level increases with the decreasing radiation dose.

Because radiologists must recognize the tumors whose size is even around 5 mm, preferably down to 2 mm. It is vital that the filtering method must be developed according to this requirement. To this end, N (the size of three sets: S^{min}, S^{med} and S^{max}) is set at 3. The parameter value of the maximum order of neighborhood n_{max} is defined as 9 through several experiments. N and n_{max} will be changed if spatial resolution of MDCT images is changed. After analyzing the image quality (noise level, contrast, definition) of low dose processed phantom images, we select a certain dose level for clinical datasets and improve its image quality by 3D AMLA and OAMLA.

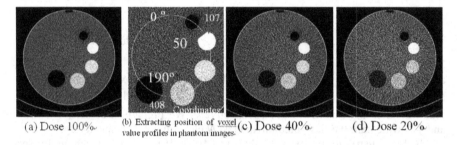

(a) Dose 100%.   (b) Extracting position of voxel (c) Dose 40%.   (d) Dose 20%.
value profiles in phantom images.

Fig. 1 Phantom images created in 40 and 20% dose level.

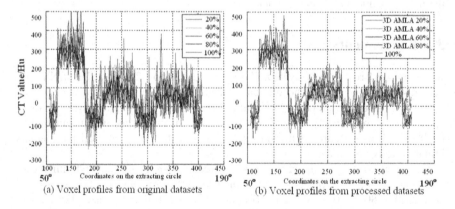

(a) Voxel profiles from original datasets   (b) Voxel profiles from processed datasets

Fig. 2 Voxel profiles from different datasets. The extracting position is shown as a white circle in Fig. 1(b).

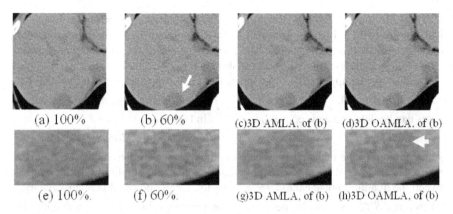

(a) 100%   (b) 60%   (c)3D AMLA, of (b)   (d)3D OAMLA, of (b)

(e) 100%.   (f) 60%.   (g)3D AMLA, of (b)   (h)3D OAMLA, of (b)

Fig. 3 Certain patient's clinical hepatic images. (a) and (b) are 100% and 60%, respectively. (c) and (d) are filtered results of 60% images by 3D AMLA and 3D OAMLA, respectively. The region indicated by the white arrow The second row is the enlarged one of the first row.

In order to evaluate the filtered results in detail, voxel value profiles in different datasets are constructed in this paper. Voxel value is extracted around a circle passing through every circle of the phantom images to show the variation of voxel value in different boundaries. The extracting position is shown by the white circle in Fig. 1(b). Only the voxel value between 50 degree and 190 degree, that is, from 107 to 408 of coordinates around the white circle is shown for simplicity.

4   Discussion and Conclusions

As shown in Fig. 1, The image quality becomes worse and the image definition become lower with the reduction of X-ray dose levels from 100% to 20%. Especially in the 20% image, the boundaries of circles are corrupted and can not be distinguished well. In Fig. 2(a) of voxel value profiles, the voxel value of 20% vibrates so strongly after 200 of coordinates that the boundaries between two circles almost can not be distinguished. Therefore, the images created in 20% dose level are not fit to be applied in low dose hepatic MDCT images in this sense.

As shown in Fig. 2, the variation of voxel value from 40% dose images (green) exceeds the envelope range of the normal dose one (purple) frequently in Fig. 2(b). On the other hand, the filtered results of 60% and 80% are close to normal dose in the aspects of the graininess, sharpness of contours, circle conspicuity, overall image quality and voxel value variation in voxel value profiles as shown in Fig. 1 and Fig. 2.

Therefore, the x-ray exposure can be reduced to 60% at least validated by the experiments on phantom images.

The AMLA and OAMLA strategies are used to enhance the 60% low dose clinical hepatic MDCT images. 3 patients who were diagnosed as hemangioma took normal and low dose scan (60%). The slice thickness of the 3D datasets is 5mm. One of these patients' filtered results is shown in Fig. 3.

Because of the application of adaptive local averaging scheme, the value interval of normal voxel, i.e., (S_{xyz}^{min,R^n}, S_{xyz}^{max,R^n}), has been shrunk, quantum noise candidates can be detected and replaced, as shown in Fig. 3(c).

However, S_{xyz}^{med,R^n} is set at the average value of S^{med} in AMLA. That blurs the subtle texture of the hepatic tumor, such as the region indicated by the white arrow in Fig. 3(d). According to the experimental results, the amplitude difference between the quantum noise candidates and the local neighborhood is relatively large. Therefore, S_{xyz}^{med,R^n} should be decided at the median value of S^{med} rather than the average value of it. This modified filtering strategy is reference as OAMLA. Furthermore, the texture of a tumor is so subtle that the iteration step should not be too many. After several tries, the iteration step of OAMLA is set at 1.

In Fig. 3, the representative filtered results are compared with normal and low dose images. The noise level is suppressed after applying AMLA and OAMLA. The subtle texture of the tumor in OAMLA can be distinguished well indicated by a white arrow in Fig. 3(h). However the results of AMLA can not show this point.

Two radiologists visually appraise the image quality of phantom datasets and evaluate the artifacts of all the clinical hepatic MDCT datasets in the hepatic subtle texture and graininess, the sharpness of liver contour, tumor conspicuity, homogeneity of the enhancement of the portal vein and overall image quality. The evaluated results show that OAMLA outperforms AMLA, and the OAMLA strategy can suppress the noise level without compromising the quality of diagnostic images based on images of 60% dose level.

Acknowledgments. The authors would like to acknowledge Prof. Hisakazu Ogura in University of Fukui for his instruction and thank Prof. Huixia Cao in General Hospital of Guangzhou Military Command for her great help.

References

1. Gray, J.E.: Safety (risk) of diagnostic radiology exposures, pp. 15–17. American College of Radiology, Reston (1996)
2. Okumura, M., Ota, T., Tsukagoshi, S., Katada, K.: New method of evaluating edge-preserving adaptive filters for computed tomography (CT): Digital phantom method. Jpn. Radiol. Technol. 62(7), 971–979 (2006)
3. Kalra, M.K., Wittram, C., Maher, M.M., Sharma, A., Avinash, G.B., Karau, K., Toth, T.L., Halpern, E., Saini, S., Shepard, J.A.: Can noise reduction filters improve low-radiation-dose chest CT images? Pilot study. Radiology 228(1), 257–264 (2003)
4. Kachelrieb, M., Watzke, O., Kalender, W.A.: Generalized multi-dimensional adaptive filtering for conventional and spiral single-slice, multi-slice, and cone-beam CT. Med. Phys. 28(4), 475–490 (2001)

5. Yasuda, N., Ishikawa, Y., Kodera, Y.: Improvement of image quality in chest MDCT using nonlinear wavelet shrinkage with trimmed-thresholding. Jpn. J. Radiol. Technol. 61(12), 1569–1599 (2005)
6. Sasaki, T., Sasaki, M., Hanari, T., Gakumazawa, H., Noshi, Y., Okumura, M.: Improvement in image quality of noncontrast head images in multidetector-row CT by volume helical scanning with a three-dimensional denoising filter. Radiat. Med. 25(7), 368–440 (2007)
7. Funama, Y., Awai, K., Miyazaki, O., Goto, T., Nakayama, Y., Shimamura, M., Hiraishi, K., Hori, S., Yamashita, Y.: Radiation dose reduction in hepatic multidetector computed tomography with a novel adaptive noise reduction filter. Radiat. Med. 26(3), 171–177 (2008)
8. Funama, Y., Awai, K., Miyazaki, O., Nakayama, Y., Goto, T., Omi, Y., Shimonobo, T., Liu, D., Yamashita, Y., Hori, S.: Improvement of low-contrast detectability in low-dose hepatic multidetector computed tomography using a novel adaptive filter: evaluation with a computer-simulated liver including tumors. Investigative Radiology 41(1), 1–7 (2006)
9. Trægde Martinsen, A.C., Kjernlie Sæther, H., Olsen, D.R., Skaane, P., Olerud, H.M.: Reduction in dose from CT examinations of liver lesions with a new postprocessing filter: a ROC phantom study. Acta Radiol. 49(3), 303–309 (2008)
10. Deng, J.H., Hiratsuka, K., Ishida, T., Shirai, H., Kuroiwa, J., Odaka, T., Ogura, H.: Improvement of low-dose MDCT images by applying a novel adaptive median filter with local averaging. International Journal of Intelligent Computing in Medical Sciences and Image Processing 3, 31–42 (2009)
11. Chan, R.H., Ho, C.W., Nikolova, M.: Salt-and-pepper noise removal by me-dian-type noise detectors and de-tail-preserving regularization. IEEE Trans. Image. Process. 14(10), 1479–1485 (2005)

Quickly Creating Illumination-Controllable Point-Based Models from Photographs

Weihua An

Abstract. Based on the theory of visual hulls, this paper presents a method to create point based models for real objects. Instead of using the expensive special equipments such as 3D laser scanners, this method deals with some silhouette images of the objects, and generates uniformly point-sampled models. We adopt a uniform-interval index table to organize the silhouette edges of each sample image, which provides much flexibility for point sampling. Moreover, combining the surface splatting technology and Layered Depth Buffers (LDB), we introduce a new algorithm to judge the visibilities of the points. The experimental results have shown the high accuracy of the visibility judgment.

Keywords: point based modeling, illumination, layered depth buffer.

1 Introduction

In traditional computer graphics, an object is constructed as a set of triangles. With the increasing requirements for realities, the triangle meshes become more and more dense[1]. To decrease the model complexity, the point based modeling and rendering technology is proposed [2, 3].

Compared with triangle meshes, the sample points have superior capability to represent sophisticated models. There is no need to maintain the connectivity between different points, and they can be easily organized into a level-of-detail (LOD) data structure [4]. Moreover, we can assign any attributes to the point sets, and achieve various rendering effects such as semi-transparence [5].

Nowadays, the direct method for point based modeling of real objects is to use 3D laser scanners to acquire their surface points. Although providing much high sampled precision and density, these equipments are too expensive to be adopted widely.

Weihua An

College of Information Science, Beijing Language and Culture University, Beijing, China

e-mail: anweihua@blcu.edu.cn

F.L. Gaol et al. (Eds.): Proc. of the 2011 2nd International Congress CACS, AISC 145, pp. 295–302.

Motivated by image based visual hulls [6], this paper presents a simple and fast image-based method for points modeling. For uniformly point sampling, we use a series of parallel rays to intersect with the visual hull from three orthogonal directions, and generate three LDIs (Layered Depth Image) [7]. And then, we combine these orthogonal LDIs into a LDC (Layered Depth Cube) [8]. In this method, the uniform-interval index table is adopted to organize the silhouette edges for each sample image, which effectively simplify the intersection computation.

Another contribution of our method is the accurate visibility judgment. we combine the surface splatting technology [5] and LDB to identify the visibilities of each point in sample images.

2   Background and Related Works

For virtual objects, Pfister [8] provides a point sampling method in image space. He uses a series of parallel rays to intersect with the virtual object from three orthogonal directions, and to generate three parallel-projective LDIs [7]. For LOD rendering, the points is usually organized as an octree based Layered Depth Cube (LDC) [4].

For real objects, we can firstly use some computer vision algorithms [9, 10] to create their corresponding virtual models, and then sample them with the above ray casting method. However, these methods usually involve much complex computation and time consumption.

Furthermore, the theory of image based visual hulls [6] can give us some inspiration from another perspective. The visual hull is an approximate representation for a 3D object [11]. As shown in Fig. 1, it is the intersection of some cones, which are formed by some view points and their corresponding silhouette images of the object. Obviously, if there are enough view points, the visual hull can be treated as the original geometry of the object.

Matusik et al [6] have provided an image based method to browse visual hulls. However, their result is a view-dependent LDI, which isn't uniformly sampled, and can not be directly treated as a point based model. In addition, the data structure of bins in that method just adapts to incremental intersection computation, and can not benefit random sampling. Afterward, Matusik substitutes parallel projections for perspective ones in his subsequent works [12], but he doesn't solve the limitation of incremental computation.

It is a very important problem to judge the visibilities of the point sets. Based on the epipolar geometry, McMillan proposes a list-priority algorithm for redisplaying projected surfaces [13, 14], which provides a back-to-front rendering order for correct visibility. Matusik also provides a similar method to judge the visibility of the LDI for each silhouette image [6]. However, any of the above methods can not guarantee the accuracy of occlusions between different points.

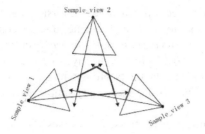

Fig. 1 The image based visual hull shown in 2D form.   **Fig. 2** The imaging system.

3   Acquiring Images and Preprocessing

Similar to the previous methods [12], the images are taken by the combination of a directional light source and a digital camera. As shown in Fig. 2, a series of light positions and camera positions around the object are selected. For any combination of a light position and camera position, we take an image for the object which is called a reflectance field image. In addition, for any camera position, another image of the object in a fixed lighting condition is taken, which is called a light field image [15]. The light fields are used for rendering the model under the fixed lighting condition, and the reflectance fields are used to control the reflectance effects.

We select a feature based method to calibrate the camera positions [16], and the light positions are recorded manually. In addition, we use Photoshop to extract the object silhouettes for each sample image.

4   Creating Point Based Models

As shown in Fig. 1, our objective is to point-sample the visual hull. For uniformly sampling, we create three parallel projective LDIs from three orthogonal directions, and combine them into a LDC with the method proposed by pfister [8]. Section 4.1 describes a fast method to create parallel projective LDIs.

4.1   Creating LDIs

In order to generate a parallel projective LDI, we firstly set a list of parallel sample rays. The LDI is exactly the intersection of all the rays and the visual hull defined by the sample images. The computation for the intersection between a ray and the visual hull can be depicted as follows.

- Firstly, we intersect the ray with the cone defined by one silhouette image, and generate a set of intersected segments.
- Secondly, all the sets of segments from different cones are intersected, and a list of depth pixels for the ray can be obtained.

We can continue to describe the ray-cone intersection as follows: Firstly, the ray is projected onto the sample image, and intersected with its silhouette edges. Secondly, the 3D intersected segments between the ray and the cone are obtained by projecting the 2D intersected segments back onto the ray. Therefore, the ray-cone intersection can be decomposed into three steps: projection, the intersection in image space, and back-projection. Considering that the first and third steps are very simple, we mainly focus on the second steps.

The silhouette edges of each sample image are expressed as a series of segments. Therefore, during the 2D intersection in image space, we need to judge which segment is intersected with the projected ray. According to the theory of vanishing point, we can construct a flexible data structure which can be accessed randomly.

For a list of parallel rays in 3D space, if their direction $\overrightarrow{d}$ is known, we can determine a unique vanishing point in each sample image. Moreover, the projections of all the rays must pass the vanishing point.

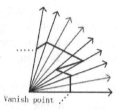

Fig. 3 Uniformly subdividing the angle intervals. (Black rays are the connections of the vanishing point and the endpoints of the silhouettes. Red rays are used to subdivide all the angle intervals)

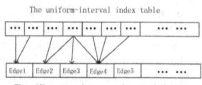

Fig. 4 The uniform-interval index table created according to the silhouette edges shown in Fig.3.

According to the characteristic of the vanishing point, we can effectively organize the silhouette edges. For each sample image, we firstly compute out its vanishing point, and connect it with the endpoints of all the silhouette edges. In this way, there is an angle for each edge to the vanishing point. And then, we subdivide all the angles uniformly according their maximum common factor. Therefore, all the sub-angles have the same interval, as shown in Fig. 3. Finally, a uniform-interval index table is used to organize all the edges. The index table is shown in Fig. 4, and each index item contains the addresses of all the edges located in the corresponding angle interval.

Since the index table is constructed based on fixed angle interval, it can be accessed randomly. For the projection of a sample ray, we can easily identify the entry of the index table according to its slope, and obtain the candidate edges. Therefore, this data structure benefits the flexibility of the intersection computation.

Finally, the procedure for creating parallel projective LDIs can be described as follows.

1) Compute the vanishing point for each sample image.
2) Construct a uniform-interval index table for the silhouette edges in each image.
3) For each sample ray, use the three-step method to compute the ray-cone intersection for each image.
4) Compute the intersection of all the ray-cone intersections, and obtain the lists of depth pixels.

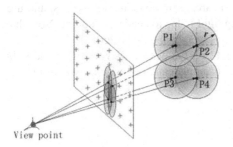

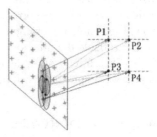

Fig. 5 Rendering the point sets with surface splatting technology. (Each point and its filter kernel are projected onto the desired view.)

Fig. 6 The LDB corresponding to Fig.5. (Each cell contains a list of 3D points which contribute to the corresponding pixel color.)

4.2 Identifying Visibilities and Shading

As we known, the depth buffer corresponds to the image buffer. Each cell in the depth buffer contains a depth value, which is the distance from the corresponding pixel in the desired view to the projected 3D point. Furthermore, the LDB can be treated as an extension of the depth buffer. Each cell in the LDB contains a list of depth values, which record the distances from the corresponding pixel to all the intersected points between the view ray and the scene. For visibility judgment, besides the list of depth values, we also store another list of values in each cell, which is the list of all the intersected points.

In this paper, the surface splatting technology [5] is used to render the points. As shown in Fig. 5, we must determine the filter kernel size for each point, and then project it onto the desired image. For the LDC is uniformly sampled, each point has the fixed filter kernel size. We set it as the interval of adjacent parallel sample rays which is denoted as r in Fig. 5. For the discreteness of the points, the points list in each cell of the LDB is the sample points which are splatted onto the corresponding pixel, instead of the accurate intersected points between the view ray and the scene. Fig. 6 shows the points list in each cell of the LDB generated by splatting the points in Fig. 5.

Combining the surface splatting technology and LDBs, we can judge the visibilities of all the points. We firstly render the point sets on each sample image, and generate a series of LDBs. Each LDB contains the depth information of all the points for the corresponding sample image. Obviously, for each cell in the LDB, the point with the minimum depth value is visible. To judge the visibilities of other points in this cell, we set a threshold δ. Only if the difference between some other depth value and the minimum are less than δ, the corresponding point is visible.

5   Rendering

For fixed illumination, we choose the n nearest sample camera positions, and use the weighted sum of their corresponding light fields to shade each point. Note that if a point is invisible for a sample image, its weight should be zero.

For controllable illumination, we should set the desired view position and light direction. Therefore, besides the n nearest sample camera positions, we also select the m nearest sample light positions.

6   Experimental Results

Our method has been examined on both virtual and real objects. For virtual objects, the OpenGL language is used to render their images. We select 30 camera positions and 30 light directions uniformly. Therefore, the amount of sample images is about 930. For real objects, we can not take the images of their bottoms due to occlusion. So, we set 20 camera positions and 20 light directions, which lead to about 420 images.

We have compared the sample images with the rendered images to illustrate the accuracy of point modeling and rendering. Fig. 7 (a) shows two sample images, and Fig. 7 (b) shows the rendered images with the same view points and light directions. Obviously, each pair of images is approximately identical. Fig. 7 (c) and Fig. 7 (d) also present the silhouette difference and color difference between them. Because of the camera calibration errors, their silhouette difference is about 2-5 pixels. The color differences just exist on the high-frequency regions of the images. It is caused by filtering, and can be avoid by increase the point sampling density for those regions.

Fig. 8 shows the same point based model with two different visibility judgment algorithms. As the right image shown, Matusik's algorithm attached the finger's colors to the trousers, and lead to the blurs. Our algorithm solves this problem effectively (see the left image in Fig. 8).

Fig. 9 shows the illumination effects. In this figure, the left, middle and right images are sample images, and other images are rendered images with lighting interpolations. We can find the continuous lighting change between each two adjacent images.

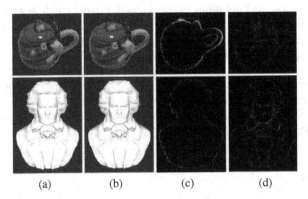

(a)   (b)   (c)   (d)

Fig. 7 Comparing the sample images with the rendered images of their point based models. (a) The sample images. (b) The rendered images. (c) The silhouette difference between (a) and (b). The green color denotes the sampled silhouettes, and the red color denotes the rendered silhouettes. (d) The color difference between (a) and (b).

Fig. 8 Comparing the results of different visibility judgment algorithms.

Fig. 9 A set of views of the point based teapot with a fixed view point and various light directions.

7 Conclusion and Future Work

An image based method for point modeling real objects has been presented. This method deals with the silhouette information of the sample images, and generates uniformly sampled point based models. For each sample image, all the silhouette edges are organized into a uniform-interval index table, which benefit the intersection computation. Moreover, we combine the surface splatting technology with LDBs, and solve the visibility judgment and shading problem.

There are also some limitations in our method. Firstly, we do not take any compression for the point sets. Secondly, we will sample high dynamic range image [17] for rendering in the future. Finally, we will explore new technologies to integrate point based models with triangular meshes.

Acknowledgment. This research was supported by the National Natural Science Foundation of P. R. China and the Microsoft Asia Research (No. 60970158). It was also funded by BLCU supported project for young researchers program (supported by "the Fundamental Research Funds for the Central Universities) (No. 10JBT02).

References

1. Deering, M.F.: Data Complexity for Virtual Reality: Where do all the Triangles Go? In: Proc. IEEE Virtual Reality Annual International Symposium (VRAIS), Singapore, pp. 357–363 (1993)
2. Alexa, M., Behr, L., Cohen-Or, D., Fleishman, S., Levin, D., Silva, C.: Point Set Surfaces. In: Proc. IEEE Visualisation, pp. 21–28 (2001)
3. Grossman, J.P., Dally, W.: Point Sample Rendering. In: Proc. Rendering Techniques 1998, pp. 181–192 (1998)
4. Pajarola, R.: efficient level-of-details for point based rendering. In: Proc. IASTED Computer Graphics and Imaging (2003)
5. Zwicker, M., Pfister, H., Baar, J.V., Gross, M.: Surface splatting. In: Proc. SIGGRAPH, pp. 371–378 (2001)
6. Matusik, W., Buehler, C., Raskar, R., Gortler, S., Mcmillan, L.: Image-based visual hulls. In: Proc. SIGGRAPH, pp. 369–374 (2000)
7. Shade, J., Gortler, S., He, L., Szeliski, R.: Layered Depth Images. In: Proc. SIGGRAPH 1998, pp. 231–242 (1998)
8. Pfister, H., Zwicker, M., van, B.J., Gross, M.: Surfels: Surface elements as rendering primitives. In: Proc. SIGGRAPH 2000, pp. 335–342 (2000)
9. Lorensen, W.E., Cline, H.E.: Marching Cubes: A High Resolution 3D Surface Construction Algorithm. In: Proc. SIGGRAPH 1987, pp. 163–169 (1987)
10. Faugeras, O.: Three-dimensional Computer Vision: A Geometric Viewpoint. MIT Press (1993)
11. Laurentini, A.: The visual hull concept for silhouette based image understanding. IEEE Trans. Pattern Anal. And Mach. Intell. 16, 150–162 (1994)
12. Matusik, W.: Image-Based 3D Photography using Opacity Hulls. In: Proc. SIGGRAPH 2002, pp. 427–437 (2002)
13. McMillan, L.: Computing Visibility Without Depth, Department of Computer Science University of North Carolina, NC 27599
14. McMillan, L.: A list-priority rendering algorithm for redisplaying projected surfaces, Technical Report 95–005, University of North Carolina (1995)
15. Wood, D., Azuma, D., Aldinger, K., Curless, B., Duchamp, T., Salesin, D., Stuetzle, W.: Surface Light Fields for 3D Photography. In: Proc. SIGGRAPH 2000, pp. 287–296 (2000)
16. Abidi, M.A., Chandra, T.: A new efficient and direct solution for pose estimation using quadrangular targets: algorithm and evaluation. IEEE Trans. Pattern Anal. And Mach. Intell. 17, 534–538 (1995)
17. Debevec, P., Malk, J.: Recovering high dynamic range radiance maps from photographs. In: Proc. SIGGRAPH 1997, pp. 369–378 (1997)

Development on Insole 3D Plantar Pressure Measurement System Based on Zigbee Technology

Yemin Guo and Lanmei Wang

Abstract. Plantar pressure measurement is now considered commonplace in the healthcare control and monitoring of normal people and patients, as well as having applications in military applications. Based on this need, we create a plan to construct a new insole plantar pressure system based on multifunction data acquisition modular, LabVIEW and Zigbee technology. Then the hardware and software parts are developed respectively. There are 3 sensors arrayed at each measurement point, that means 3 sensors are assembled in 3 different directions of X,Y and Z . The piezoelectric ceramic type sensors are designed, manufactured and calibrated according to scientific methods. Meanwhile, the DAQ card is selected carefully. Of course, the software part is developed based on LabVIEW. The singnal transmission and inception are performed based on wireless communication technology of Zigbee. A series of tests are performed in order to validate the function of the measurement system. The results satisfy the anticipated design requirements. The existing problems and application trend of the measurement system are predicted.

Keywords: virtual instrument, insole, piezoelectric ceramic, plantar pressure, labview, Zigbee.

1 Introduction

As we know, feet are important organs of human body. The human foot and ankle is a strong and complex mechanical structure containing 26 bones (some people have more), 33 joints (20 of which are actively articulated), and more than a hundred muscles, tendons, and ligaments. The foot can be subdivided into the hind foot, the midfoot, and the forefoot. They are closed linked with the health of

Yemin Guo · Lanmei Wang
Shandong University of Technology, Zibo China
e-mail: {gym,wanglanmei}@sdut.edu.cn

F.L. Gaol et al. (Eds.): Proc. of the 2011 2nd International Congress CACS, AISC 145, pp. 303–308.

human beings. According to the statistics released by American Podiatric Medical Association, the walking distance of one whole life is about 2.5 times as long as that of the earth's perimeter. The force that the feet bore can reach 1.5 times as big as that of the gravity of human body. During running, the force that the feet bore can reach 2-3 times as big as that of the gravity of human body.[1]

How to keep the feet in good health and prevent from being hurt and evaluate the health status of the feet become concerns of medical circles and the related circles. The test and analysis of plantar pressure can acquire many effective parameters that human body is under different conditions. The data have great importance to clinical diagnosis, evaluation of diseases, and evaluation of curative effect after surgical operation, biomechanics, anthropology, space navigation, ergonomics, industry and so on.[2,3]. It is very necessary to construct a new plantar measurement system to present the pressure distribution that can conduct accurate, real-time and continuous measurement.

ZigBee is a specification for a suite of high level communication protocols using small, low-power digital radios based on the IEEE 802.15.4-2003 standard for Low-Rate Wireless Personal Area Networks(LR-WPANs), such as wireless light switches with lamps, electrical meters with in-home-displays, consumer electronics equipment via short-range radio needing low rates of data transfer. The technology defined by the ZigBee specification is intended to be simpler and less expensive than other WPANs, such as Bluetooth. ZigBee is targeted at radio-frequency (RF) applications that require a low data rate, long battery life, and secure networking.

2 Structure of the System

The system can be divided into hardware and software. Of course, the structure of the system is presented as the following figure in detail.

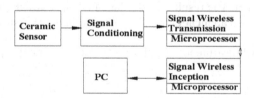

Fig. 1 Structure of the system

3 Design of Hardware

3.1 Design of Piezoelectric Ceramic Sensor

The piezoelectric ceramic sensors have the following strongpoints---high sensitivity, wide working frequency, high SNR(signal to noise ratio),low mass, good stability of temperature and so on. It is this type sensor that has many advantages over other type sensors. So the piezoelectric ceramic type sensor is chosen.

3.2 Layout of Sensors

Each foot can be divided into several zones as follows.

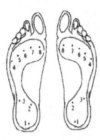

Fig. 2 Plantar Anatomy Zone[2]

The entire sensors are inserted into two insoles equally. Each insole has 8 measurement points. There are 3 sensors that are placed onto each specified measurement points specified as the anatomy zone. That means that there will be 48 sensors will be deployed totally. The accurate position of each sensor depends upon the size of foot(see the following figure).

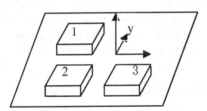

Fig. 3 Position of Each Sensor at One Measurement Point

3.4 Calibration of Sensors and Measurement of Dynamic Characteristic

Sine comparison calibration method[4-7] is adopted to determine the sensitivity and response characteristic curve of amplitude &frequency.

Calibration of sensors is presented as the specified program.

3.5 Determination of Response Characteristic Curve of Amplitude & Frequency

The following chart shows the determination process of response characteristic curve of amplitude &frequency.

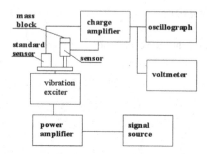

Fig. 4 Determination of Response Characteristic

4 Development of Software

4.1 Introduction to LabVIEW

Virtual Instrument is one of important new technologies in the CAT field. Lab-VIEW (short for Laboratory Virtual Instrumentation Engineering Workbench) is a platform and development environment for a visual programming language from National Instruments. Originally released for the Apple Macintosh in 1986, Lab-VIEW is commonly used for data acquisition, instrument control, and industrial automation on a variety of platforms including Microsoft Windows, various flavors of UNIX, Linux, and Mac OS X.The programming language used in Lab-VIEW, also referred to as G, is a dataflow programming language.

4.2 Development of Software

According to the requirement of the measurement system, the development is designed carefully. After much amendment and adjustment, the software can satisfy the requirement of the system. The appearances of the finished program are as follows.

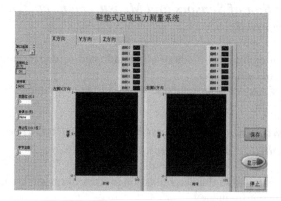

Fig. 5 Interface of the Program

5   Conclusion

With the help of the insole measurement system, many useful data have been acquired that satisfy the conclusions presented on some literatures [8-10]. Though the system has some advantages, there exist some problems[11-14] As we know, the study on plantar pressure is a systematic engineering. There still exists something needed to be improved such as the following aspects.

Though many sensors have been used, there still exist gap between the ideal numbers and the finished system. If the numbers of the sensors embedded into the insole can cover the whole feet, the more satisfied results can be acquired.

On how to integrate the technology of GPS and wireless communications. If the related technology can be integrated into this system, the function can be greatly improved.

On how to integrate the related development software such as Visual C++, Visual Basic, C++ Builder and so on.

On how to realize the internet measurement.

If the above-mentioned aspects are improved gradually, the system will have more functions.

References

1. Wu, J.: Study on biomechanics of foot &ankle and related assistance tools, vol. 18 (April 2002), http://www.chinese-ortho.com
2. Eason, G., Noble, B., Sneddon, I.N.: On certain integrals of Lipschitz-Hankel type involving products of Bessel functions. Phil. Trans. Roy. Soc. London A247, 529–551 (1955)
3. Wei, Q., Lu, W., Fu, Z.: A New System For Foot Pressure Measurement and Gait Analysis. Chinese Journal of Biomedical Engineering 19, 32–40 (2000)
4. Yuan, G., Zhang, M., et al.: The Measurement System for Dynamic Foot Pressure and Its Clinical Application. Chinese Journal of Rehabilitation 18 (February 2003)
5. Yuan, X.: Manual of Sensors. China National Defense Press, Beijing (1986)
6. Liu, Y., Ye, X.: Principle Design and Application of Sensors. China National Defense University Press, Changsha (1997)
7. Ma, M., Zhou, C.: Data Acquisition and Treatment. Xi'an Jiaotong University Press, Xi'an (1998)
8. Jiang, H., Zhang, Q.: Theory and application of Dynamic measurement. Southeast University Press, Nanjing (1998)
9. Li, W.: The application of Testpoint. Electrnic Technology 3, 37–38 (2002)
10. RongGuang, T., Wallace, W.A.: Static and Dynamic measurement of normal foot pressure. Chinese Journal of Biomedical Engineering 13, 175–177 (1994)
11. Yuan, G., Zhang, M., Wang, Z., et al.: The distribution of foot pressure and its influence factors in Chinese people. Chinese Journal of Physical Medicine and Rehabilitation 26, 156–159 (2004)

12. Jiménez, F.J., De Frutos, J.: Virtual instrument for measurement, processing data, and visualization of vibration patterns of piezoelectric devices. Computer Standards & Interfaces 27, 653–663 (2005)
13. Torán, F., Ramírez, D., Navarro, A.E., Casans, S., Pelegrí, J., Espí, J.M.: Design of a virtual in-strument for water quality monitoring across the Internet. Sensors and Actuators B: Chemical 76, 281–285 (2001)
14. Nuccio, S., Spataro, C.: Assessment of virtual instruments measurement uncertainty. Computer Standards & Interfaces 23, 39–46 (2001)

A Study of Software Architecture for Real-Time Image and Graphic Processing for Time-Sequenced 3-D CT Images

Yoshio Yanagihara

Abstract. Software architecture is studied, in which real time image and Graphic processing are executed for time-sequenced three-dimensional computer tomogram images, to observe the organs with movement, such as heart or lung. Because such system must process so huge number of images, its observer might wait a while for processing images and then generating graphic images. In developing such system, it might be efficient to utilize GPU (graphic processor unit) for processing. The experimental result shows that the processing times on GPU are almost 1/4 of the times on CPU including memory transfer. Furthermore some suitable software architecture must be investigated in consideration of the size of targeted images and the processing time for image process and graphic process. In this paper, software architecture in the system will be proposed, which continues to display animated graphic images naturally as heart movement even while huge image process executes.

1 Introduction

Recently, GPU (graphic processor unit) makes rapid progress to get higher performance. At the earlier stage, GPU is utilized only for generating graphic images and displaying them on a monitor. At the next, GPU obtained programmable ability of pixel shading on its first progress. Now, GPU obtained more computational ability for calculation of general purposes. GPU produced by NVIDIA Corporation has more than one hundred processing cores for calculation inside of one graphic chip. Its performance of calculation may become more 10X-100X than CPU (Central Processing Unit). So, GPU can have high performance especially for applications of determinant calculation such as array processing. CUDA

Yoshio Yanagihara
3-3-138 Sugimoto Sumiyoshi Osaka Japan
e-mail: yanagihara@info.eng.osaka-cu.ac.jp

F.L. Gaol et al. (Eds.): Proc. of the 2011 2nd International Congress CACS, AISC 145, pp. 309–317.
springerlink.com

(Compute Unified Device Architecture) produced by NVIDIA Corporation is a parallel computing environment to get high performance on such GPU [1].

In the field of manipulating huge data such as retrieval of data base, CUDA is one of very useful techniques [2, 3, 4]. Also, in medical applications, many images must be maintained and processed by calculation in general purpose. Especially, multi-sliced CT (computer tomography) equipment generates huge number of images both on volumes and on time phases. The photographed tomogram images could be displayed as animated graphic images. So, such system for observation of multi-sliced CT images must consume large resources of computer and long time for processing and, then generating graphic images. In this paper, software architecture is investigated, which is proper both for image process and for graphic process.

2   Module of Image Process

Here, the methods of processing images for the extraction of organs are briefly described in [5, 6]. In this experiment, a set of data has 10 time phases presenting for the period of one heart beat. At each time phase, 116 CT images are obtained and thinned out to its half because of data reduction for utilization in the experiment. One image has 512 x 512 pixels, each of which has 16 bits density. The size of one pixel is almost 0.5 mm x 0.5 mm and the distance of neighboring slices is almost 0.5 mm.

2.1   Processing Images

First, several threshold values are utilized for discrimination of blood regions, bone regions and other regions on a simple tomogram plane image, where the threshold values were selected by the preliminary experiments. And then, the neighboring connectivity on the same plane is checked for area selection and eliminates small noises.

Second, the expanded bone regions on each time-sequenced plane at the same slice position are gathered and eliminated. Using the all time-sequenced tomogram images at each slice position, bone regions on the all plane images are added to get a mask image. And then, the positions of the pixels discriminated as bone on the entire mask images along time-sequence are decided as true bone and the mask plane is obtained. This mask plane is utilized as bone templates to eliminate small bone regions.

Last, for the more precise region extraction, blood regions and some small bone regions at one time phase are selected. Each of the pixels on selected regions is checked if it is overlapped on the same selected regions in images at the two neighboring slices. If a pixel is not overlapped on the same selected regions in any image at its neighboring slice position, then the pixel is eliminated. This process is repeated for all combination of three neighboring sliced plane images.

After extraction of organs, the process of preparing and setting textures is executed for display by using the graphic library of OpenGL. Here, animation images can be obtained by generating graphic images at all sequenced time phase. A set of textures at each time phase is corresponding to each time-sequential volume. Each processed tomogram image (on whose the pixels are segmented) is mapped to the corresponding texture. The sequential number for using the library is assigned to a group of textures about each volume according to its time phase. Finally, graphic images are displayed in the loop process of the library. Fig.1 shows one of generated graphic images in animation.

As huge processing time consumed, the module of image process is apart from graphic process (including animated display). As supposing that Human's heart beat is 60 counts per minute, the period of one beat is almost one second and a set of data for one beat has 10 time phases and the interval of one time phase becomes 100ms. On the other side, image process needs 800 to 1000 ms on the two PCs, (one has AMD phenom (2.50GHz) and another has Intel Core2 Duo (2.26GHz)) averagely for each time phase. The presented evaluated time excludes memory transfer time.

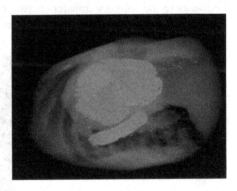

Fig. 1 Example of generated CG images [6].

2.2   Replacement by CUDA Program

For using CUDA program, the proposed processes could be classified as follows: 1) Processing for each individual pixel (i.e., threshold), 2) Processing for each pixel of interest and its neighbors on the same image, 3) Processing for each pixel of interest and the neighbors on its neighboring images, and 4) Processing for each pixel of interest and the neighbors of the pixel at the same position on the images neighboring along time sequence. Furthermore, each pixel is presented as a set of density value and transparency value. Then processes are arranged for the set of the values. The first and the second classified processes are replaced by CUDA program.

Original data (a[]) presents the density (16 bits) of each element. After processing, each element is represented as the 2-tuple of density value (8 bits) and transparency value (8 bits). In the data (b[]) obtained in these process, b[2*n] means the density value and b[2*n+1] means the transparency value of the n-th element, respectively. The notation (c[]) is also represented as the same structure of data.

1) Threshold process: For each element of original data, its density value of b[i] after processing is determined with an arbitrary translation function (f()) applied for a[i] and the range of the calculated value is restricted to 0 to 255. And Its transparency value will be set to 255 (means not transparent) when its density value is over than the previously assigned value. Otherwise, the transparency value is set to a small value (means almost transparent). This process is classified as the first category.

2) Selection process: For any of data (b[i]), if b[i] is transparent and one more pixels of its neighbors are not transparent, then its transparency value is reset to the value meaning to half transparency. Otherwise, its transparency value still remains. This process is classified as the second category.

3) Combined-in-system process: This process applies threshold process and then selection process for one volume (including 58 images) at one time phase. That is, the output data obtained by threshold process is used as input for selection process. Furthermore the memory transfers between host (PC side) and device (GPU side) are also launched. After then, obtained data is used in the graphic process.

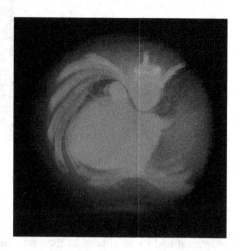

Fig. 2 Example of images by ray-tracing.

The processing times on GTX240 are presented in table 1. The values on the rows of "threshold" and "selection" present the measurements for one image. The values of the row of "combined-in-system" present the measurement for one volume including memory transfer. In both of the threshold and selection processes, the processing times on GPU are almost 1/10 of the times on CPU. In combined-in-system process, the processing time on GPU is almost 1/4 of the time on CPU, which means that the time demanded for data transfer between host and device must be taken into consideration.

Table 1 Average process time on GPU and CPU.

Process	process time on GPU	process time on CPU
"Threshold"	0.13 ms	1.30 ms
"Selection"	0.20 ms	2.20 ms
"combined-in-system"	95 ms	400 ms

3   Module of Graphic Process

As limitation of memory size of the GPU board, it is difficult to store all image data in the memory. That is, the memory transfer between host and device are needed on each change of image process and graphic process, which means expensive cost. Its reason is mainly that the module of graphic process has all image data. If some sophisticated method of ray-casting is utilized instead of using the library for generating graphic images, the memory transfer might be reduced because the sophisticated method can process the images in image process without memory transfer.

A set of input images at one time phase is considered as volume data including voxels. Some technique of ray-casting is utilized for graphic generation [7, 8, 9]. The basic method is described here. First, a position of the pixel of a graphic image to generate is selected and converted to a space position including the volume data. A ray from the position is radiated for the volume data. When the ray passes any voxel, a cumulated value is calculated with the density value and the transparency value of the voxel. This process is repeated for all pixels of the graphic image. Fig.2 shows an example image obtained by ray casting.

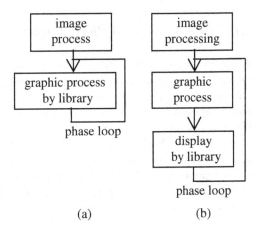

Fig. 3 Construction of phase loop.(a) the construction used for Fig.1, (b) the construction with CG-generation implemented.

(a)   (b)

The figure 3 shows the concepts of software construction. Fig.3 (a) shows the flow diagram for the modules of image process and graphic process using the library for all tomogram images. Fig3 (b) shows the flow diagram of the modules of image process, graphic process, and display of one generated graphic image by the library. As the small memory utilized for display in the later case, the large memory transfer between host and device is not needed. Furthermore, its transfer is executed on the memory of the GPU board by using a memory transfer function (called Buffer Object in the library). As result, for graphic process, it takes about 54ms to 89ms by using global memory and about 14ms to 17ms by using texture memory on GTX480 for one whole volume (116 images).

4   System Construction for Natural Display

By considering that one period of Heart's beat may need one second, the redrawing time of 100ms for each time phase (described before) must be needed for natural display. It means animated display for the same interval as the real world. It is proper that this condition is satisfied in the case that any image processing is not executed in the redrawing time for natural display (100ms). But, in actual, image process must be executed before the graphic process starts. On the case of observing another effects for data (which means such that different organs is displayed or extracted with varying thresholds as an examples), the observer must waits until image process terminated   (8-10 seconds on CPU) and graphic image is then displayed, which means not-natural display. The process flow diagram of such not-natural display is illustrated in Fig.3 (a) and (b), where the module of image process is still apart from graphic process.

One idea for natural display is the usage of multi-thread programming by which different threads for image process and for graphic process run. But the present GPU is executable on single thread. In the next section, the process flow applicable to natural display in one thread programming is described.

4.1   Improvement of Phase Loop

A method for natural display is considered, in which image process is implemented as one component of phase loop shown in Fig.4 (a). In this case, images only at one time phase are processed for each redrawing, so that the waiting time until image process terminated is reduced according to the number of time phases. But, if the processing time of image process is over than the redraw time for natural display, the observer must wait for a while and feel unnaturally. In this paper, the processing time takes almost 800-1000 ms on CPU needed for a time phase corresponding to each volume, which is over than the natural redraw time of 100ms. This method might be improved, but the observer must still wait for almost a second at every time phase.

For more suitability to natural display, it is needed that image process is divided to several small procedures, each of which becomes terminated in the natural redraw time. The additional process to administrate these small procedures must be implemented as some framework. The process flow diagram of the refined method is illustrated in Fig.4 (b). In this figure, the image process has the post-component for saving the procedure status utilized to keep which procedures terminated in a capable time. And the display by the library has the post-component of restoring the procedure state for the subsequent procedures in the image process. By using this architecture, image process can be implemented without avoiding natural display.

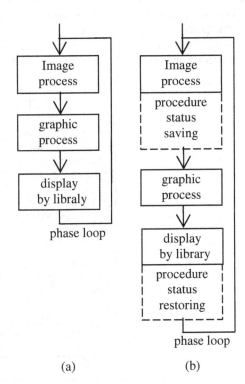

(a)   (b)

Fig. 4 Construction of phase loop (a) the construction with also image processing, (b) the construction for natural display.

4.2   Discussion

The method is described for software construction applicable for natural display on animation of time-sequenced 3D CT images. Its features are discussed here.

1) The proposed method is rather proper for single thread computing, which is suitable for a single GPU device and easily implemented on it. On using parallel mechanism, the multi GPU environment must be necessary and some sophisticated technique of memory transfer must be implemented.

2) In the proposed method, the component of image processing must be divided to several small procedures based on some units of time phase, procedure, or to-mogram image suitable for one of the sequential procedures. Each of the small procedures must terminate in consideration of realizing natural display. On using parallel mechanism, almost the same small procedures in image process can be utilized. That means the architecture might be a simple one.

3) The proposed method must include the process of management of sequential small procedures, which means that the state instance of State Pattern shown in Fig.5 might become the management data. Another method (called doStrategy as pattern) might be implemented for main part of management mechanism of image process. Furthermore, the management data is treated as a class (called CareTaker pattern). On using parallel mechanism, this mechanism is not necessary, but the process of managing buffer between both components of image process and graphic process must be necessary.

Totally, the proposed method might be preferred as its simple architecture on construction and suitable for single thread and a single GPU.

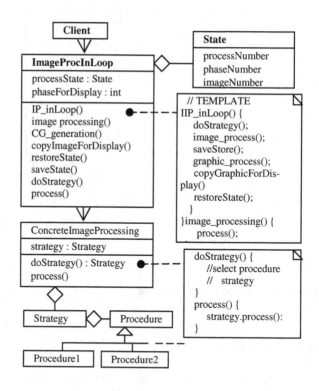

Fig. 5 Class diagram of the improvement method.

5   Conclusion

In this paper, system architecture is studied for applicable to development of an observation system which can display generated graphic images naturally with executing image processing in consideration of time-balance. The subject of the system is a time-sequenced data of 3-D CT images which has huge amount of data, so that its observer must wait a while for redrawing of graphic images until the image process completely terminate. In this case, GPU is actually important for fast processing. By utilizing GPU, several of the procedures become fast and the experimental result shows that its processing speed becomes almost 4X faster than by CPU including memory transfer. Suitable software architecture in consideration of both of image process and graphic process is discussed. The method implementing the architecture needs to divide the component to small procedures, each of which can be executed in the time needed for natural display. As so, it is necessary to utilize some management mechanism. Next, the observation system will be implemented by the proposed architecture.

References

1. Sanders, J., Kandrot, E.: Cuda By Example. Addison-Wesley (2010)
2. Munekawa, Y., Ino, F., Hagihara, K.: Acceleration Smith-Waterman Algorithm for Biological Database Search on CUDA-Compatible GPUs. IEICE Trans. Inf. & Syst. E93-D(6m) (2010)
3. Chen, G., Li, G., Pei, S., Wu, B.: Gpgpu supported cooperative acceleration in molecular dynamics. In: Proceeding of 13th Conf. of Computer Supported Cooperative Work in Design, pp. 113–118 (2009)
4. Nukada, S.M.A.: Auto-tuning 3-d fft library for cuda gpu. In: Proc. High Performance Computing Networking (2009)
5. Yanagihara, Y.: A Study of Region Extraction and System Model on an Observation System of Time-Sequenced 3-D CT Images. In: Proc. ISITA, M-TA-4 (2008)
6. Yanagihara, Y.: A study about software architecture for realtime processing and smoothed presentation on an observation system of time-sequenced 3-D CT images. CARS 5,1, S341 (2010)
7. Smelyanskiv, M., Holmes, D., Chhugani, J., et al.: Mapping High-Fidelity Volume Rendering for Medical Imaging to CPU, GPU, and Many-Core Architectures. IEEE Trans. Visualization and Computer Graphics 15, 6 (2009)
8. Patidar, S., Narayaman, P.J.: Ray Casting Deformable Models on the GPU. In: ICVGIP 2008 (2008)
9. Popov, S., Gunther, J., Sedel, H., et al.: Stackless KD-Tree Traversal for High Performance GPU Ray Tracing. Computer Graphics Forum 26(3), 415–424 (2007)

Entropy Application in Partial Discharge Analysis with Non-intrusive Measurement

Guomin Luo and Daming Zhang

Abstract. Partial discharge (PD) occurs when insulation deterioration happens in electrical apparatus. It is often detected in order to evaluate the state of insulation. For metal-clad equipments, external sensors which are easy to install and interruption-free on operations are preferred. However, their performances are compromised by heavy noise. Although time-frequency (TF) spectrum provides much information to discriminate PDs and noises, automatic selection remains a tough issue in field application. Entropy, a measure of disorder, is applied in this paper to extract PD pulses automatically. This entropy-based algorithm is implemented and examined by two field-collected datasets. Practical results show that true PDs can be identified and extracted effectively.

1 Introduction

For metal-clad equipment, partial discharge (PD) measurement is often employed as a means of insulation evaluation. Traditionally, the PD sensors are mounted inside the metallic tank which shields external interferences. But for operational apparatus, arranging a shutdown specifically to fit such couplers rarely can be justified[3]. Then external non-intrusive sensors which are easy to install and interruption-free on operations are preferred. But without the shielding of metallic enclosure, the surrounding interference to the measured PD signal output from such non-intrusive sensors is a major problem for obtaining the true PD pulses. Time-frequency (TF) spectrum that reveals the energy variation with both time and frequency is an effective de-noising tools in PD analysis[8]. But there still exist some difficulties, for example, PD extraction from heavy noisy background and the need of samples for machine learning.

This paper proposes an effective PD extracting tool that extracts PD pulse when signal-to-noise ratio (SNR) is very low and the number of samples is limited. This paper starts with the introduction of non-intrusive PD measurement. The TF analyses of PDs and major noises are followed. Then an entropy based method is provided to select the frequency bands that contain possible PD energies and to locate the time regions of PDs. The method is then used to analyze field data and its effectiveness is demonstrated.

F.L. Gaol et al. (Eds.): Proc. of the 2011 2nd International Congress CACS, AISC 145, pp. 319–324.
springerlink.com   © Springer-Verlag Berlin Heidelberg 2012

2   Fundamentals of Non-intrusive PD Measurement

Partial discharges inside metal enclosure generate electromagnetic waves and a part of them radiate out from slits at insulated parts, gasket joints and cable insulation terminals[5]. These escaped electromagnetic waves are captured by sensors. Fig.1(a) shows the drawing of a self-developed coaxial sensor for non-intrusive PD measurement, where part 1 is the female BNC interface and it is integrated with part 2. Its frequency response ranges from 80Hz to 9MHz.

A laboratory test is set up as in Fig.1(b),(c) to illustrate the mechanism of non-intrusive measurement. A PD generator is placed inside the metallic enclosure. Two self-developed sensors are placed at inner bottom and outer top of the cabinet respectively. Besides the PD generator, there is also a commercial HFCT sensor.

Fig. 1 Coaxial PD sensor and the laboratory test: (a) coaxial sensor, (b) enclosure with its cover open, (c) enclosure with its cover close

The captured PD signals are shown in Fig.2 for two different durations. The measured signals are almost the same from the two self-made sensors placed inside and outside the enclosure. This test provides fundamental basis for field test of metal-clad equipments using non-intrusive PD sensing technique.

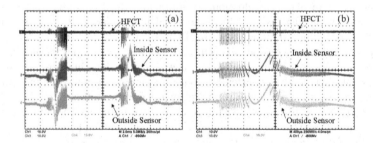

Fig. 2 Measured PD pulses by different sensors (a) PDs of a power frequency cycle (20 milliseconds), (b) PDs of 4 milliseconds

3   Time-Frequency Features of PD and Noises

The TF spectrums of all signals in this paper are generated by short time Fourier transform (STFT). The sliding window has a size of 20 microseconds.

The PD pulse often has quite short duration. If the frequency response range of the sensor is wide enough, frequency-axis-paralleling energy strips can be found in the TF spectrum of PD signals, as in Fig.3(d).

The TF spectrums of two typical kinds of noises: harmonics and communication signals are portrayed in Fig.3(b),(c). With similar frequency components in all sliding windows, the regular signals appear to be some time-axis-paralleling strips as in Fig.3(e). The disturbances from communication systems, such as cell phone, have narrow frequency ranges which appear to be some dots in TF spectrum.

The difference between PDs and noises in Fig.3 is quite pronounced. But such pronounced difference becomes smaller due to heavy noises in field tests. To automate the PD pulse extraction when analyzing field data, TF analysis is insufficient. This motivates us to develop an entropy based technique as described in the following sections to extract true PD pulse from field datasets automatically.

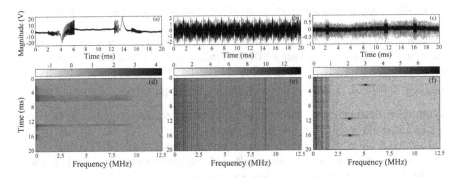

Fig. 3 PD, noises and their TF spectrum, (a) original PDs in Fig.2(a), (b) harmonics, (c) telecommunication signal from cell phone, from (d) to (f) logarithm of TF spectrum of signals above.

4   Entropy Based PD Extraction Algorithm

Entropy is a measure of disorder. Shannon Entropy defined in [6] is employed in this algorithm. The entropy and time-frequency analysis illustrated in Fig.4 suggest the following PD extraction procedure:

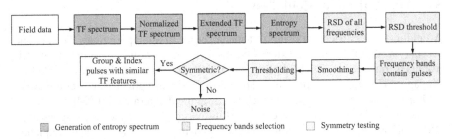

Fig. 4 Flow chart of entropy based PD extraction method

Generation of TF Entropy Spectrum: Perform the TF transform of signal and generate its entropy spectrum.

To produce an entropy spectrum unaffected by magnitude, the TF spectrum is normalized by dividing the maximum value of each frequency. Then, the maximum value in whole TF spectrum is 1. Then a sliding window with odd dimensions slides along rows and columns of TF spectrum to form an entropy spectrum. One window produces one entropy value. To keep the TF spectrum and entropy spectrum being in the same dimension, all four sides and corners of the TF spectrum are symmetrically extended.

Frequency Bands Selection: Select the frequency bands that contain possible PDs according to the entropy parameters.

For a particular frequency, larger differences between TF energies suggest the existence of pulses and cause larger variations. Relative standard deviation (RSD) which equals the standard deviation σ divided by mean μ, is a suitable yardstick to describe this variation.

Then a threshold generated by using minimax estimation is employed to select the frequency bands with larger RSD[7]. Here, an approximate minimax estimator is used[4]. The threshold λ equals $\varepsilon * (0.3936 + 0.1829 * \log_2 N)$, where N is the length of time axis, ε is the median of RSD of all frequencies.

Symmetry Testing: Test the symmetry of phase region of pulses in two half cycles and remove the pulse-like disturbances.

Usually in oil-insulated systems, there are four typical types of PD patterns that occur on both positive and negative half cycles at similar phase regions[1]. Then, testing the symmetry of phase regions in entropy spectrum can help find the possible PDs. Similarity coefficient δ between the phase regions of two half cycles is calculated as $\delta = \sum (T_1 \cap T_2) / (\sum (T_1 + T_2) / 2)$. Here, T_1 and T_2 are the phase regions of pulses in two half cycles. They are extracted by thresholding with mean value. An empirical threshold, for instance 0.5, is adopted to distinguish PDs and noises.

Pulse Extraction: Group the pulses with similar characteristics and index them in original TF spectrum. The universal threshold is employed to remove the noises in those PD-containing frequency bands[2].

5   Application in Field Test

A field test was carried out using self-made sensor in Singapore Shaw Tower. The sensors were placed on the external surface of the metallic enclosure of an oil-insulated transformer. The sampling rate is 50MHz. Another dataset provided by Singapore Hoestar Inspection International Pte Ltd was collected by a commercial air-borne sensor on the same equipment and sampled at a rate of 100MHz.

The original signals of one cycle (20 milliseconds), their entropy spectrums, RSD and selected phase regions are displayed in Fig.5. The selected frequency bands that have greater RSD are emphasized by dashed lines and numbered in series from low frequency to high frequency. One can see that for the frequency bands with greater RSD than threshold, the corresponding entropy spectrum at two half cycles are very similar and their centers are apart from each other approximately 180^0. For example, the 5th frequency bands (from 6MHz to 6.6MHz) in Fig.5(h). The similarity coefficients δ of the three bands in Fig.5(g) are 0.81081, 0.70189, and 0.7509 in series. The δ of selected bands in Fig.5(h), from first to fifth, are 0.91444, 0.96154, 0.72137, 0.7992, and 0.96709, respectively. As all of them are greater than 0.5, all these selected frequency bands are considered to contain PDs. Next, the frequency bands with similar entropy distribution and occurring within almost the same time intervals at two half cycles are grouped. For instance, the second and fifth bands in Fig.5(h) with similar phase regions are grouped. Then, the universal threshold described in step 4 in section 4 is employed to examine TF spectrum within selected frequency bands. After thresholding, the left points on TF spectrum of each group are indexed by time to form the extracted pulses.

The extracted pulses of two signals are portrayed in Fig.6. The magnitude of pulses from self-made sensor is larger than those of data from commercial sensor. Besides the influence from amplitude response of different sensors, the greater magnitude from self-made sensor may partly result from its electrical contact with the cabinet surface.

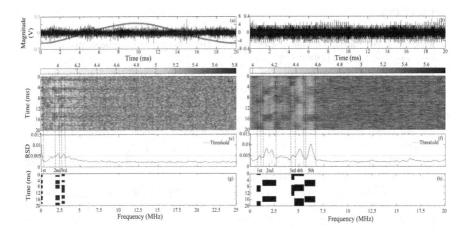

Fig. 5 Data from self-developed sensor and commercial sensor, (a) original signal from self-made sensor, on the left is the measure of sinusoidal, and on the right is the measure of PD signal, (b) data from commercial sensor, (c) the entropy spectrum of Fig.5(a), (d) the entropy spectrum of Fig.5(b), (e) the RSD of entropy of Fig.5(c), (f) the RSD of entropy of Fig.5(d), (g) the selected phase region of Fig.5(c), (h) the selected phase region of Fig.5(d).

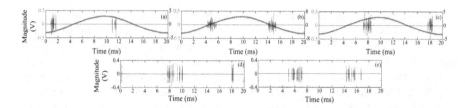

Fig. 6 Extracted pulses, (a) to (c) are the extracted pulses of data from self-made sensor (three groups). On the left is the measure of sinusoidal, and on the right is the measure of PD signal, (d) to (e) are the extracted pulses of data from commercial sensor (two groups)

6   Conclusion

To solve PD extraction problem of non-intrusive PD measurement, entropy based time-frequency analysis is presented in this paper as an efficient tool to identify real PD pulses in the presence of high levels of noises. The efficiency and adaptability of this proposed method are demonstrated by analyzing two field datasets as well as some noise data and laboratory-created PD data. This article shows that entropy based time-frequency analysis can successfully complement other existing PD de-noising methods in solving the problem of pulse extraction.

References

1. Kreuger, F.H.: Partial Discharge Detection In High-Voltage Equipment, pp. 129–142. Butterworths, London (1989)
2. Luo, G., Zhang, D.: Application of wavelet transform to study partial discharge in XLPE sample. In: 19th Australasian Universities Power Engineering Conf.: Sustainable Energy Technologies and Systems, AUPEC 2009, Adelaide, Australia, pp. 1–6 (2009)
3. Judd, M.D., Farish, O., Pearson, J.S., Hampton, B.F.: Dielectric windows for UHF partial discharge detection. IEEE Trans. Dielectrics and Electrical Insulation 8, 953–958 (2001)
4. Misiti, M., Misiti, Y., Oppenheim, G., Poggi, J.M.: Wavelet Toolbox for Use with MATLAB: User's Guide, 1st edn. (1996)
5. Davies, N., Yuan, T., Tang, J.C.Y., Shiel, P.: Non-intrusive partial discharge measurements of MV switchgears. In: International Conference on Condition Monitoring and Diagnosis, CMD 2008, pp. 385–388 (2008)
6. Verdú, S., McLaughlin, S.: Fifty Years of Shannon Theory. In: Information Theory: 50 Years of Discovery, pp. 13–15 (1999)
7. Mallat, S.G.: A Wavelet Tour of Signal Processing: The Sparse Way, pp. 535–545. Elsevier /Academic Press, Sparse, Amsterdam; Boston (2009)
8. Jia, Z., Hao, Y., Xie, H.: The Degradation Assessment of Epoxy/Mica Insulation Under Multi-Stresses Aging. IEEE Transactions on Dielectrics and Electrical Insulation 13, 415–422 (2006)

Three-Dimensional Imaging of a Human Body Using an Array of Ultrasonic Sensors and a Camera

Hideo Furuhashi, Yuta Kuzuya, Chen Gal, and Masatoshi Shimizu

Abstract. This paper describes the three-dimensional (3D) imaging of a human body using an array of ultrasonic sensors and a camera. We have previously reported on a system that measures the shape of an object (including a human body) using a camera and combines this information with 3D position data obtained using an ultrasonic array; however, this system only detects the center position of the object. In the present research, a prototype system was constructed using an ultrasonic sensor array and a camera. The 3D positions of each body part were measured using the ultrasonic array and combined with the image of the human body. Some difficulties in combining the 3D positions and the camera image are discussed. One problem is the lack of 3D position data (partially due to the poor signals of the reflected ultrasonic waves), which can be resolved by using the neighboring position data.

1 Introduction

Many imaging sensor systems have been developed that are capable of measuring the three-dimensional (3D) positions and shapes of various objects, such as human bodies; an ultrasonic sensor system is one such system [1, 2, 3, 4, 5, 6]. We have previously reported on a system that measures the shape of an object using a camera and combines this information with 3D position data obtained using an ultrasonic array; however, this system only detects the center position of the object. In

Hideo Furuhashi · Yuta Kuzuya · Chen Gal
Aichi Institute of Technology, Dep. of Electrical and Electronics Engineering,
Aichi Institute of Technology, 1247 Yachigusa, Yakusa-cho, Toyota, Aichi, Japan

Masatoshi Shimizu
Power Engineering R&D Center, Kansai Electric Power Co., Inc., 3-11-20 Nakoji, Amagasaki
661-0947, Japan

F.L. Gaol et al. (Eds.): Proc. of the 2011 2nd International Congress CACS, AISC 145, pp. 325–330.
springerlink.com   © Springer-Verlag Berlin Heidelberg 2012

many applications, it is required to know the position of each part of the body. Therefore, this paper discusses a system that measures the 3D positions of each body part using an ultrasonic sensor array and combines this data with a camera image of the entire body.

2 Theory

Figure 1 shows a schematic of the sensor array system, which consists of an ultrasonic transmitter and 16 ultrasonic receivers. Ultrasonic waves are emitted by the transmitter, and delay-and-sum operations are used to process the signals detected by the receivers; this determines the 3D positions of the target objects. The camera captures an image of the human body, and the computer processes the image, detecting the body using the background difference method. Finally, the shape and 3D position information are combined and displayed.

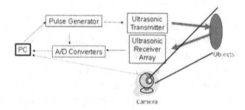

Fig. 1 Schematic of the ultrasonic sensor array and camera system.

Figure 2 shows the coordinate system of the camera's imaging system. If a point at position $P(x,y,z)$ is projected on the camera, the position of the camera's pixel $Q(i,j)$ corresponding to that point is given by [6]

$$x = z \cdot k_x \cdot i , \tag{1a}$$

$$y = z \cdot k_y \cdot j . \tag{1b}$$

Here,

$$k_x = \frac{2}{M} \tan \frac{R}{2} , \tag{2a}$$

$$k_y = \frac{2}{N} \tan \frac{S}{2} , \tag{2b}$$

where the angular field of view of the camera is $R{\times}S$, there are pixels, and $M{\times}N$ the center position of the pixel array is $Q(0,0)$.

Similarly, the relationship between the reflection point $P(x,y,z)$ and the position $Q'(i',j')$ of the pixel on the ultrasonic sensor array is

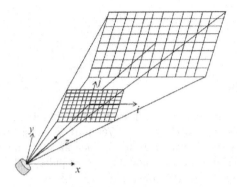

Fig. 2 Coordinate system of the camera.

$$x = z \cdot k_x' \cdot i', \tag{3a}$$

$$y = z \cdot k_y' \cdot j'. \tag{3b}$$

Here

$$k_x' = \frac{2}{M'} \tan \frac{R'}{2}, \tag{4a}$$

$$k_y' = \frac{2}{N'} \tan \frac{S'}{2}, \tag{4b}$$

where the angular field of view of the ultrasonic sensor array is $R' \times S'$, there are $M' \times N'$ pixels, and the center position of the pixel array is $Q'(0,0)$.

If the angular field of view of the ultrasonic sensor array is equal to that of the camera, then

$$i' = \frac{M'}{M} \cdot i, \tag{5a}$$

$$j' = \frac{N'}{N} \cdot j. \tag{5b}$$

In this ultrasonic sensor array, delay-and-sum operations are performed. By delaying each signal detected by the ultrasonic receivers and applying sum operations, pixel signals from the ultrasonic sensor array can be obtained. The range $d(i',j')$ can calculated from the time $t_{max}(i',j')$ at which the summed signal is maximum using

$$d(i', j') = \frac{t_{max}(i', j')}{2} \upsilon \qquad (6)$$

3 Experimental

Figure 3 shows the ultrasonic receiver array, which has 16 receivers (Knowledge Acoustics, SP0103NC3-3; 6.15 mm × 3.76 mm). A single transmitter (Murata Manufacturing, MA40S4S) was used to produce an ultrasonic wave with a frequency of 40 kHz. The wave was modulated by a pulse with a half-width of 0.4 ms. The received signal was A/D converted every 2.5 μs.

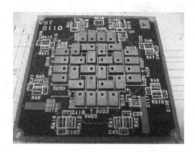

Fig. 3 Ultrasonic receiver array.

A 160×120 pixel CCD camera was placed 4 cm below the center of the ultrasonic array. It has an angular field of view of 71° diagonally (nominal value) and an angular field of view of 58° (horizontally) × 45° (vertically). The angular field of view of the ultrasonic sensor is 58° × 45°, which is the same as that of the camera. Additionally, the ultrasonic sensor has 16×12 pixels, making its angular resolution 3.6° × 3.8°/pixel. Using (5a) and (5b), the coordinates of the camera pixels and the ultrasonic sensor array are given by

$$i' = \frac{1}{10} \cdot i, \qquad (7a)$$

$$j' = \frac{1}{10} \cdot j. \qquad (7b)$$

Figure 4 shows the imaging and 3D measurements of a person standing in an office. The right arm is extended in front of the torso, as shown in Fig. 4(a). The background image was subtracted from the image, and an object extraction was applied (Fig. 4(b)), which contains no information about the range. The range of each point was obtained from the time of the maximum t_{max} of the ultrasonic sensor array (Fig. 4(c)). The ultrasonic sensor array is unable to determine the detailed shapes of the object. The range data within the imaged area shown in Fig. 4(b) were applied to each pixel, resulting in the 3D image shown in Fig. 4(d). The 3D image clearly shows that the right arm is closer than the torso, while the left arm is well over 2 m away, behind the torso.

(a) Camera image   (b) Human shape extracted from the image

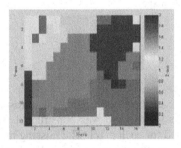
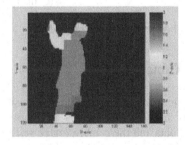

(c) 3D image obtained by the ultrasonic sensor   (d) 3D image of the human body

Fig. 4 3D measurements of a human body.

Figure 5 shows another case; Figure 5(a) shows the camera image, and Figure 5(b) shows the 3D image. In this case, some 3D position data are missing for the right hand due to poor signals of the reflected ultrasonic waves. If the maximum value of the ultrasonic sensor array signal is smaller than the threshold value, that value is ignored, resulting in incomplete range data. The threshold value is determined from the noise level. Therefore, the range data of these pixels were obtained by interpolation using the neighboring range data, and the results are shown in Fig. 5(c)

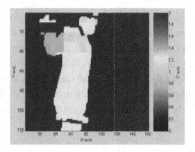

(a) Camera image   (b) 3D image obtained by the ultrasonic sensor

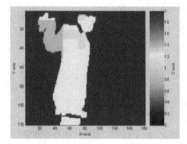

(c) Interpolated 3D image

Fig. 5 Interpolation of the range data.

4 Conclusion

A system that measures the 3D positions of each body part using an ultrasonic sensor array and combines this data with a camera image was constructed. The range data and the camera image were combined for each pixel. The missing range data due to poor signals of the reflected ultrasonic waves were interpolated using the neighboring range data.

References

1. Keating, P.N., Sawatari, T., Zilinskas, G.: Proceedings of the IEEE 67, 496–510 (1979)
2. Griffiths, L.J., Buckley, K.M.: IEEE Trans. Acoustic, Speech and Signal Processing ASSP-35, 917–926 (1987)
3. Furuhashi, H., Uchida, Y., Shimizu, M.: In: Proceedings of ICCAS 2007. FEP-35 (2007)
4. Furuhashi, H., Uchida, Y., Shimizu, M.: In: Proceedings of IECON 2008, vol. 85 (2008)
5. Furuhashi, H., Uchida, Y., Shimizu, M.: In: Proceedings of IEEE Sensors 2009, pp. 1467–1472 (2009)
6. Furuhashi, H., Kuzuya, Y., Uchida, Y., Shimizu, M.: In: Proceedings of IEEE Sensors 2010, pp. 713–718 (2010)

A Parallel Adaptive Block FSAI Preconditioner for Finite Element Geomechanical Models

Carlo Janna, Massimiliano Ferronato, and Giuseppe Gambolati

Abstract. Efficient ad hoc preconditioners are a key factor for a successful implementation of linear solvers in a parallel computing environment. The class of Factorized Sparse Approximate Inverses (FSAI), although originally developed for scalar machines, has proven extremely promising in multicore hardware. A recent evolution of FSAI is Block FSAI (BFSAI) which clusters the largest coefficients of the preconditioned matrix in a number of diagonal blocks defined in advance. A further improvement of BFSAI is the adaptive BFSAI (labelled ABF) where the non zero pattern of the BFSAI preconditioner is not prescribed a priori but computed automatically and adaptively by a suitable algorithm. Numerical results from large finite element (FE) geomechanical models show that ABF coupled with an incomplete Cholesky factorization of each individual diagonal block, i.e. ABF-IC, may outperform BFSAI-IC by up to a factor 4 while exhibiting an excellent degree of parallelization on any multiprocessor computer.

1 Introduction

Finite elements (FE) in structural mechanics yield large sparse symmetric positive definite (SPD) systems that call for efficient solvers with the solution typically addressed in a parallel computing environment. This is even truer for nonlinear problems that often require the solution of a sequence of large sets of linear equations. Unstructured FE grids lead generally to sparse matrices with the sparsity pattern dictated by both the geometry of the grid and the elements type. The availability of massively parallel machines [14] enhance the development of parallel solvers that must ideally possess a high parallelization degree in addition to a good scalability as is required by the increasingly finer granulation of the very hard problems

Carlo Janna · Massimiliano Ferronato · Giuseppe Gambolati
Department of Mathematical Methods and Models for Scientific Applications,
University of Padova, Italy
e-mail: {janna,ferronat,gambo}@dmsa.unipd.it

F.L. Gaol et al. (Eds.): Proc. of the 2011 2nd International Congress CACS, AISC 145, pp. 331–338.
springerlink.com   © Springer-Verlag Berlin Heidelberg 2012

expected to be solved in a near future in the most advanced areas of engineering, physics and biosciences. The solution algorithms which are successfully applicable to large sparse systems belong to the class of either direct [13, 1] or iterative solvers [15, 5]. Among the latter Krylov projection methods [6] deserve a special mention. To be effective any projection solver has to be cost-effectively preconditioned [2]. Preconditioning based on sparse approximate inverses has attracted much attention in recent years for two major reasons: 1- relative to incomplete Cholesky (IC) factorizations the approximate inverses does not suffer from pivotal breakdown or numerical instabilities, and 2- can be very efficiently implemented on parallel computers as their application requires only matrix-vector products. The basic idea underlying the construction of an approximate inverse of the matrix A relies on finding a matrix M, with prescribed non zero pattern, that minimizes the Frobenius norm of $(I - AM)$ or $(T - AM)$ with I and T the identity and a targeted matrix, respectively [4, 7]. For the special case of SPD matrices, a very effective approximate inverse of A can be found also in a factorized form (FSAI) by minimizing the Frobenius norm of $(I - GL)$ [11], with L the exact lower triangular factor of A (that need not be known explicitly). Generally, unless A possesses some special properties such as the diagonal dominance or a banded structure, the selection of a suitable pattern for M is in no way a trivial task. A most widespread strategy is to select the non zero pattern of a power of A that may nevertheless give rise to a preconditioner expensive to compute and apply. This inconvenience is obviated to by sparsifying the preconditioner via a pre- [3] or a post- filtration [12] technique.

Recently a novel FSAI, denoted as block FSAI (i. e. BFSAI), has been developed by Janna et al. [8]. This is a generalization of the FSAI concept of [11] with the aim at defining a preconditioned matrix under the form FAF^T which resembles as far as possible to a block diagonal matrix (with the off block coefficients relatively unimportant). F is computed by minimizing the Frobenius norm of the matrix $(D - FL)$ with L again the exact lower triangular factor of A and D an arbitrary block diagonal matrix. An IC block diagonal preconditioner is then applied to FAF^T thus giving rise to the new preconditioner BFSAI-IC that lends itself to a high degree of parallelism with BFSAI-IC in general superior to FSAI for any number of processors [8].

The actual computational performance of BFSAI-IC depends to a large extent on the a priori selected pattern of F (generally equal to that of A^2 or A^3). To improve the latter an algorithm has been derived [9] by which an optimal F pattern is obtained via an automatic and adaptive technique that is based on the minimization of an upper bound of the Kaporin number of the preconditioned matrix. Starting from an initial guess, typically the identity matrix, F is recursively improved extending its non zero pattern. The corresponding preconditioner is denoted by ABF-IC (for the sake of brevity the letters SAI are dropped). The present communication addresses the construction of ABF-IC and its application on a multicore environment. ABF-IC is then used with large sparse FE matrices arising in geomechanics and compared to BFSAI-IC and standard FSAI showing its superior performance on both.

2 The Adaptive Block FSAI Algorithm

We first recall how the BFSAI preconditioner is computed. Let $\mathscr{S}_L$ and $\mathscr{S}_{BD}$ be a sparse lower triangular and a dense block diagonal non-zero pattern, respectively, for a $n \times n$ matrix. For the sake of simplicity, assume that $\mathscr{S}_{BD}$ consists of n_p diagonal blocks with size $m = n/n_p$, n_p being the number of available processors. The $\mathscr{S}_L$ pattern is prescribed a priori. The BFSAI preconditioner for a symmetric matrix A is defined by minimizing the Frobenius norm [8]:

$$\|D - FL\|_F \to \min \tag{1}$$

over the set of matrices F with non-zero pattern $\mathscr{S}_{BL} = \mathscr{S}_{BD} \cup \mathscr{S}_L$. As said above, in Equation (1) D is an arbitrary full-rank block diagonal matrix with pattern $\mathscr{S}_{BD}$, F is the lower block triangular factor of BFSAI and L is the exact lower triangular factor of A. The differentiation of (1) with respect to the non-zero F entries $[F]_{ij}$, $(i, j) \in \mathscr{S}_{BL}$, leads to the equation:

$$[FA]_{ij} = [DL^T]_{ij} \qquad \forall (i, j) \in \mathscr{S}_{BL} \tag{2}$$

As DL^T is block upper triangular, the vector $\mathbf{f}_i$ of the nz_i non-zeroes of the i-th row of F is found by solving the dense linear system:

$$A[\mathscr{J}_i, \mathscr{J}_i]\mathbf{f}_i = \mathbf{v} \tag{3}$$

where $A[\mathscr{J}_i, \mathscr{J}_i]$ is the submatrix of A made of the coefficients $[A]_{ij}$ such that $i, j \in \mathscr{J}_i$, $\mathscr{J}_i$ is the set of indexes k such that $(i, k) \in \mathscr{S}_{BL}$, and $\mathbf{v}$ is a vector with the first $(nz_i - m)$ components equal to zero and the last m equal to the corresponding $[DL^T]_{ij}$. As D is arbitrary, the coefficients of DL^T, hence $\mathbf{v}$, are also. Therefore, Equation (3) can be rewritten in a more convenient block form by defining the set:

$$\overline{\mathscr{P}}_i = \{k : (i, k) \in \mathscr{S}_{BD}\} \tag{4}$$

and its complement set:

$$\mathscr{P}_i = \mathscr{J}_i \setminus \overline{\mathscr{P}}_i \tag{5}$$

Separating accordingly the unknown components of $\mathbf{f}_i$, Equation (3) becomes:

$$\begin{bmatrix} A[\mathscr{P}_i, \mathscr{P}_i] & A[\mathscr{P}_i, \overline{\mathscr{P}}_i] \\ A[\overline{\mathscr{P}}_i, \mathscr{P}_i] & A[\overline{\mathscr{P}}_i, \overline{\mathscr{P}}_i] \end{bmatrix} \begin{Bmatrix} \mathbf{f}_{i1} \\ \mathbf{f}_{i2} \end{Bmatrix} = \begin{Bmatrix} \mathbf{0} \\ \mathbf{v}_2 \end{Bmatrix} \tag{6}$$

Since the vector $\mathbf{v}_2 \in \mathbb{R}^m$ is arbitrary but certainly not null because of the non-singularity of D, m components of $\mathbf{f}_i$ can be also set arbitrarily. A practical choice is to take $[F]_{ii}$ equal to 1 and all the remaining components of $\mathbf{f}_{i2}$ equal to 0. This is equivalent to prescribing F to be a unit lower triangular matrix with the form:

$$
\begin{bmatrix}
I & 0 & \cdots & 0 \\
F_{21} & I & & \vdots \\
\vdots & & \ddots & \vdots \\
F_{n_p 1} & \cdots & \cdots & I
\end{bmatrix}
\tag{7}
$$

With such an assumption Equation (6) reads:

$$
A[\mathscr{P}_i, \mathscr{P}_i]\mathbf{f}_{i1} = -A[\mathscr{P}_i, i]
\tag{8}
$$

where $A[\mathscr{P}_i, i]$ is the $(nz_i - m) \times 1$ matrix with the coefficients of the i-th column of A with row index in $\mathscr{P}_i$. The algorithm for the computation of F can be efficiently implemented into a parallel environment as it requires the solution of n independent dense subsystems in the form (8). Moreover, if A is symmetric and positive definite the existence and uniqueness of F are always theoretically guaranteed [8].

The effect of a BFSAI application over A is to concentrate the largest entries of FAF^T into n_p diagonal blocks. However, as D is arbitrary, there is still no a priori reason why FAF^T should be better than A in a conjugate gradient iteration. Hence, to accelerate convergence, each diagonal block B_{i_p} of FAF^T is computed via a sparse matrix-matrix product and its incomplete Cholesky decomposition with partial fill-in found:

$$
B_{i_p} \simeq L_{i_p} L_{i_p}^T
\tag{9}
$$

Collecting all matrices L_{i_p}, $i_p = 1, \ldots, n_p$, provides an incomplete Block Jacobi pre-conditioner for FAF^T:

$$
J = J_L J_L^T
\tag{10}
$$

with:

$$
J_L =
\begin{bmatrix}
L_1 & & & \\
& L_2 & & \\
& & \ddots & \\
& & & L_{n_p}
\end{bmatrix}
\tag{11}
$$

The resulting final preconditioned matrix is:

$$
J_L^{-1} FAF^T J_L^{-T} = WAW^T
\tag{12}
$$

i.e. a sequential application of F and J_L to A. The "inner" preconditioner F is intended to reduce the interaction between the diagonal blocks of A, while the "outer" precondi-tioner J^{-1} improves the conditioning index of each block. The overall preconditioner WW^T is denoted as BFSAI-IC as it is a combination of BFSAI and IC. Note that for $n_p = 1$ F is the identity matrix so BFSAI-IC incompletely factorize A, only.

Once n_p is set, the quality of M^{-1} as a preconditioner for A mainly depends on the choice of the non-zero pattern $\mathscr{S}_{BL}$, whose optimal a-priori selection is still an open issue. In [8] $\mathscr{S}_{BL}$ is generally selected as the block lower pattern of small powers of A which is perhaps the most popular choice in approximate inverse preconditioning.

In this communicatiom we propose an iterative algorithm that adptively constructs the pattern of F with the aim to improve the overall preconditioner performance. We need first to recall some theory. The Kaporin conditioning number κ of a matrix A is defined as [10]:

$$\kappa(A) = \frac{\frac{1}{n}\mathrm{tr}A}{\det(A)^{1/n}} \tag{13}$$

and gives some indication as to the Conjugate Gradient (CG) rate of convergence in solving a linear system with A. Quite obviously $\kappa(A) \geq 1$ and the closer $\kappa(A)$ to 1, the faster the CG convergence. What we ideally need is to find a non-zero pattern for F such that the Kaporin number of the preconditioned matrix $\kappa(WAW^T)$ is close to 1 as much as possible. Unfortunately a direct minimization of $\kappa(WAW^T)$ is too expensive and we need to use a weaker result. It can be shown that the following inequality holds:

$$1 \leq \kappa(WAW^T) \leq [\det(A)\Psi_F(A)]^{\frac{1}{n}} \tag{14}$$

with $\Psi_F(A) = \prod_{i=1}^{n}[FAF^T]_{ii}$. The proposed algorithm is designed so as to iteratively reduce $\Psi_F(A)$ and works as follows. Let us assume $\mathscr{S}_{BL}^0$ to be an arbitrary initial guess for the non-zero pattern of the factor F. We aim at finding an augmented pattern $\mathscr{S}_{BL}^1$ providing a reduction of $\Psi_F(A)$. If $F[i,:]$ denotes the i-th row of F, we note that $[FAF^T]_{ii}$ depends on $F[i,:]$ only, and the gradient of $\Psi_F(A)$ with respect to $F[i,:]$, denoted as $\nabla_{F[i,:]}\Psi_F(A)$, is proportional to:

$$\mathbf{g}_i = \nabla_{F[i,:]}[FAF^T]_{ii} \tag{15}$$

The j-th component of $\mathbf{g}_i$ can be computed quite inexpensively as:

$$[\mathbf{g}_i]_j = 2\left(\sum_{r=1}^{n}[A]_{jr}[F]_{ir} + [A]_{ji}\right) \qquad \forall j = 1,\ldots,n \tag{16}$$

and we can obtain $\mathscr{S}_{BL}^1$ by adding to $\mathscr{S}_{BL}^0$ the positions (i,j) with the column indices corresponding to the largest components of $\mathbf{g}_i$. After computing F over the augmented pattern $\mathscr{S}_{BL}^1$, the procedure above can be iterated to find $\mathscr{S}_{BL}^2$, $\mathscr{S}_{BL}^3$, and so on. For the i-th F row, the procedure can be stopped when a satisfactorily small tolerance is satisfied, i. e.:

$$\frac{[FAF^T]_{ii}^k - [FAF^T]_{ii}^{k-1}}{[FAF^T]_{ii}^0} \leq \varepsilon \tag{17}$$

where the apex k stands for the iteration count and ε is the desired exit tolerance.

Once F is obtained, following the same approach as with BFSAI-IC, we compute an incomplete Block Jacobi preconditioner of FAF^T giving rise to the final ABF-IC preconditioner.

3   Numerical Experiments

The ABF-IC performance is compared with that of FSAI and BFSAI-IC on a large-size 3D geomechanical example. The system matrix arises from the discretization of a deep reservoir with tetrahedral Finite Elements. The geological structure of the reservoir is characterized by a sequence of thin layers of porous material that are quite hard to represent with a regular grid. Moreover, from a mechanical viewpoint the medium is strongly heterogeous. For these reasons, the resulting linear system, totaling 923,136 equations and 41,005,206 non-zeroes, has a large bandwidth and is ill-conditioned thus representing a challenging problem for any iterative solver. The parameters used for the preconditioners set-up have been chosen so as to minimize the total wall-clock time required for the system solution. In particular, FSAI and BFSAI-IC have been computed using the non-zero pattern of A^2 while an exit tolerance $\varepsilon = 0.001$, eq. (17), has been set for the ABF-IC pattern search procedure. Both the IC decompositions combined with BFSAI and ABF allow for a small fill-in degree of about 10 entries in addition to the original non-zeroes of each row. Table 1 compares the performance of the three preconditioners vs the number of computing processors from 1 to 32. It can be noticed from the table that BFSAI-IC largely outperforms FSAI for small n_p, say $n_p = 1 \div 8$, and this is mainly due to a drastic reduction in the number of iterations to convergence. However, for larger n_p, the performance of BFSAI-IC approaches that of FSAI with a similar iteration count.

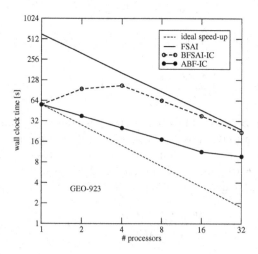

Fig. 1 Wall-clock time vs. the number of processors for the optimal FSAI, BFSAI-IC and ABF-IC. The ideal speed-up is also shown

Table 1 Number of iterations and wall-clock time of *ABF-IC*, *BFSAI-IC* and *FSAI* for the proposed geomechanical problem with $n_p = 1 \div 32$

n_p	ABF-IC		BFSAI-IC		FSAI	
	# iter.	time	# iter.	time	# iter.	time
1	121	55.76	121	55.76	3383	599.40
2	149	37.96	441	94.82	3378	312.63
4	195	25.24	1292	105.94	3379	161.19
8	292	17.18	1487	63.84	3381	86.25
16	347	11.33	2065	47.21	3381	46.17
32	398	9.72	2622	25.77	3381	23.80

By contrast, the use of ABF-IC allows for a further improvement relative to BFSAI-IC for small n_p ($1 \div 4$) proving more than 8 time faster than FSAI. ABF-IC exhibits a very good performance for $n_p = 32$ as well with a speed-up of 2.5 with respect to FSAI. The same result is evindenced also in figure 1 showing the total wall-clock time versus n_p required by CG preconditioned with ABF-IC, BFSAI-IC and FSAI to solve the linear system. For the sake of comparison, the ideal profile obtained from scaling the outcome of one processor with n_p is also reported. Again note that, increasing n_p, ABF-IC is closer to the ideal speed-up line, while BFSAI-IC quickly approaches the FSAI performance.

4 Conclusion

BFSAI is a factored approximate inverse of a SPD matrix A which concentrates (and only computes) the most important coefficients of FAF^T into the diagonal blocks of an arbitrary block diagonal matrix D. F possesses a given pattern (generally equal to that of A^2 or A^3) and its coefficients are provided by minimizing the Frobenius norm of the matrix $D - FL$, with L the exact lower triangular factor of A. FAF^T is then incompletely block factorized by an incomplete block Jacobi algorithm with the final outcome giving rise to the BFSAI-IC hybrid preconditioner that can be efficiently computed and implemented into a multiprocessor hardware. A further evolution of BFASI-IC is ABF-IC where the pattern of F is not prescribed a priori but is automatically computed with the aid of an adaptive cost-effective procedure. ABF-IC has been compared with both FSAI and BFSAI on SPD systems arising from realistic large size geomecanical FE applications. The results show that ABF-IC drastically reduces the iteration count and is always superior to FSAI and BFSA-IC in terms of the overall wall-clock time up to a factor 8 and 4, respectively.

Acknowledgements. This study has been supported by the CINECA project "PARPSEA: Parallel preconditioners for large scale engineering applications".

References

1. Agullo, E., Guermouche, A., L'Excellent, J.Y.: A parallel out-of-core multifrontal method: storage of factors on disk and analysis of models for an out-of-core active memory. Parall. Comp. 34, 296–317 (2008)
2. Benzi, M.: Preconditioning techniques for large linear systems: A survey. J. Comp. Phys. 182, 418–477 (2002)
3. Chow, E.: A priori sparsity patterns for parallel sparse approximate inverse preconditioners. SIAM J. Sci. Comput. 21, 1804–1822 (2000)
4. Chow, E., Saad, Y.: Approximate inverse preconditioners via sparse-sparse iterations. SIAM J. Sci. Comput. 19, 995–1023 (1998)
5. Couturier, R., Denis, C., Jezequel, F.: GREMLINS: A large sparse linear solver for grid environment. Par. Comput. 34, 380–391 (2008)
6. Erhel, J.: Some properties of Krylov projection methods for large linear systems. In: Ivany, P., Topping, B.H.V. (eds.) Computational Technology Reviews, vol. III, pp. 41–70. Saxe-Coburg Publications, Stirlingshire (2011)
7. Holland, R.M., Wathen, A.J., Shaw, G.J.: Sparse approximate inverses and target matrices. SIAM J. Sci. Comput. 26, 1000–1011 (2005)
8. Janna, C., Ferronato, M., Gambolati, G.: A block FSAI-ILU parallel preconditioner for symmetric positive definite linear systems. SIAM J. Sci. Comput. 32, 2468–2484 (2010)
9. Janna, C., Ferronato, M.: Adaptive pattern research for block FSAI preconditioning. SIAM J. Sci. Comput. (to appear)
10. Kaporin, I.E.: New convergence results and preconditioning strategies for the conjugate gradient method. Numer. Lin. Alg. App. 1, 179–210 (1994)
11. Kolotilina, L.Y., Yeremin, A.Y.: Factorized sparse approximate inverse preconditionings, 1. Theory. SIAM J. Matrix Anal. Appl. 14, 45–58 (1993)
12. Kolotilina, L.Y., Yeremin, A.Y.: Factorized sparse approximate inverse preconditionings, 4. Sipmple approaches to rising efficiency. Numer. Lin. Alg. Appl. 6, 515–531 (1999)
13. Li, X.Y.S.: An overview of SuperLU: algorithms, implementation and user interface. ACM Trans. on Math. Software 31, 302–325 (2005)
14. Resh, M.M.: High performance computer simulation for engineering: A review. In: Ivany, P., Topping, B.H.V. (eds.) Trends in parallel, distributed, grid and cloud computing for engineering, pp. 177–186. Saxe-Coburg Publications, UK (2011)
15. Saad, Y.: Iterative methods for sparse linear systems. SIAM, Philadelphia (2003)

An Effective Inductive Learning Algorithm for Extracting Rules

Rein Kuusik and Grete Lind

Abstract. In this paper we present a new inductive learning algorithm named MONSAMAX for extracting rules. It has some advantages compared to several machine learning algorithms: it uses several new pruning techniques which guarantee great effectiveness of the algorithm; it extracts overlapping rules; as a result it finds determinative set of rules that we can use for post-analysis of extracted rules. Compared to a former algorithm MONSIL it is much less labor-consuming.

1 Introduction

In the domain of inductive learning there are many different algorithms in use. Such a variety of algorithms shows that there are different problems which are hard to solve by one specific algorithm.

In the inductive learning (IL) environment we have the problem how to manage large example sets with an unknown complicated structure. From our experience we can say that monotone systems algorithms have been very efficient in ordering large data tables for finding regularities from these [2, 3, 5-7]. We try to use techniques that have been successful in data analysis in the inductive learning field.

This paper defines a new term "Determinative set of rules", how to use it and presents a new IL algorithm for finding it. We mainly originate from the notions of the article [1]. The inductive learning algorithms have to allow us to find descriptions that are at the same time both consistent and complete.

2 A New Approach

Next we present a new approach for IL which bases on a new term "Determinative set of rules" (DSR).

Rein Kuusik · Grete Lind
Department of Informatics, Tallinn University of Technology
Raja 15, Tallinn 12618, Estonia
e-mail: kuusik@cc.ttu.ee, grete@staff.ttu.ee

F.L. Gaol et al. (Eds.): Proc. of the 2011 2nd International Congress CACS, AISC 145, pp. 339–344.
springerlink.com   © Springer-Verlag Berlin Heidelberg 2012

2.1 Basis of the New Approach

Let be given a data table X(N,M) and a set B of all possible rules and each rule in B is presented only once.

The *determinative set of rules (DSR)* consists of all rules which are not contained in other rules of B.

B = {Ri}, i=1, 2,..., K where K is a number of all possible rules. Ri ≠ Rj, i ≠ j. Ri ∈ DSR if there /∃ Rt ∈ B, Ri ⊂ Rt, t ≠ i. DSR ⊆ B

It means that DSR does not contain the sub-rules of its rules. In order to get DSR from B we have to throw out all the sub-rules of the rules in B. We call this process „rule set compression".

Example. Let B contain 4 rules:

r1: IF T1=1 & T2=1 THEN CLASS=1
r2: IF T1=1 & T3=2 THEN CLASS=2
r3: IF T2=1 THEN CLASS=1
r4: IF T3=2 THEN CLASS=2

As we see, the rule r1 is contained in r3 and r2 is contained in r4. According to the definition DSR_B = {r3, r4}.

The main features of DSR are:

1. there are no redundant attributes (zero factors) in rules,
2. the same object in class Y can be described by several rules.

On the basis of DSR we can formulate and solve next tasks, for example, to find

1. the shortest rules (by the number of attributes (selectors) in the rule),
2. the longest rules (by the number of attributes in the rule),
3. the rules with specific features (for example, all rules with r selectors),
4. the shortest rule system (i.e. the rule system with the smallest number of rules),
5. the rule system which consists of rules with minimal number of selectors,
6. all rule systems we can form on the basis of DSR.

These tasks are important for post-analysis of rules.

2.2 Description of the Algorithm

Here we describe the algorithm realizing the new IL approach. The findable set of rules is DSR together with some redundant rules which are eliminated afterwards (rule set compression).

Algorithm MONSAMAX is given in Fig.1.

This is a depth-first-search algorithm that makes subsequent extracts of objects containing certain factors (i.e. attribute with certain value). At each level first the rules (of that extract) are detected and then the factors for making extracts of the next level are selected one by one. The algorithm uses frequency tables for all objects of the current extract and for each class of the current extract. If a factor has equal frequencies for all objects and in some class then this factor completes a rule. The rule includes also the factors chosen on the way to that extract.

The selection criteria for choosing the next factor are based on frequencies, the maximal frequency for all objects (of extract). If only one attribute (of the extract) has free (unused) value(s) (indicated by frequencies over zero) then it is not practical to make a next (further) extract because there would be no free factors to distinguish objects of different classes in that extract. If there are no free factors (i.e. no frequencies over zero) then obviously it is not possible to make a next extract. In both cases the algorithm backtracks to the previous level.

Each factor that has been used for making an extract or completing a rule is set to zero in the corresponding frequency table. Each frequency table (except for the initial level) inherits all zeroes of previous level (we call it "bringing zeroes down").

These zeroing techniques prevent many redundant extracts and rules.

Algorithm MONSAMAX

S0. t:=0; U_t:=$\varnothing$

S1. Find frequencies in whole dataset and each class
    If t>0 then Bring zeroes down

S2. For each factor A such that its frequency in some class C is equal to its frequency in whole set
        output rule $\{U_i\}$&A$\rightarrow$C, i=0,...,t
        A$\leftarrow$0

S3. If not enough free factors for making an extract then
        If t=0 then Goto End
        Else t:=t-1; Goto S3

S4. Choose a new (free) factor U_t
    $U_t \leftarrow$0; t:=t+1;
    extract subtable of objects containing U_t;
    Goto S1

End. Rules are found

Fig. 1 Algorithm MONSAMAX

2.3 Example

In the following example coded data from [4] are used. For given data the frequencies are found across all data and across each class (see Table 1). We call them "3D frequency tables". If frequencies of some factor are equal in the whole dataset and some class then we can complete the rule. In the given dataset/extract this factor determines the class. From the initial frequency tables (Table 1) 3 rules are found this way: R1: T2.1 → Class 1, R2: T3.2 → Class 1, R3: T2.2 → Class 2.

The frequencies of those factors (T2.1, T2.2, T3.1) are set to zero in the current frequency table (see Table 2). Now the factor with the biggest frequency is selected for making an extract. Chosen factor is T3.1 with frequency 5. Extract by T3.1 and the corresponding frequencies are given in Table 3.

Table 1 Initial data and frequencies

Object	1	2	3	4	5	6	7	8
T1	2	1	2	2	2	1	1	2
T2	1	1	3	2	3	3	3	1
T3	1	1	1	1	2	1	2	2
Class	1	1	2	2	1	2	1	1

Value	1	2	3	1	2	3	1	2	3
T1	3	5		2	3		1	2	
T2	3	1	4	3	0	2	0	1	2
T3	5	3		2	3		3	0	
Class		all			1			2	

Table 2 Frequencies after extracting 3 rules

Value	1	2	3	1	2	3	1	2	3
T1	3	5		2	3		1	2	
T2	0	0	4	0	0	2	0	0	2
T3	5	0		2	0		3	0	
Class		all			1			2	

Table 3 Extract by T3.1=5 and corresponding frequencies

Object	1	2	3	4	6	Value	1	2	3	1	2	3	1	2	3
T1	2	1	2	2	1	T1	2	3		1	1		1	2	
T2	1	1	3	2	3	T2	0	0	2	0	0	0	0	0	2
Class	1	1	2	2	2	Class		all			1			2	

The cells with grey background are prohibited factors that have zeroed frequencies at the previous level.

At current level (i.e. extract by T3.1) factor T2.3 has equal frequencies in Class 2 and the whole extract which gives the next rule: R4: T3.1&T2.3 → Class 2.

After completing a rule, the frequency of T2.3 is set to zero. Now only one attribute (T1) has frequencies over zero (i.e. allowed factors). In such situation further extract is not made (because this extract cannot give any rule). Instead of it the algorithm backtracks to the previous level.

The current state of frequencies at the initial level is in Table 4. The frequency of T3.1 which was the basis for making an extract is set to zero. The next extract is made by T1.2=5 (see Table 5).

Table 4 Frequencies at the initial level

Value	1	2	3	1	2	3	1	2	3
T1	3	5		2	3		1	2	
T2	0	0	4	0	0	2	0	0	2
T3	0	0		0	0		0	0	
Class		all			1			2	

Table 5 Extract by T1.2=5 and corresponding frequencies

Object	1	3	4	5	8	Value	1	2	3	1	2	3	1	2	3
T2	1	3	2	3	1	T2	0	0	2	0	0	1	0	0	1
T3	1	1	1	2	2	T3	0	0		0	0		0	0	
Class	1	2	2	1	1	Class		all			1			2	

This extract gives no rules. As there is only one frequency over zero in the frequency table (T2.3), the algorithm backtracks to the initial level again. The basis for the next extract is T2.3 with frequency 4.

This extract consists of objects 3, 5, 6 and 7. After bringing zeroes down (for T1.2, T3.1 and T3.2) there is only one factor with non-zero frequency (T1.1 with frequency 2). It does not give a rule. Also there are not enough free factors to make a subsequent extract, therefore the algorithm returns to the initial level.

Now there is only one usable (non-zero) frequency in the frequency table of the initial level (T1.1 with frequency 4). This factor (T1.1) does not give a rule. Also it makes no sense to make an extract by it. At the initial level it is not possible to backtrack also. The work is finished.

So, we have extracted 4 rules: R1: T2.1 → Class 1, R2: T3.2 → Class 1, R3: T2.2 → Class 2, R4: T3.1&T2.3 → Class 2.

After using compression it occurs that the extracted rule set is DSR.

As we see there is no need to find all possible rules. In our example as a result of the algorithm's work the only rule set which is DSR was found. All other rules are sub-rules of this rule set.

3 Conclusion

This paper presents the algorithm MONSAMAX for finding a determinative set of overlapping rules. The algorithm is based on frequency tables and some new pruning techniques which make it easy to detect a potential DSR rule. We can use DSR for post-analysis of rules for finding rule sets with several features.

After compression the algorithm MONSAMAX presented here gives the same result as the algorithm MONSIL (described in [3, 5]) after compression. In order to determine the belonging of extracted objects to the same class in the process of extracting of rules, MONSIL must make an extract and usually the objects do not belong to the same class. It means that we have made a superfluous work. MONSAMAX works so effectively because we have data to check the belonging of objects to the same class using 3D frequency tables, there is no need to make these extracts. And second, for MONSAMAX the amount of extracted rules is always smaller than for MONSIL because of extracting shorter rules first. MONSIL always extracts longer rules first.

References

1. Gams, M., Lavrac, N.: Review of Five Empirical Learning Systems within a Proposed Schemata. In: Bratko, I., Lavrac, N. (eds.) Progress in Machine Learning, Proceedings of EWSL 1987: 2nd European Working Session on Learning, Bled, Yugoslavia, pp. 46–66. Sigma Press, Wilmslow (1987)
2. Kuusik, R.: The Super-Fast Algorithm of Hierarchical Clustering and the Theory of Monotone Systems. Transactions of Tallinn Technical University 734, 37–62 (1993)
3. Kuusik, R., Treier, T., Lind, G., Roosmann, P.: Machine Learning Task as a Diclique Extracting Task. In: 2009 Sixth International Conference on Fuzzy Systems and Knowledge Discovery: FSKD 2009, Tianjin, China, August 14-16, pp. 555–560. Conference Publishing Service, California (2009)
4. Quinlan, J.R.: Learning efficient classification procedures and their application to chess end games. In: Carbonell, J.G., Michalski, R.S., Mitchell, T.M. (eds.) Machine Learning. An Artificial Intelligence Approach, pp. 463–482. Springer, Heidelberg (1984)
5. Roosmann, P., Võhandu, L., Kuusik, R., Treier, T., Lind, G.: Monotone Systems approach in Inductive Learning. International Journal of Applied Mathematics and Informatics 2(2), 47–56 (2008)
6. Võhandu, L.: Fast Methods in Exploratory Data Analysis. Transactions of Tallinn Technical University (705), 3–13 (1989)
7. Võhandu, L., Kuusik, R., Torim, A., Aab, E., Lind, G.: Some Monotone Systems Algorithms for Data Mining. WSEAS Transactions on Information Science and Applications 3(4), 802–809 (2006)

Computational Knowledge Modeling in Cardiovascular Clinical Information Systems

Nan-Chen Hsieh, Jui-Fa Chen, Hsin-Che Tsai, and Fan Su

Abstract. Cardiac surgery is a complex surgical operation that is performed on patients with a severe insufficiency in their cardiac function. In this study, we present a CIS (Clinical Information System) with knowledge modeling that combines information extracted from the heterogeneous data sources in order to assess the evolution of the Cardiac surgery after the intervention and data extracted from Follow-Up Record. Once the integrating of data, the homogeneous data could be useful in answering important clinical questions and could help optimize cardiac methodologies in the clinical decision field. The results show that the system proposed by this approach yields valuable information and knowledge for Cardiac surgery patients.

1 Introduction

Cardiac surgery is a complex surgical operation that is performed on patients with a severe insufficiency in their cardiac function. During the operation, in the postoperative ICU and on the nursing wards, there is considerable morbidity for cardiac surgery patients with postoperative complications, which results in increased surgical mortality. A cardiovascular CIS (clinical information system) is helpful prior to and during cardiac surgery [1].

The information collected can be used by surgeons and patients to evaluate whether or not the surgical procedures are likely to be successful. The development of a robust CIS can therefore both assist vascular surgeons in

Nan-Chen Hsieh · Fan Su
Department of Information Management,
National Taipei University of Nursing and Health Sciences, Taiwan, ROC
e-mail: {nchsieh,fiona7714}@gmail.com

Jui-Fa Chen · Hsin-Che Tsai
Department of Computer Science and Information Engineering, Tamkang University,
Taiwan, ROC
e-mail: alpha@mail.tku.edu.tw, shin@live.com

F.L. Gaol et al. (Eds.): Proc. of the 2011 2nd International Congress CACS, AISC 145, pp. 345–350.
springerlink.com © Springer-Verlag Berlin Heidelberg 2012

evaluating the expected outcome for a given patient as well as facilitate counseling and operative decision-making. Due to the potential usefulness of these data (contains data on patients' preoperative characteristics, risk factors, details of the surgical procedure, physical characteristics of the heart and postoperative physiological and laboratory findings.) in monitoring trends, quality, and outcomes, it is valuable to translate these data to useful knowledge.

This study explores the development of a web-based cardiovascular CIS, ontologies based of EMR (electronic medical record), electronic registries, automatic feature surveillance schemes, and decision support systems. The framework of the CIS has three layers: monitoring, surveillance and model construction. The monitoring layer provides a visual user interface. At the surveillance and model construction layers, we also explored the application of model construction and intelligent prognosis to aid in making postoperative predictions and follow-up. The accumulation of data is quickly accelerating the process of transforming clinical information into clinical knowledge.

As shown in Fig. 1, the research aims of this study include the following:

- Pre-operative risk evaluation: the statistical significance of certain intraoperative parameters can be estimated using the information from previous patients [2].
- Ontology-based data modeling: this modeling ability used to describe concepts in a standardized manner and define existing interrelationships of cardiovascular surgery information.
- Follow-up risk evaluation: this mechanism could provide additional information to the risk prediction of a given cardiac surgery patient. It takes into account past events' outcome and follow-up information of the current patient [2].

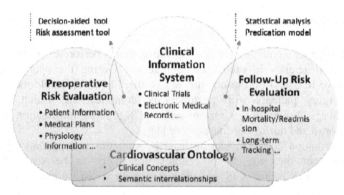

Fig. 1 A conceptual diagram of the research.

2 The Architecture of the CIS

The WWW is a standardized, cross-platform environment were used of CIS. WWW applications can be effective in creating virtual working platforms, which provide easy ways to collaborate and communicate with co-workers. This platform allowed the surgeons, lab technicians and nurses to work together through the use of the cardiovascular CIS [3, 4]. The architecture we used is described in detail below:

- The Monitoring Layer: This layer is primarily provided for user interfaces. The medical plan and the patient's case management system are developed at this layer and the relative information is stored.
- The Surveillance Layer: The relative decision models are implemented at this layer. The clinical decision model is designed using the surgeons' knowledge. The success rate of the test methods is calculated, and this information is fed back into the model to assess its suitability. This feature helps surgeons to make decisions regarding treatment.
- The Model Construction Layer: The basic aim of this layer is knowledge modeling. Cardiovascular ontologies are a means of describing concepts, things or entities in the cardiovascular diseases, while defining the relationships that exist between them. The ontologies play a role in standardizing the data, meaningful the terminologies and connecting it and making it machine-readable [5, 6]. As this knowledge is established, it is helpful to analyze the preoperative and postoperative details of each case to estimate the patients' quality of life.

This study uses statistics and data mining to assist surgeons' research and understanding of the factors than can affect cardiovascular treatment methods, as well as to establish a knowledge base for the comparison of biochemical data of patients at the preoperative versus Follow-up stages.

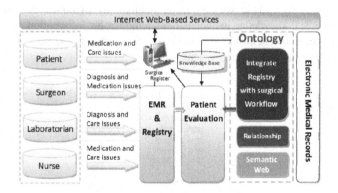

Fig. 2 The cardiovascular CIS architecture

3   Cardiovascular Ontology

The development of cardiovascular ontologies is for the purposes of controlling terminologies with advantages such as standardization of data and enabling inference at the granular level. This mechanism not only provides a uniform format to describe the concepts but also relates the concepts together into hierarchies and networks which provide the basis for inference and automated reasoning. Importantly, domain experts are consulted to explain the meaning of domain-specific concepts.

The CIS facilitated the linkage of patient and EMR data required for statistical, management and research objectives. The cardiovascular ontologies consists of a diagnostic ontology of the cardiovascular diseases, a meta-database for connecting EMRs information, classification models based on data mining techniques, the process of extracting patterns of data to convert data into knowledge, and an interactive web-based interface for managing the interaction between surgeons, patients. Application of ontologies to this study ranges from building computational ontology for clinical practice, clinical practice guidelines and computer-aided decision.

Medical ontology is a model of the knowledge from a clinical domain such as the heart disease. It contains all of the relevant concepts related to the diagnostics, treatment, operational procedures and patient data. The point to design ontology is to allow knowledge inference and reasoning. They are different from terminologies which are static structures used for knowledge reference.

4 Cardiovascular Ontology Creation Process

4.1 Research Scope

The creation of cardiovascular ontology is based on the surgical operation stages. As shown in Fig. 3, all the relevant terminology and data from the clinical guidelines were represented in a systematic way by using a hierarchy of concepts. Other related sources of medical knowledge include patient demographic information, medical plan, physiology and biochemistry informatics, and other medical ontologies or terminologies related to the cardiovascular disease.

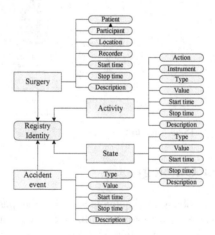

Fig. 3 Medical ontology in a fact hierrchy

4.2   Ontology Tool Selection

After determining the research scopes, the next step is to decide which tool will be used to design the ontology. In this proposal, Protégé (2000) developed by Stanford Medical Informatics, was chosen as the ontology design tool (available from(http://protege.stanford.edu/overview/). Protégé (2000) is an open-source, java-based ontology tool with a pleasant editing environment containing several third-party plugins. Protégé implements a rich set of knowledge modeling structures and actions that support the creation, visualization, and manipulation of ontologies. This study focuses on the development of descriptive knowledge.

4.3   Connecting Surgical Workflow Models with Ontology

Surgical workflow provides high amount of potentially useful information. By surgical workflow ontology, the choice of suitable sampling of information restricts, and various surgical questions could be answered by the help of surgical workflow ontology.

In this study, the surgical work domain that we try to model is related to the Cardiac surgical treatment performed in different surgical timeline for the individual patient. This necessary information, which includes the data shall be recorded, major surgical steps and their sequencing, was collected by interviews with surgeons and studies the different surgical timeline's reports. That is, the development of ontology is a crucial and required step in modeling the cardiac surgical domain. We provide a method for describing objects and their relationships within this domain. Ontology provides a language for describing objects and relationships within this domain. We develop the ontology of cardiovascular work domain by identifying the main concepts, relationships and cardinalities of from the descriptions of surgical cases, and will formalize the main concepts and relationships by UML class diagram.

5   Results and Discussions

Analyzing surgical workflow or procedures is a promising research topic of medical informatics that aimed at improving surgical assist systems. The terminology of surgical workflows is used for describing the methodological framework for acquiring formal descriptions of surgical interventions. This study designs a surgical workflow ontology which optimized the recording of surgical workflows information. This model categorizes the data as facts with associated numerical measures and descriptive dimensions characterizing the surgical-relative facts that is useful in recording data. The development of surgical workflow models with ontology can support preoperative planning by retrieving similar retrospective cases, and support postoperative data exploration analysis.

Acknowledgment. This research was partially supported by National Science Council of Taiwan (NSC99-2410-H-227-002-MY2).

References

[1] Wright, J.G., Bieniewski, C.L., Pifarre, R., Gunnar, R.M., Scanlon, P.J.: A database management system for cardiovascular disease. Computer methods and Programs in Bio-medicine 20(1), 117–121 (1985)

[2] Legarreta, J.H., Boto, F., Macía, I., Maiora, J., García, G., Paloc, C., Graña, M., de Blas, M.: Hybrid Decision Support System for Endovascular Aortic Aneurysm Repair Follow-Up. In: Graña Romay, M., Corchado, E., Garcia Sebastian, M.T. (eds.) HAIS 2010. LNCS, vol. 6076, pp. 500–507. Springer, Heidelberg (2010)

[3] Bellazzi, R., Montani, S., Riva, A., Stefanelli, M.: Web-based telemedicine systems for home-care: technical issues and experiences. Computer Methods and Programs in Biomedicine 64(3), 175–187 (2001)

[4] Vanoirbeek, C., Rekika, Y.A., Karacapilidis, N., Aboukhaleda, O., Ebel, N., Vader, J.-P.: A web-based information and decision support system for appropriateness in medicine. Knowledge-Based Systems 13(1), 11–19 (2000)

[5] Noy, N.F., Rubin, D., Musen, M.A.: Making Biomedical Ontologies and Ontology Repositories Work. IEEE Intelligent Systems 19(6), 78–81 (2004)

[6] Rubin, D.L., Dameron, O., Bashir, Y., Grossman, D., Dev, P., Musen, M.A.: Using ontologies linked with geometric models to reason about penetrating injuries. Artificial Intelligence in Medicine 37, 167–176 (2006)

Automatic Extraction and Categorization of Lung Abnormalities from HRCT Data in MDR/XDR TB Patients

Saher Lahouar, Clifton E. Barry, 3rd, Praveen Paripati, Sandeep Somaiya,
Yentram Huyen, Alexander Rosenthal, and Michael Tartakovsky

Abstract. An ancient disease, tuberculosis (TB) remains one of the major causes
of disability and death worldwide. In 2006, 9.2 million new cases of TB emerged
and killed 1.7 million people. We report on the development of tools to help in the
detection of lesions and nodules from High Resolution Computed Tomography
(HRCT) scans and changes in total lesion volumes across a study. These
automated tools are designed to assist radiologists, clinicians and scientists assess
patients' responses to therapies during clinical studies. The tools are centered upon
a rule-based system that initially segments the lung from HRCT scans and then categorizes the different components of the lung as normal or abnormal. A layered
segmentation process, utilizing a combination of adaptive thresholding, three-
dimensional region growing and component labeling is used to successively peel
off outside entities, isolating lung and trachea voxels. Locating the Carina allows
logical labeling of the trachea and left/right lungs. Shape and texture analysis are
used to validate and label normal vascular tree voxels. Remaining abnormal voxels are clustered on density, gradient and texture-based criteria. Several practical
problems that arise due to large changes in lung morphology due to TB and

Saher Lahouar · Praveen Paripati · Sandeep Somaiya
NET ESOLUTIONS CORPORATION, McLean, VA
e-mail: {saher,Praveen,sandeep}@nete.com

Yentram Huyen · Alexander Rosenthal · Michael Tartakovsky
Office of Cyber Infrastructure and Computational Biology (OCICB), NIAID, NIH,
Bethesda, MD
e-mail: {huyeny,alexr,mtartakovs}@niaid.nih.gov

Clifton E. Barry, 3rd
Tuberculosis Research Section, Laboratory of Clinical Infectious Diseases, Division of
Intramural Research, National Institute of Allergy and Infectious Diseases (NIAID),
National Institutes of Health (NIH), Bethesda, MD
e-mail: cbarry@niaid.nih.gov

F.L. Gaol et al. (Eds.): Proc. of the 2011 2nd International Congress CACS, AISC 145, pp. 351–360.
springerlink.com   © Springer-Verlag Berlin Heidelberg 2012

patients' inability to hold their breath during scan operations need to be addressed to provide a viable computational solution. Comparisons of total common volumes of lesions by size for a given patient across multiple visits are in concordance with expert radiologist's manual measurements.

Keywords: Lung Abnormalities, HRCT, Pulmonary Tuberculosis, MDR/XDR TB, Automatic Extraction, Lung Nodules.

1 Introduction

Tuberculosis (TB) is one of the leading infectious diseases in the world, with approximately one-third of the world's population harboring the causative agent, Mycobacterium tuberculosis (Mtb). Though previously a disease associated with aristocratic societies, TB is now predominantly a third-world disease, particularly affecting Asian communities and sub-Saharan Africa. Mtb isolates are increasingly resistant to drug therapies: multidrug-resistant TB (MDR TB), or more severely, extensively drug-resistant TB (XDR TB). An ancient disease, TB remains one of the major causes of disability and death worldwide. In 2006, 9.2 million new cases of TB emerged and killed 1.7 million people [10]. We report on the development of tools to help in the detection of lesions and nodules from High Resolution Computed Tomography (HRCT) scans of TB patients and changes in total lesion volumes across visits. These automated tools are meant to assist radiologists, clinicians and scientists assess patients' responses to therapies during clinical studies. The tools employ a rule-based system that initially segments the lung from HRCT scans and then categorizes the different components of the lung as normal or abnormal. The practical problem of patients' inability to hold their breath and other movements during HRCT scanning introduces challenges that need to be addressed in order to enable proper computations as rescanning is not a viable option most of the time.

2 Review of Existing Work

The past decade has seen dramatic improvements in CT scanner technology, especially with the introduction of multi-detector-row scanners which can acquire up to 64 1-mm slices simultaneously per rotation and perform each rotation in less than a second. The resulting data explosion [11] has been followed by a sharp increase in research on computer analysis of chest CT scans [12]. An excellent summary of relevant work on this subject up to 2004 has been reported by Sluimer et al. [12]. More recently, significant work continues to be reported on various aspects of computer analysis of CT scans, especially on the following broad subjects: Lung Segmentation; Bronchial / Vascular Tree Segmentation; Lesion / Nodule Detection.

The difficulty in lung segmentation in CT scans is attributable to several factors such as partial volume effects, beam hardening, motion artifacts and low-radiation induced noise [1]. Additionally, the contrast between lung tissue and surrounding tissue significantly decreases in the presence of lung pathology: as parts of the lung harden or soften in response to disease, the affected areas can become indiscernible from surrounding tissue, especially in the juxtapleural regions on the lung boundaries. Moreover, lesions inside the lung may start occupying the same Hounsfield Unit (HU) range as the normal vascular or bronchial structures. This is especially pronounced in TB affliction where lung lobes may partially or entirely collapse, or harden.

Lung segmentation approaches generally use two broad categories of methods: threshold-based and pattern classification based (or a hybrid of the two). They traditionally operate on 2-dimensional slices, or 3-dimensional volume data, or use what are dubbed 2 ½-dimensional approaches where a subsequent pass reasons with results obtained from 2-dimensional slice analysis. Semi-automatic to fully automatic methods are largely replacing approaches that used to require some form of manual intervention such as region of interest (ROI) or seed selection. In what follows, we give a synopsis of methods attempted and their reported successes and failures.

Hu et al. [4] use Gray Level Thresholding to separate the low density lung tissue from the denser surrounding tissue. A second step aims at separating the left and right lungs at the anterior and posterior junction lines. This is followed by morphological closing to account for the harder vessel and nodules inside the lung and a smoothing operation of the boundary. The method works reasonably well, and is used as a basis by several authors, as long as no moderate or severe edge hardening occurs. Similarly, Gao et al. [2] use a parallel 3-step approach in which they select an optimal threshold for initial lung segmentation, followed by a tracking algorithm to separate the left and right lung, then a rolling ball filter to smooth the edges and fill the inside holes in the lung. Tong et al. [13] compute an effective average gradient to determine the optimal gray level threshold as the first lung segmentation step in their nodule detection approach. Guo et al. [3] use an iterative algorithm in which the parenchyma and non-parenchyma voxels are successively clustered in two groups based on a threshold that is updated by halving the distance between the average HU values in the two clusters. This is followed by contour tracking to collect all the lung voxels in a given slice.

The work described in the literature on nodule detection and classification is geared towards cancer and consequently categorization of a nodule as benign or malignant and in a few cases on detecting ground glass opacity [7][18][6]. The HRCT scans used for this study were collected with informed consent from subjects enrolled in an IRB-approved Phase II clinical trial entitled "A Randomized, Double-blind, Placebo-controlled Phase II Study of Metronidazole Combined with Antituberculous Chemotherapy vs. Antituberculous Chemotherapy with Placebo in Subjects with Multi-drug Resistant Pulmonary Tuberculosis" (NIAID 07-I-N041, ClinicalTrials.gov identifier NCT00425113). The amount of infection and degeneracy in the lung is substantially higher in patients affected with TB, as illustrated in Fig. 1.

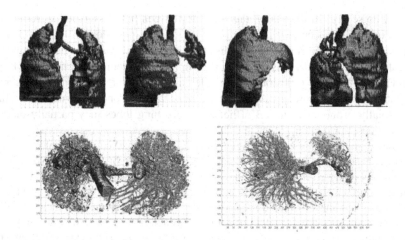

Fig. 1 Deformations of the Lung due to TB infection

3 Approach

3.1 Initial Lung Segmentation

The prototypical lung is fairly symmetric and atlases can be used to help segment, as well as label, the components [17] [9]. In lungs affected by TB, the structure of the lungs can be significantly altered, reducing the utility of atlases. Our approach consequently involves analyzing the structural components that constitute the lungs, and developing the space and labeling components into normal and abnormal categories. This architecture provides the flexibility to deal with substantial changes to the normal lung structure that occurs in a patient affected by TB.

To isolate the lung voxels, a first pass "layered segmentation" is performed. In this pass, we use a combination of adaptive thresholding, 3-dimensional region growing and component labeling to first separate the patient's body from the background and any outside objects (such as articles of clothing and scanner table), then the skin, muscles and fat, skeletal system and other organs are successively peeled off, leaving the voxels that belong to the lung and trachea. This is illustrated in Fig. 2.

3.2 Locating the Carina

In a healthy lung, one can take advantage of the symmetry inherent in the anatomy of the lung in order to predict the location of certain entities. Under TB stress, when one of the lungs is showing signs of collapse, the other lung oftentimes expands to accomodate these changes resulting in gross deformations to the position and layout of the lungs and trachea. In order to locate the carina under such conditions, several techniques are used including contouring, shape analysis, fuzzy reasoning and geometric sweeps.

The first step is to identify a valid seed point in the trachea from which to start the carina tracking process. Relying on the fact that the trachea is a nearly vertical tube filled with air (relatively low HU values) with an expected size range (nominally 2 to 4 cm in diameter), we perform contour analysis on the various CT slices and extract the location and size of ellipsoidal shapes. The individual ellipses are then stacked into tubular shapes upon which length, roundness and regularity metrics are computed for each resulting tube. Using fuzzy metrics, a score is assigned to each tube and the one corresponding to the trachea is thus selected.

Once a good seed is selected for the trachea, directed cross-sectional analysis is used to track the trachea until the first bifurcation. The center of the bifurcation is recorded as the carina location. Using the carina location information, region growing, and component labeling, low density lung voxels are further labeled as belonging to the trachea, left or right lung. Sample results of this step are shown in Fig. 2.

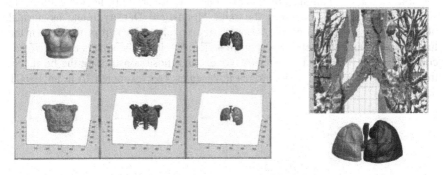

Fig. 2 Layered Segmentation, Carina, Trachea and Right/Left Lung Voxel Labeling

An additional step is performed to attempt to recover juxta-pleural abnormalities that might have been clustered with surrounding muscle/fat tissue. A directional search is performed from every edge voxel belonging to either lung for a predetermined distance through non-lung voxels. If the search reaches a voxel belonging to the same lung, then all voxels along that path are marked as belonging to the same lung. Islands that may get formed due to this operation are also labeled as belonging to the same lung through a morphological operation.

3.3 Lung Sextants

To help with diagnosis, radiologists subdivide the lung into smaller components and analyze each of the components and report on its condition. These components are either the five lobes of the lung or sextants. The two lung components above the horizontal plane through the carina are the upper left and right sextants and the rest of the lungs are divided into middle and lower left and right sextants respectively. While a score can be assigned to each sextant, we have found that direct comparison of scores needs to be done with care as the sextant volume and geometry can change significantly between visits. To help with the comparison across visits, the sextant partitions are primarily computed for the first visit and corresponding regions are used in subsequent visits.

3.4 Characterizing Abnormalities in the Lungs

The voxels inside the lung belong to one of many physical entities including lung surface, lung parenchyma, airway tubes (bronchial tree), blood vessels (vascular tree), interlobular fissures, and abnormalities. Abnormalities cover the entire spectrum of densities, from air or liquid filled cavities, to fibroid nodules. The approach we've taken in detecting abnormalities is to err on the side of caution, and possibly have false positives which can later be identified and reclassified.

For this purpose, we try to label as much of the known "normal" voxels as possible, and all remaining voxels are subject to abnormality tests.A significant class of abnormalities occupies the same Hounsfield Density range as normal blood vessels, so separating the vascular tree voxels (VTV) from the rest of the lung voxels is an extremely important first step. We label the voxels recognized as constituting the vascular tree as VTV voxels and those constituting the bronchial tree as BTV voxels. Fig. 3 illustrates a Vascular Tree extracted from an HRCT data set.

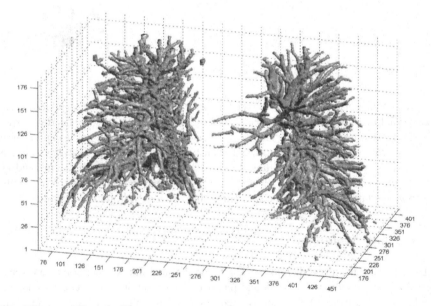

Fig. 3 Extracted Vascular Tree

Voxels with higher HU intensity than -400 and not previously labeled as a normal lung voxel are grouped into abnormal voxels. These are subject to further classification into lesions of different types.

4 Application in Human Clinical Studies

Given that study patients suffer from TB, their ability to hold their breath in maximum inhalation position is not as robust as that of the general population. As such, these patients are typically not able to hold their breath long enough for a complete HRCT scan in one pass. While protocols exist for the scans to be stopped, allowing the patient to catch their breath and then resume scanning, the lung degeneracy caused by TB makes it difficult for the patients to hold their breath steadily. This leads to significant, artificial movements in the features of interest - lung boundaries, tree segments and lesions, causing incorrect reconstruction and categorization. Fig. 4(A) illustrates this problem, Fig. 4(B) provides a zoomed-in view and Fig. 4(C) shows its resolution. The reconnected branch segments allow for proper labeling. One cannot attempt to re-scan in these situations as the patient's exposure to radiation needs to be limited.

We have developed capabilities to address these real world problems and are testing their effectiveness in current clinical studies. The algorithms developed to segment the lungs and isolate lesions is being applied in the TB studies conducted by NIAID scientists, led by second author, Dr. Clifton Barry, 3rd. The presence of large numbers of lesions in TB patients can make it difficult for radiologists to compute the volumes and their changes manually. This tool can assist radiologists in performing these tasks. Detecting global changes in a lung could also be complicated by imaging constraints - if only a portion of the lung is imaged in a particular visit, and a larger volume is imaged in a previous visit, the system needs to be able to register lungs across visits and use equivalent lung volumes. The currently reported tool incorporates this capability. The example chart in Fig. 5

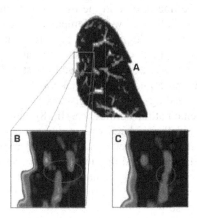

Fig. 4 (A) A portion of lung affected by breathing (B) Zoomed-in view of the section, (C) Motion adjusted view.

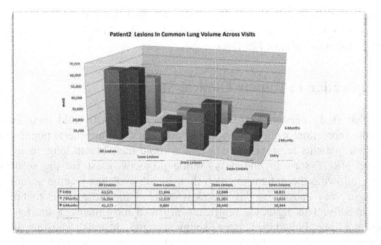

Fig. 5 Lesion Volumes across visits

represents the computations of global changes in lesion volumes, broken down by size. The charts and the graphs summarize the lesion volumes at entry into the study, after 2 months and 6 months into the study. The categorizations and calculations are based on HRCT scan images, using the automated processes described in the previous sections.

5   Results

Fig. 6 illustrates correlation between three independent radiologists' findings and volumes automatically computed by our method on nodules 4 to 10 mm in range. Similar results were obtained for nodules in the < 2mm range. The highest correlation, at 0.81, was between radiologist R_b and our automated tool (the subscript b represents the line in blue in the figure). It should be noted that when findings of the three radiologists were compared to each other, the highest correlation was between radiologists R_b and R_y at 0.82. An additional interesting observation is that the highest correlation with the automated evaluation was with the most experienced radiologist. These techniques have proven viable in correlating Standardized Uptake Value (SUV) results from PET scans with HRCT scan readings by radiologists as well. Additional results along these lines will be presented at the conference. In [8], human notions of similarity of lesions were not very well correlated with traditional machine learning methods of automatically derived content based similarity. The strong correlations between automated analysis and human readings in our experiments here indicate that a generic similarity metric spanning human and automated analysis is indeed feasible and a shorter path to it may lie in choosing to start with the vocabulary that radiologists already use.

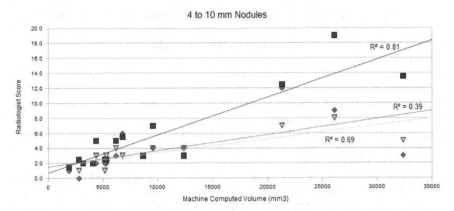

Fig. 6 Correlation between three independent radiologists' scores and computed nodule volumes

References

1. Ceresa, M., Artaechevarria, X., Munoz-Barrutia, A., Ortiz-de-Solorzano, C.: Automatic leakage detection and recovery for airway tree extraction in chest CT images. In: 2010 IEEE International Symposium on Biomedical Imaging: From Nano to Macro, April 14-17, pp. 568–571 (2010)
2. Gao, Q., Wang, S., Zhao, D., Liu, J.: Accurate Lung Segmentation For X-ray CT Images. In: Third International Conference on Natural Computation, ICNC 2007, August 24-27, vol. 2, pp. 275–279 (2007)
3. Guo, S., Wu, X.: Automatic Segmentation and Quantitative Diagnosis of Pulmonary Parenchyma in Thoracic CT. In: The 1st International Conference on Bioinformatics and Biomedical Engineering, ICBBE 2007, July 6-8, pp. 668–670 (2007)
4. Hu, S., Hoffman, E.A., Reinhardt, J.M.: Automatic lung segmentation for accurate quantitation of volumetric X-ray CT images. IEEE Transactions Medical Imaging 20(6), 490–498 (2001)
5. Kim, H.S., Yoon, H.-S., Trung, K.N., Lee, G.S.: Automatic Lung Segmentation in CT Images Using Anisotropic Diffusion and Morphology Operation. In: 7th IEEE International Conference on Computer and Information Technology, CIT 2007, October 16-19, pp. 557–561 (2007)
6. Kim, H., Maekado, M., Tan, J.K., Ishikawa, S., Tsukuda, M.: Automatic extraction of ground-glass opacity shadows on CT images of the thorax by correlation between successive slices. In: 17th IEEE International Conference on Tools with Artificial Intelligence, ICTAI 2005, November 16, p. 607– 612 (2005)
7. Kim, H., Nakashima, T., Itai, Y., Maeda, S., Tan, J.K., Ishikawa, S.: Automatic detection of ground glass opacity from the thoracic MDCT images by using density features. In: International Conference on Control, Automation and Systems, ICCAS 2007, October 17-20, pp. 1274–1277 (2007)
8. Kim, R., Dasovich, G., Bhaumik, R., Brock, R., Furst, J.D., Raicu, D.S.: An investigation into the relationship between semantic and content based similarity using LIDC. In: Proceedings of the International Conference on Multimedia Information Retrieval (MIR 2010), pp. 185–192. ACM, New York (2010)

9. Li, Q., Li, F., Suzuki, K., Shiraishi, J., Abe, H., Engelmann, R., Nie, Y., MacMahon, H., Doi, K.: Computer-Aided Diagnosis in Thoracic CT. Seminars in Ultrasound, CT, and MRI 26(5), 357–363 (2005); Update of Chest Imaging-Part I

10. NIAID, (2010) weblink,
    `http://www.niaid.nih.gov/topics/tuberculosis/Pages/`
    `Default.aspx` (accessed August 31, 2010)

11. Rubin, G.D.: Data explosion: the challenge of multidetector-row CT. European Journal of Radiology 36(2), 74–80 (2000)

12. Sluimer, I., Schilham, A., Prokop, M., van Ginneken, B.: Computer analysis of computed tomography scans of the lung: a survey. IEEE Transactions on Medical Imaging 25(4), 385–405 (2006)

13. Tong, J., Zhao, D.-Z., Wei, Y., Zhu, X.-H., Wang, X.: Computer-Aided Lung Nodule Detection Based on CT Images. In: IEEE/ICME International Conference on Complex Medical Engineering, CME 2007, May 23-27, pp. 816–819 (2007)

14. Tong, J., Zhao, D.-Z., Yang, J.-Z., Wang, X.: Automated Detection of Pulmonary Nodules in HRCT Images. In: The 1st International Conference on Bioinformatics and Biomedical Engineering, ICBBE 2007, July 6-8, pp. 833–836 (2007)

15. van Rikxoort, E.M., de Hoop, B., van de Vorst, S., Prokop, M., van Ginneken, B.: Automatic Segmentation of Pulmonary Segments From Volumetric Chest CT Scans. IEEE Transactions on Medical Imaging 28(4), 621–630 (2009)

16. van Rikxoort, E.M., Prokop, M., de Hoop, B., Viergever, M.A., Pluim, J., van Ginneken, B.: Automatic Segmentation of Pulmonary Lobes Robust Against Incomplete Fissures. IEEE Transactions on Medical Imaging 29(6), 1286–1296 (2010)

17. van Rikxoort, E.M., Goldin, J.G., van Ginneken, B., Galperin-Aizenberg, M., Ni, C., Brown, M.S.: Interactively learning a patient specific k-nearest neighbor classifier based on confidence weighted samples. In: 2010 IEEE International Symposium on Biomedical Imaging: From Nano to Macro, April 14-17, pp. 556–559 (2010)

18. Zhou, J., Chang, S., Metaxas, D.N., Zhao, B., Ginsberg, M.S., Schwartz, L.H.: An Automatic Method for Ground Glass Opacity Nodule Detection and Segmentation from CT Studies. In: 28th Annual International Conference of the IEEE Engineering in Medicine and Biology Society, EMBS 2006, August 30-September 3, pp. 3062–3065 (2006)

19. This research was partially supported by the Intramural Research Program of the National Institutes of Health, National Institute of Allergy and Infectious Diseases, and the Bill & Melinda Gates Foundation and Wellcome Trust through the Grand Challenges in Global Health Initiative (PI, Douglas Young, Imperial College, London)

20. This research was primarily supported by the National Institutes of Health, National Institute of Allergy and Infectious Diseases, Office of Cyber Informatics and Computational Biology funding to NET ESOLUTIONS CORPORATION (NETE)

21. This research utilizes proprietary computational algorithms and software tools developed by the NET ESOLUTIONS CORPORATION (NETE) team

DNA Algorithm Based on K-Armed Molecule and Sticker Model for Shortest Path Problem

Hong Zheng, Zhili Pei, Qing'an Yao, QingHu Wang, and Yanchun Liang[*]

Abstract. DNA computing is a computing paradigm based on biochemical reactions by using DNA molecule. Because of its advantages, such as mass storage and parallelism, it has been used to solve many complex problems. Shortest path problem is a frequently used problem in practical application. In this paper, a DNA algorithm based on k-armd molecule and sticker model to solve shortest path problem for weighted graph is proposed. The encoding method for the vertex, edges and weights is described in detail. This method shorten the length of the DNA strands for the weight code and may decrease the error during the biochemical reaction in a certain degree. The solving process for shortest path problem is also described in this paper.

Keywords: DNA computing, k-armed molecule, sticker model, shortest path problem.

1 Introduction

Since Adleman proposed a method to solve Hamilton Path Problem by using DNA molecule and biological operating, research on DNA computing are carried out in

Hong Zheng · Yanchun Liang
College of Computer Science and Technology, Jilin University, Changchun, China

Hong Zheng · Qing'an Yao
College of Computer Science and Engineering, Changchun University of Technology

Zhili Pei · QingHu Wang
College of Computer Science and Technology,
Inner Mongolia University for the Nationalities

Zhili Pei
Key Laboratory of Symbolic Computation and Knowledge
Engineering of Ministry of Education, Jilin University

[*] Corresponding author.

F.L. Gaol et al. (Eds.): Proc. of the 2011 2nd International Congress CACS, AISC 145, pp. 361–366.
springerlink.com

many fields. DNAcomputing is a new computing paradigm based on DNA strands and biochemical reaction. The central idea of DNA computing is using the mass storage of the DNA molecule as well as the mass parallelism of genetic code. So, DNA computing model must have mass storage and high calculating speed. Because of this advantagement, DNA algoritms are used to solve NP-hard problems. Many encouraging results have been obtained. Such as Hamilton Paths [1], 3-SAT problem[2], maximal clique problem [3] , and Chinese postman problem [4].

Shortest path problem for weighted graph is an important problem and has many applications in practice. It is a kind of NP hard problem. Conventional algorithms need massive calculation. Especially when the problem scale is large, it need a long time to find the probable solution. In this paper, we propose a DNA algorithm based on k-armed molecule and sticker operation to find the shortest path of a weighted graph.

2 K-Armed Molecule and Sticker Operation

In DNA computing, the process to solve problem is the biochemical reactions process of DNA strands, such as ligation, denaturation, renaturation, cut, and so on. DNA is composed of nucleotides distinguished by four bases adenine, guanine, cytosine and thymine(abbreviated as A, G, C, T). DNA strands is formed by end-to-end linked nucleotides. The nucleotides C and G are complementary, so do A and T(called Watson-Crick Complementary principle). Two complementary single-stranded DNA sequences will join together to form a double-strand called double helix. The reverse process is such a process that a doubler helix coming apart to two complementary single strands when the temperature is high enough.

N.Jonoska, *et al.* proposed a method with DNA molecules by constructing 3-dimensional structures[5]. Fig. 1 indicates *k*-armed DNA molecule. The molecule in the structure has partly double strands and *k*-armed single-strand branches. Because of its *k* single ends, the molecule is called *k*-armed molecule. Researches show that *k*-armed DNA molecule are exist in nature. 3-armed and 4 armed molecule are much stable[6].

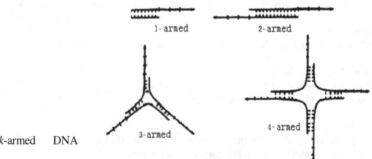

Fig. 1 *k*-armed DNA Molecules

This character can be used for linking among vertexes in a graph. A vertex with k-degree can link to the other k vertex by a k-armed molecule. Just set the single part as complementary with the neighborhood vertexes.

Sticker operation is used when building the graph. Sticker operations are indicates in Fig. 2. Single ends of two incomplete molecule such as x and y join together to double-strand according to Watson-Crick Complementary principle.

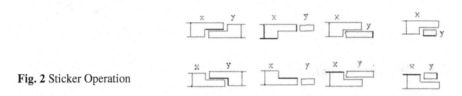

Fig. 2 Sticker Operation

3   Encoding Scheme

We will describe our encoding method of the vertexes, the weights and the edges for a weighted graph given in Fig. 3 as an instance.

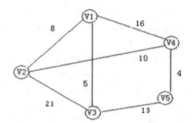

Fig. 3 A Weighted Graph

3.1   *Vertex Encoding*

We use single-strand to encode vertex. Here we adopt the length of single-strand is 10 bp. Each encoding of the vertex is unique.

For the instance shown in Fig. 3, the encoding of vertex are as follows:

$VC(V_1)$= TGACGGTTCA
$VC(V_2)$= GACTCTGTAC
$VC(V_3)$= ATGCGAGGTC
$VC(V_4)$= GATCGGTAAC
$VC(V_5)$= TCGATTGAAA

3.2   *K-Armed Molecules*

For a k-degree vertex, a k-armed molecule is used for joining part with other vertexes. The k single ends are complmentary strands of the neighborhood. For the instance, vertex V_1 connect to V_2, V_3 and V_4, so we use a 3-armed molecule, the 3 single strands are respectively complementary with V_2, V_3 and V_4, denoted as V_2 ′, V_3 ′ and V_4 ′. Fig. 4 shows the encoding way. Similarly, we can encode for other vertexes.

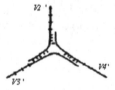

Fig. 4 3-armed molecule for V_1

3.3 *Weight Encoding*

Encoding method for weight is an important problem in weighted graph when using DNA algorithm. Many methods have been proposed by researchers. S Y Shin, B T Zhang, and S S Jun used the number of bonds to express weights[4]. M Yamamoto, Y Hiroto, and T Matoba used the contents of C/G pairs for weight encoding[7]. J Y Lee, S Y Shin, and T H Park used the melting temperature of double strands with certain length[8]. The boichemical reaction of these methods   are more complex than detecting the length of a double strand. So we use the length of double-strand to denote weight. Long strand is easy to break during reaction and then cause errors, in order to decrease errors, we shorten the length by the following method.

We translate the value of weight to binary string. The highest bit should be 1. We use G and C pair to express 1, while A and T pair to express 0. So the weight is encoded as a double-strand. In order to shorten the length, we set w_m as the minimum of the weight of all edges, and *ave* as the average of the difference between the weights. Let *MIN* is the minimum of w_m and *ave* , the weights are encoded based on the value of *MIN*. For the instance in Fig. 3, w_m=4, *ave*=8, so *MIN*=4.

The binary encoding of 4 is "100", so 5 can be encoded as "1100", 8 can be encoded as "100100", and 21 can be encoded as "1100100100100100", and so on. Thus its DNA molecules can be expressed as shown in Table 1.

Use this encoding method, to detect the length of a strand is easy to operate and a much shorter strand can decrease the error during the biochemical reaction in a certain degree.

3.4 *Edge Encoding*

The edge between vertexes is incomplete molecule including three parts. The two single-stranded ends are codes of the two vertexes. The middle of the sequence is double-strand of the weight, shown as Fig. 5.

Fig. 5 Encoding for E_{ij}

By encoding the edges like discussing above, we can see, the single-strand(V_i) of the edge is complementary with the single-strand of the joining part(V_i '). So by sticker operation, the edges can link with the k-armed joining parts. When all the molecules are offered, the graph can be built in biochemical reaction. All the codes of the instance are listed in Table.1.

Table 1 Edge Encoding

Edge	Code of Weight	Code of Edge
E_{45}	CAA GTT	CAA GATCGGTAAC GTT TCGATTGAAA
E_{13}	GGAA CCTT	GGAA TGACGGTTCA CCTT ATGCGAGGTC
E_{12}	CAACTT GTTGAA	CAACTT TGACGGTTCA GTTGAA GACTCTGTAC
E_{24}	GTCAACTT CAGTTGAA	GTCAACTT GACTCTGTAC CAGTTGAA GATCGGTAAC
E_{35}	CGTACAACTT GCATGTTGAA	CGTACAACTT ATGCGAGGTC GCATGTTGAA TCGATTGAAA
E_{14}	CATGTACAACTT GTACATGTTGAA	CATGTACAACTT TGACGGTTCA GTACATGTTGAA GATCGGTAAC
E_{23}	CGTTCATGTACAACTT GCAAGTACATGTTGAA	CGTTCATGTACAACTT GACTCTGTAC GCAAGTACATGTTGAA ATGCGAGGTC

4   Algorithm for Shortest Path Problem

In this section, we describe the algorithm of our solution for shortest path problem. For a given graph G=(V,E,W) which has n vertexes, propose that V(G)={V_0, V_1,...,V_{n-1}}. We want to find the shortest path from Vi to Vj and pass each vertex only once. The algorithm could be given as follows.

Step1   Encode for the vertex, edge and weights.
Step2   Randomly generating the paths.
Step3   Retain all the paths begin from Vi and end to Vj.
Step4   Retain all the paths not including the same vertex.
Step5   Retain the paths in step 4 with the shortest path.
Step6   Detection solutions.

Biotechnological operations of the algorithm described as follows:

1. Encode:Encode the vertexes, k-armed joining part and edges using the method illustrated in section 3. Each code is identifiable and unique. we put different coloured radioactivity label at each vertex.

2. Building graph: Put enough copies of moleules in the test tube, controll the temperature of reaction to make the complementary single strand annealing.

3. Select beginning vertex:Select a certain vertex Vi as a beginning vertex to get the path.

4. Select paths: Because each vetex is marked by different coloured radioactivity label, we can filter the paths do not include a certain vertex and the paths that include a certain vertex more than once.

5. Select the shortest from all the paths got from 4. All the joining part and vertexes have the same length, so they can be left out.

5 Conclusion

In this paper, k-armed molecule and sticker operation are used to solve shortest path problem. The encoding method and the algorithm are described. We use incomplete molecules to encode the edges, including two single ends as the corresponding vertexes and a double-strand in the middle as the weight. The length of the double-strand denotes the weight. The encodiing method of weight is easy operating and may decrease the error in a certain degree.

Acknowledgments. This work was supported by the Open Project Program of Key Laboratory of Symbolic Computation and Knowledge Engineering of Ministry of Education, Jilin University under Grant 93K172011K07, and the Scientific Research Project of the Inner Mongolian Colleges and Universities under Grant NJ10118.

References

1. Adleman Leonard, M.: Molecular Computation of Solution to Combinatorial Problems. Science 66, 1021–1024 (1994)
2. Lipton Richard, J.: DNA Solution of Hard Computational Problems. Science 268, 542–545 (1995)
3. Ouyang, Q., Kaplan, P.D., Liu, S., et al.: DNA solution of the maximal clique problem. Science 278, 446–449 (1997)
4. Shin, S.Y., Zhang, B.T., Jun, S.S., et al.: Solving traveling salesman problems using molecula programming. In: Proc of the 1999 Congress on Evolutionary Computation, pp. 994–1000. IEEE Press (1999)
5. Jonoska, N., Karl, S.A., Saito, M.: Three dimensional DNA structures in computing. Biosystems 52, 143–153 (1999)
6. Xu, J., Tan, G.-J., Fan, Y.-K., Guo, Y.-A.: DNA Computing Principle, Progress and Difficulty(IV)—Discussion on DNA Computing Model. Chinese Journal of Computers 36, 881–893 (2007)
7. Yamamoto, M., Matsuura, N., Shiba, T., Kawazoe, Y., Ohuchi, A.: Solutions of Shortest Path Problems by Concentration Control. In: Jonoska, N., Seeman, N.C. (eds.) DNA 2001. LNCS, vol. 2340, pp. 203–240. Springer, Heidelberg (2002)
8. Lee, J.Y., Shin, S.Y., Park, T.H., et al.: Solving traveling salesman problems with DNA molecules encoding numerical values. BioSystems 78, 39–47 (2004)

Smith Normal Form Using Scaled Extended Integer ABS Algorithms

Effat Golpar-Raboky and Nezam Mahdavi-Amiri*

Abstract. Classes of integer ABS methods have recently been introduced for solving linear systems of Diophantine equations. The Smith normal form of a general integer matrix is a diagonal integer matrix, obtained by elementary nonsingular (unimodular) operations. Such a form may conveniently be used in solving integer systems of equations and integer linear programming problems. Here, we present a class of algorithms for computing the Smith normal form of an integer matrix. In doing this, we propose new ideas to develop a new class of extended integer ABS algorithms generating an integer basis for the integer null space of the matrix. Finally, we test our algorithms and report the obtained numerical results on randomly generated test problems.

1 Introduction

Smith [10] proved that any integer matrix A with rank r can be transformed by elementary row and column operations into the Smith normal form, a diagonal matrix with r nonzero integer diagonal elements $\lambda_1, \cdots, \lambda_r$ so that $\lambda_1 \mid \cdots \mid \lambda_r$ [8]. ABS methods constitute a large class of methods, first introduced by Abaffy, Broyden and Spedicato [1], for solving linear algebraic systems, and later extended to solve least squares problems, nonlinear algebraic equations, and optimization problems [1, 2]. Esmaeili, Mahdavi-Amiri and Spedicato [5, 6] presented a classes of integer ABS (IABS) algorithms based on the basic ABS algorithms and later proposed the scaled ABS algorithms [11, 12] for solving linear Diophantine systems of equations. Each integer ABS algorithm decides if the Diophantine system has an integer solution, and if so, obtains a particular solution along with an integer matrix with possibly dependent rows generating the integer null space of the equations. Let A be

Effat Golpar-Raboky · Nezam Mahdavi-Amiri
Faculty of Mathematical Sciences, Sharif University of Technology, Tehran, Iran
e-mail: g_raboky@math.sharif.edu, nezamm@sharif.edu

* Corresponding author.

F.L. Gaol et al. (Eds.): Proc. of the 2011 2nd International Congress CACS, AISC 145, pp. 367–372.
springerlink.com   © Springer-Verlag Berlin Heidelberg 2012

an $m \times n$ integer matrix and b be an integer vector with m components. Consider the following linear system,

$$Ax = b. \tag{1}$$

The system (1) is equivalent to the scaled system, $V^T A x = V^T b$, where V is an arbitrary $m \times m$ unimodular matrix (an integer matrix with determinant equal to $+1$ or -1). A scaled integer ABS method starts with an arbitrary initial vector $x_1 \in Z^n$, a nonzero vector $v_1 \in Z^m$ and a unimodular matrix $H_1 \in Z^{n \times n}$. Given x_i, an integer solution of the first $(i-1)$ equations, and H_i, a matrix with rows generating the integer null space of the first $(i-1)$ rows of the coefficient matrix, a scaled integer ABS algorithm computes x_{i+1} as an integer solution of the first i equations and H_{i+1}, with rows generating the integer null space of the first i rows of the coefficient matrix, by performing the following steps:

1. Compute $v_i \in Z^m$ independent of $v_1, \cdots, v_{i-1}$ and let $s_i = H_i^T A v_i$.
2. Determine $z_i \in Z^n$ such that $z_i^T H_i A^T v_i = gcd(s_i)$ and set $p_i = H_i^T z_i$.
3. Update the solution by $x_{i+1} = x_i + \alpha_i p_i$, where $\alpha_i = \frac{v_i^T b - v_i^T A x_i}{v_i^T A p_i}$.
4. Update the Abaffian matrix by

$$H_{i+1} = H_i - \frac{H_i A^T v_i w_i^T H_i}{w_i^T H_i A^T v_i}, \tag{2}$$

where, $w_i \in Z^n$ is an arbitrary integer vector satisfying $w_i^T H_i A^T v_i = gcd(s_i)$.

The scaled integer ABS method produces a lower triangular factorization. Assume that A has rank r. Define $P = (p_1, \cdots, p_r)$ as the search vector matrix. Then, P is a full rank matrix and the factorization $V^T A P = L$, holds, where, $L \in Z^{m \times r}$ is a lower triangular matrix.

Chen, Dang and Xue [3] gave a generalization of the ABS algorithms, called extended ABS (EABS) class of algorithms for the real case, which differs from the ABS algorithms only in updating the Abaffian matrices H_i. In the EABS algorithms, the Abaffian matrices H_i are updated as follows:

1. $H_{i+1} = G_i H_i$, where, $G_i \in R^{j_{i+1} \times j_i}$ is such that we have $G_i x = 0$ if and only if $x = \lambda H_i a_i$, for some $\lambda \in R$.

Using an integer extended ABS algorithm, we can produce H_{i+1} with full row rank so that the columns of H_{i+1}^T forms a basis for the integer null space of the first i rows of A. Let H have full rank and the columns of H^T be a basis for the integer null space of A. Then, $U = (p_1, \cdots, p_r, H^T)$ is a unimodular matrix [6]. In a recent work, Khorramizadeh and Mahdavi-Amiri [7] have also presented a new class of extended integer ABS algorithms for solving linear Diophantine systems, controlling the growth of intermediate results. Here, we present a new class of algorithms for computing the Smith normal form of an integer matrix making use of our scaled extended integer ABS algorithms (SEIABS). We also present a new approach for designing extended integer ABS class of algorithms to compute a basis for the integer null space of A. In doing this, we need to solve a quadratic Diophantine system $x^T A y = b$. We present two algorithms for solving such quadratic equations. Finally, we give a brief description of the randomly generated test problems used to test various given algorithms.

2 Smith Normal Form Using Scaled Extended Integer ABS Algorithms

Here, we describe our approach briefly.

2.1 A New Class of Extended Integer ABS Algorithms

We have designed a new approach for computing G_i in an SEIABS algorithm so that the columns of G_i^T form an integer basis for the integer null space of $v_i^T A H_i^T$ and the rows of $H_{i+1} = G_i H_i$ constitute an integer basis for the integer null space of $v_1^T A, \cdots, v_i^T A$.

2.2 The Smith Normal Form Based on ABS Algorithms

We have developed a new class of SEIABS algorithms for computing the Smith normal form of an integer matrix A with rank r. The process is inductive, simultaneously computing v_i and p_i, for each i, $1 \le i \le r$, to form the first r columns of V and U. In doing this, we generate two sequences of full rank matrices $\{H_i\}$ and $\{R_i\}$, compute $v_i = R_i^T y_i$ and $p_i = H_i^T x_i$, for some $y_i \in Z^m$ and $x_i \in Z^n$, so that $v_1^T A p_1 \mid \cdots \mid v_r^T A p_r$. We have also shown $V = (v_1, \cdots, v_r, R^T)$ and $U = (p_1, \cdots, p_r, H^T)$ to be unimodular matrices and $V^T A U$ to be a Smith normal form of A.

2.3 Solving a Quadratic Diaphantine Equation

In our proposed algorithm, we need to solve quadratic Diaphantine equation of the form $x^T A y = gcd(A)$. We have proposed two algorithms for solving such equations, as outlined below.

2.3.1 Solution of the Quadratic Equation by Divisibility Sequence (QEDS) Algorithm

The algorithm uses an integer basis for the integer row space of A and computes y so that $gcd(Ay) = gcd(A)$. After finding y, x is found by Rosser's algorithm [9].

2.3.2   Solution of the Quadratic Equation by Integer ABS (QEIABS) Approach Based on a Conjecture

In the second algorithm, we give a conjecture for finding an integer vector d so that $gcd(d) = gcd(A)$ and $Ay = d$ has an integer solution. We make use of Dirichlet's theorem [4] for computing d, and solve the system $Ay = d$ using a recently proposed integer ABS algorithm [7] with the intention of controlling the growth of intermediate results.

Our numerical results on randomly generated test problems show a better performance of the second algorithm, QEIABS algorithm, in controlling the size of the solution. Comparative results of the QEIABS algorithm and ismith of Maple show a more balanced distribution of the components of V and U obtained by our algorithm.

3   Numerical Experiments

Next, we compute the Smith normal form of 10 integer matrices and compare the QEDS and QEIABS algorithms. Let m be the number of rows and n be the number of columns of A, and BL be the maximum bit length of the components of A. The random integer matrices are generated by Maple's

$$RandomMatrix(m, n, generator = -2^{BL}..2^{BL}).$$

Tables 1 and 2 show the numerical results on finding the Smith normal form of A for the randomly generated problems of increasing size. In these tables, TN refers to the test number, m, n and BL are as defined above, x-Mean and y-Mean are the ratio of the means of the Euclidean norms of the computed solution vectors x and y corresponding to the quadratic equation $x^T Ay = gcd(A)$, respectively, obtained by dividing the one from the QEIABS algorithm to the one from the QEDS algorithm, BL-QEIABS and BL-QEDS give the maximum bit lengths of the vectors x and y computed by the QEIABS and QEDS algorithms, respectively.

We implemented the QEIABS algorithm, starting with y_1 as the zero vector and H_1 as the identity matrix. The algorithms were implemented using Maple 9.5 and the programs were executed on a Pentium 4 having 2.4 Ghz processor and 5.12 Mb storage. The results in Table 1 show that in most of the test problems, the QEIABS algorithm significantly outperforms the QEDS algorithm in having smaller values of the mean and bit length of the obtained results. In the few instances to the contrary, however, the outperformance of the QEDS algorithm over the QEIABS algorithm is not as significant.

The results on the Smith normal forms obtained for the same problems 1-10 of Table 1, using Algorithm QEIABS and the procedure *ismith* from Maple are given in Table 2. The headings $\|V\|_2 - SNF$ and $\|U\|_2 - SNF$ are Euclidean norms of V and U, respectively, obtained by the QEIABS algorithm and $\|V\|_2 - Maple$ and $\|U\|_2 - Maple$, are Euclidean norms of V and U, respectively, obtained by *ismith*. The numerical results show that the QEIABS algorithm generates a more balanced U and V as compared to the ones obtained by *ismith*.

Table 1 Comparative results for QEIABS and QEDS algorithms.

TN	m	n	BL	x-Mean	y-Mean	BL-QEIABS	BL-QEDS
1	3	3	27	0.1878	0.0104	13	20
2	4	6	14	0.003959	0.09055	10	13
3	5	5	13	0.0121	0.4555	7	13
4	8	8	7	15.8362	0.04163	11	13
5	8	16	4	3	2	2	1
6	9	9	11	4.4444	0.00010	7	12
7	10	12	11	2.3333	1.5	3	2
8	12	17	3	0.6667	4.5	4	3
9	17	19	4	0.01695	0.16185	7	9
10	18	18	1	0.1034	0.6533	5	8

Table 2 Euclidean norms of unimodular matrices V and U generated by QEIABS and the *ismith* procedure of Maple.

TN	m	n	BL	$\|V\|_2 - SNF$	$\|U\|_2 - SNF$	$\|V\|_2 - Maple$	$\|U\|_2 - Maple$
1	3	3	27	3.8460E+005	6.1535E+008	0.7433E+007	0.6201E+007
2	4	6	14	1.4701E+006	2.04988E+005	5.1288	0.1187E+010
3	5	5	13	3.1361E+006	2.3604E+006	0.3138E+009	0.2912E+014
4	8	8	7	4.0347E+008	3.0311E+006	0.1260E+010	0.5692E+013
5	8	16	4	899.8250	135.4543	67.4070	4799.9348
6	9	9	11	2.0802E+005	3.3391E+004	0.6753E+008	0.1823E+008
7	10	12	11	119.8925	66.4436	130.596	1365.5846
8	12	17	3	1.1157E+005	179.9770	967.5534	28673.7232
9	17	19	4	1.5038E+007	1.8384E+006	28262.9158	0.3090E+008
10	18	18	1	1	0.1199	0.1389	0.1226

4 Conclusion

We presented a new class of algorithms for computing unimodular matrices V and U so that $V^T A U$ be the Smith normal form of an integer matrix with arbitrary rank, making use of scaled extended integer (ABS) algorithms. For the Smith normal form, we needed to solve quadratic Diophantine equations. We presented two algorithms for solving such equations. The first algorithm, solving the quadratic equation by the divisibility sequence (QEDS) approach, made use of a divisibility sequence basis for the row space of A, and the second algorithm, solving the quadratic equation by integer ABS (QEIABS) algorithms with the intention of controlling the growth of intermediate results, made use of our given conjecture and devised a method based on recently proposed integer ABS algorithms. Numerical results showed that more often the QEIABS algorithm produce solutions having smaller size as compared to the QEDS algorithm. Our test results on the computed Smith normal form using our given QEIABS algorithm and Maple's *ismith* procedure showed that the unimodular matrices produced by our algorithm to be more balanced.

References

1. Abafy, J., Spedicato, E.: ABS Projection Algorithms. Mathematical Techniques for Linear and Nonlinear Equations. Ellis Horwood, Chichester (1989)
2. Abaffy, J., Broyden, C.G., Spedicato, E.: A class of direct methods for linear equations. Numer. Math. 45, 361–378 (1984)
3. Chen, Z., Dang, N.Y., Xue, Y.: A general algorithm for underdetermined linear systems. In: The Proceedings of the First International Conference on ABS Algorithms, Luoyang, China, pp. 1–13 (1992)
4. Dirichlet, P.G.L.: Beweis des Satzes, dass jede unbegrenzte arithmetische progression, deren erstes glied und differenz ganze zahlen ohne gemeinschaftlichen factor sind, unendlich viele primzahlen enthalt. Mathematische Werke 1, 313–342 (1837)
5. Esmaeili, H., Mahdavi-Amiri, N., Spedicato, E.: A class of ABS algorithms for Diophantine linear systems. Numer. Math. 90, 101–115 (2001)
6. Esmaeili, H., Mahdavi-Amiri, N., Spedicato, E.: Generationg the integer null space and conditions for determination of an integer basis using the ABS algorithms. Bulletin of the Iranian Mathematical Society 27, 1–18 (2001)
7. Khorramizadeh, M., Mahdavi-Amiri, N.: Integer extended ABS algorithms and possible control of intermediate results for linear Diophantine systems. 4OR 7, 145–167 (2009)
8. Pohst, M., Zassenhaus, H.: Algorithmic Algebraic Number Theory. Cambridge University Press, New york (1989)
9. Rosser, J.B.: A note on the linear Diophantine equation. Amer. Math. Monthly 48, 662–666 (1941)
10. Smith, H.J.S.: On systems of linear indeterminate equations and congruences. Phil. Trans. Roy. Soc. London 151, 293–326 (1861)
11. Spedicato, E., Bodon, E., Del Popolo, A., Mahdavi-Amiri, N.: ABS methods and ABSPACK for linear systems and optimization. A review. 4OR 1, 51–66 (2003)
12. Spedicato, E., Bodon, E., Zunquan, X., Mahdavi-Amiri, N.: ABS methods for continous and integer linear equations and optimization. CEJOR 18, 73–95 (2010)

The Characteristics of Flat-Topped and Pinnacle Building on SAR Image

Wang Min, Zhou Shu-dao, Liu Zhi-hua, Huang Feng, and Bai Heng

Abstract. Because of the characteristics of SAR side-looking imaging, the buildings in SAR images showing the geometric characteristics obviously, and the different roof structure buildings have the different features in the SAR images, these characteristics play an important role in the city monitoring of building three-dimensional reconstruction. First, discusses the SAR imaging principle and general characteristics, including layerover, shadow, perspective contraction, the dihedral angle reflection characteristics, and gives the different imaging characteristics of flat-topped and pinnacle building which have variety of different geometries, provides information for future city buildings identification and three-dimensional reconstruction.

1 Introduction

SAR images have side-view imaging characteristics are different from the optical image, compared with the optical image, SAR visual readability is poor, but because of this unique side-view imaging characteristics make it better reflect the goals of the three-dimensional information, when the mountain or a building area observed in SAR image, there is a three-dimensional, because the mountains or buildings in the radar images with perspective contraction, shadow, and even the formation of layerover and other features. These characteristics for the urban environment (including urban classification of the typical urban structure extraction, change detection, etc.) and building 3D structure of inversion [1] provides a valuable data source, and for the promotion of radar remote sensing technology in the city of great significance.

This paper summarize and analysis the imaging principle and imaging characteristics of SAR images, and study characteristics of the flat-topped and pinnacle buildings in the simple single background SAR image, provide reference basis for future buildings information extraction.

Wang Min · Zhou Shu-dao · Liu Zhi-hua · Huang Feng · Bai Heng
Institute of Mcteorology, PLA University of Science and Technology, Nanjing, China
e-mail: yu0801@163.com, zhousd70131@sina.com

F.L. Gaol et al. (Eds.): Proc. of the 2011 2nd International Congress CACS, AISC 145, pp. 373–379.
springerlink.com   © Springer-Verlag Berlin Heidelberg 2012

2 SAR Imaging Principles and Imaging Features

2.1 SAR Imaging Principle

Synthetic aperture radar is a high-resolution coherent imaging radar, which using pulse compression to achieve high distance resolution at distance direction, and using record the Doppler frequency shift to achieve aperture synthesis to improve the orientation resolution, aperture synthesis is the synthesis of pulse compression in principle [2]. Fig1 shows SAR geometry imaging, it towards laterally underside launches coherent electromagnetic waves, and according to the sequence of time (distance) records the intensity and phase information of the surface echo-wave. Flying along the orientation at a distance, we can get a SAR image. Because the SAR is a side- looking imaging, rather than the optical image as the vertical projection or scanning imaging, so its image reflect the distance information from the target to the radar is better, extracted from a single SAR image target height information from this principle is possible.

2.2 SAR Imaging Characteristics[3]

1) Slant range and Ground range image

The target distance in the SAR image along the distance-direction has two displays: one for the slant range, the other is the ground range, as shown in Fig2. The relative distance between objects in the slant range image is proportional to the corresponding ground range, and scale is a constant, while the slant range is entirely showing the distance from the map to the radar antenna that the scale is not constant . For the slant range image, can be converted to ground range to remeasure[4].

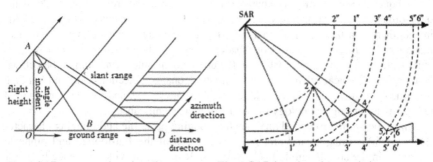

Fig. 1 SAR geometry imaging diagram   **Fig. 2** SAR imaging characteristics

2) Perspective contraction

When the slope of local area is less than the incident angle, the ramp-up region in the slant range image will be reduced, this is the perspective contraction, as shown 3 to 4 point area in Fig 2. This will make the energy to concentrate on the smaller region within images, that reflected from the ramp side facing the radar, so the performance of the brightness of the region brighter than the surrounding areas.

3) Layerover

When the slope of local area is more than the incident angle, the signal will return to the radar receiver from peak is earlier than from at the end point, resulting in the phenomenon of layerover. As shown in Fig2, 1 and 2 points will appear upside down each other's situation in the SAR image. The target echo from multiple locations in the layerover areas may return to the radar receiver in the same time, the performance of the SAR image will be brighter.

4) Shadow

When the slope which back to the radar is more than the incident angle, there isn't signal return to the radar receiver, resulting in the phenomenon of shadow. As shown in Fig 2, the area from 2 to 3 point, because the radar beam can not irradiat to the area, showed a dark area.

2.3   Pinnacle Single Reflection Effect

As the roof of the pinnacle building is perpendicular to the radar beam or with the radar beam incidence angle within 20 ° in the vertical approximation, shown in Fig3, the exposure to this plane is reflected back to the radar beam to form a single reflection effect, but also because this roof is nearest to the sensor, therefore, its echo was the first return to the sensor and recorded . Radar get a strong reflection energy, the brightness is very high, so form a bright line parallel to the azimuth direction [4].

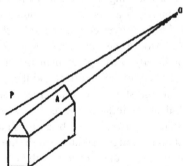
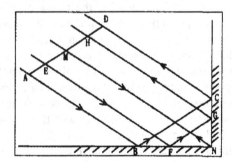

Fig. 3 Pinnacle single reflection effect   Fig. 4 Double scattering of buildings

2.4   Double Scattering of Buildings

Double scattering is defined as the incident wave radiation to the walls which perpendicular to the ground, some directly reflected back, the other part first reflected
to the ground by a wall, and then reflected to the SAR from the ground. If the dihedral angle facing the radar beam, the radar beam after a few reflecting, then reflecting back to the SAR sensor. As there is a very long distance from SAR sensor
to the ground, that the incident wave of the angle reflector is parallel to the echo,
the dihedral angle reflection diagram shown in Fig4, according to the geometric
relationship from the echo, the incident wave of the dihedral angle reflector different parts is reflected back to the sensor, which the distance is equal and reach to
the sensor at the same time, so they formed a very high backscatter coefficient in
SAR image [5-6]. Therefore, the wall in SAR image parallel to the direction and
have a certain length to perpendicular to the ground will have the effect of a bright
line parallel to the direction resulting from double scattering.

3   SAR Imaging Properties of the Two Roof Structures

In general, ordinary walls perpendicular to the ground, the radar beam imaging on
the building will have several phenomen as above. This paper focuses on the two
imaging features of flat roof and pinnacle buildings. SAR flight direction is parallel to the bottom edge of building.

3.1   Imaging Characteristics of Flat-Topped Buildings [7-8]

Analysis of imaging properties of the flat-topped building, needs to study two situation which have different length/height, as shown the two rectangles ABCD in
Fig 5 (a), 5 (b). Slant range image for the building above the schematic diagram
and below is the ground range image, the backscattering component from the
building target consisting the following main parts: ① as the ordinary building
wall slope angle is 90 °, so it meet the production layerover conditions, the echo
from roof, walls and floors can reach the SAR sensor at the same time, layerover
later than the pinnacle echo, the brightness of layerover is lighter than the surrounding environment, weaker than the single surface reflection effect and the angle of reflection effect. layerover area of radar image shown the gray area on he
left side in Fig5 (a), 5 (b); ② next is dihedral angle reflection effect between
building wall facing the radar with floors, the images showed a high brightness parallel to the azimuth direction, Fig5 (a), 5 (b) shall be a high-brightness light lines
in the right side followed layerover ; ③ to Fig 5(a), the height is greater than distance length, the radar beam to the back slope of the wall can not reach to the SAR
sensor, therefore, there can be no echo signal, the corresponding location on the
image appear dark, black areas in the figure is the shadow area for the building;

④ For Fig 5 (b), the distance length is greater than the height and incident cross-section is wider, after the dihedral angle effect is the roof area which have the middle brightness, this part only have of part echo of the roof, the right half roof backscattering is weak as the mirror reflection, the SAR image is usually darker, and finally is the dark shadow areas similar to Fig5 (a).

3.2   Imaging Characteristics of Pinnacle Buildings [9-10]

Analysis of imaging properties of the pinnacle building, needs to study several situation which have different length/height and angle of the pinnacle ridge, according to the following two conditions analysis, as shown the two five- polygon ABCDE in Fig 6(a), 6 (b). Slant range image for the building above the schematic diagram and below is the ground range image, the backscattering component from the building target consisting the following main parts: ① in the 6 (a), as the pinnacle is nearly perpendicular to the radar beam forming a single scattering, the image appears a high-brightness light cable parallel to the azimuth direction. But 6 (b) does not exist this highlight tape, because the angle with roof surface and the SAR incident more than 20 °, can't forming single scattering; ② We can think similar to the flat-topped buildings SAR images, walls, ground, roof is also produced to meet the conditions of the layerover, shown as Fig6 (a), 6 (b) of the gray area; ③ after layerover there both appear highlighted line parallel to the azimuth direction based on the dihedral angle reflection effect; ④ as the angle of roof to the horizontal plane is small in 6 (b), similar to flat-topped buildings considered as specular reflection occurs, little echo received, it was dark black; ⑤ Finally, there are shadow in the side of the building away from SAR sensors, some ground covered by buildings, forming a black shadows.

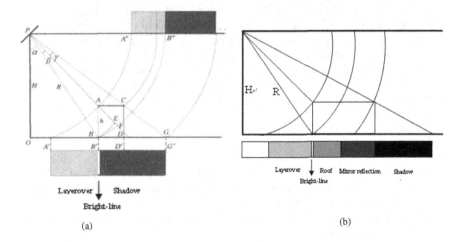

(a)   (b)

Fig. 5 SAR characteristics of flat-topped building

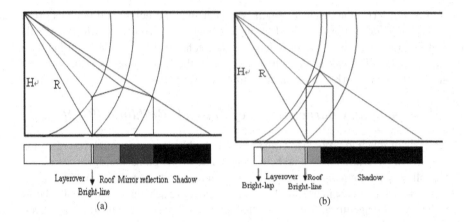

Fig. 6 SAR characteristics of pinnacle building

4 Conclusion

Based on the analysis of the imaging principle and imaging characteristics of SAR images, this paper discussed several imaging characteristics of the flat-topped and pinnacle building, both in comparison, all with overlapping cover, shadows, dihedral angle reflections and other characteristics. However, due to the special geometry of pinnacle roof, there was a high brightness straight line as single reflection. SAR image features sequence of two buildings can be summarized as follows: a) For the length / height> 1 flat-topped buildings, there are layerover - bright line - shadow; b) For the length / height< 1 flat-topped buildings, there are layerover - bright line - roof - mirror reflection - shadow; c) For the length / height< 1 and smaller ridge angle pinnacle building, there are bright line lap- layerover - bright line - shadow; d) For the length / height > 1 and bigger ridge angle pinnacle building, there are layerover - bright line - mirror reflection - shadow. Based on the above analysis, we can facilitate to the further application of the building identification and 3D reconstruction from SAR image.

References

1. Zhu, J.J., Ding, C.B., You, H.J.: 3D reconstruction of building based on high-resolution SAR and optical images. In: Geoscience and Remote Sensing Symposium, pp. 1215–1218. IEEE, Washington DC (2006)
2. Ulaby, F.T., Moore, R.K., Fung, A.K.: Microwave Remote Sensing, pp. 100–108. Science Press, Beijing (1987)
3. Zou, B., Xu, K., Zhang, L.-M., et al.: Study on Extraction Methods of Three-Dimension Information of Buildings in SAR Images. Radar Science and Technology 4, 95–101 (2009)

4. Ning, S.: Principles of Microwave Remote Sensing, pp. 130–142. Wuhan University Press, Wuhan (2000)
5. Franceschetti, G., Iodice, A., Riccio, D.: A Canonical Problem in Elect romagnetic Backscattering from Building. IEEE Trans. on Geoscience and Remote Sensing 1(8), 1787–1801 (2002)
6. Yamaguchi, Y., Imura, Y.K.: ALOS-PALSAR Image Simulation Invarious Polarization Bases. IEEE Trans. on Geoscience and Remote Sensing 1(6), 381–383 (2002)
7. Tupin, F., Roux, M.: Markov Random Field on Region Adjacency Graph for the Fusion of SAR and Optical Data in Radar-grammetric Applications. IEEE Transactions on Geoscience and Remote Sensing 43(8), 1920–1928 (2005)
8. Soergel, U., Schulz, K., Thoennessen, U., et al.: Integration of 3D Data in SAR Mission Planning and Image Interpretation in Urban Areas. Information Fusion 6(4), 301–310 (2005)
9. Zhang, F.-L., Yun, S.: Urban Target Monitoring Using High Resolution SAR Data. Remote Sensing Technology and Application 25(3), 415–422 (2010)
10. Fu, X.-Y., You, H.-J.: 3D Geometrical Feature Analysis of Buildings on High-resolution SAR Image. Remote Sensing Technology and Application 25(4), 469–473 (2010)

Haar Wavelet Method for Solving Two-Dimensional Burgers' Equation

Miaomiao Wang and Fengqun Zhao

Abstract. In the present paper, An novel and efficient combination of two-dimensional Haar wavelet functions for solving a two-dimensional Burger problem with the aid of tensorial products. The numerical results demonstrate that making use of Haar wavelets to solve two-dimensional Burgers equation could reach higher accuracy and calculate easily.

1 Introduction

At present, many different orthogonal basis functions (such as Fourier function, Chebyshev polynomial and Wavelet function) have been developed and applied to the collocation method and Galerkin method to solve differential equations, integral equations. All of these functions in terms of application have their own advantages and disadvantages. However, wavelet basis is the most attractive method in these functions, which is due to good approximation and fast convergence of the wavelet sequence.

In recent year, Haar wavelet method is extensively applied to solve differential equations and integral equations. Hsiao adapted single term Haar wavelet series (STHWS) to solve stiff differential equations[1]. Lepik have deeply discussed with respect to solution of differential equations by using Haar wavelet. In 2005, the solution of ordinary differential equation by segmentation method is proposed [2], which convent second-order ODE into two first-order ODEs, and use Haar wavelet method to solve these ODEs respectively. In 2006, Haar wavelet collocation method is applied for nonlinear integral-differential equation[3], the proposed method is based on the collocation technology; and the calculation of wavelets coefficients could be made use of the Newton iteration. In 2007, the solution of evolution equation by the aid of the Haar wavelet method is proposed[4]. And it turned out that the accuracy of the Haar wavelet solution is quite high even in the case of a small number of grid points, and this method is

Miaomiao Wang · Fengqun Zhao
School of sciences, Xi'an University of Technology, Xi'an, Shaanxi
e-mail: mmjenny@126.com, zhaofq@xaut.edu.cn

F.L. Gaol et al. (Eds.): Proc. of the 2011 2nd International Congress CACS, AISC 145, pp. 381–387.
springerlink.com   © Springer-Verlag Berlin Heidelberg 2012

extended to high order nonlinear ordinary differential equation[5]. Shi and Deng make use of Haar wavelet to solve convention-diffusion equation[6]. which the partial differential equation with constant coefficient and satisfied with initial and boundary conditions are converted into algebraic equations to solve, and pointed out that this method is applied to general differential (integral) equation. Hariharan, Kannan and Sharma use Haar wavelet method to solve Fisher's equation[7]. Which used Haar wavelet to be discrete the space variables, and time variables t is viewed as constant function in each small period Δt, then conversion differential equations into algebraic equations were solved.

Above research is mostly based on one-dimensional problems, but the studies with respect to two-dimensional problem haven't ever been involved.

In this paper, based on the literature[7], the paper presents a novel, combination of one-dimensional Haar wavelet functions for solving a two-dimensional Burger problem with the aid of tensorial products. Thus one-dimensional numerical solution of Haar wavelet method is extended to two-dimensional problem, which format two-dimensional Burgers' equation for solving Haar wavelet collocation method. Numerical experiments indicate that the method is reliable theory, algorithm is simple, and easy to calculate, which are successfully obtained Haar wavelet method for two-dimensional Partial differential equation..

2   Haar Wavelet and Integral Operator Matrix

Haar function $h_i(x)$ is an orthogonal set, for $x \in [0,1]$, the Haar wavelet family is defined as follows:

$$h_i(x) = \begin{cases} 1, & x \in \left[\dfrac{k}{m}, \dfrac{k+0.5}{m}\right) \\ -1, & x \in \left[\dfrac{k+0.5}{m}, \dfrac{k+1}{m}\right) \\ 0, & elsewhere \end{cases} \tag{1}$$

where integral $m = 2^j (j = 0,1,2,...,J)$ is the level of the wavelet, J is the scale factor; $k = 0,1,2,...,m-1$ is the translation parameter, and the index i is calculated according to the formula $i = m + k + 1$. Particularly, the value $i = 1$ corresponds to the scaling function $h_1(x) = 1, x \in [0,1]$. Let $m = 1, k = 0$, we have $i = 2$, which corresponds to the Haar mother wavelet function.

We define integration

$$p_i(x) = \int_0^x h_i(x)dx, q_i(x) = \int_0^x p_i(x)dx \quad i = 1,...,2M, M = 2^J \tag{2}$$

and introduce vector function:

$$H_x(x) = [h_1(x), h_2(x),...,h_{2M}(x)]^T$$

$$P_x(x) = [p_1(x), p_2(x),..., p_{2M}(x)]^T , Q_x(x) = [q_1(x), q_2(x),..., q_{2M}(x)]^T$$

When choosing the collocation points as

$$x_l = (l - 0.5)/2M \qquad l = 1,2,...,2M \ , M = 2^J$$

we get the Haar coefficient matrix

$$H_x = [H_x(x_1), H_x(x_2),..., H_x(x_{2M})] \tag{3}$$

and first-order integral operator matrix   and second-order integral operator matrix are respectively as follows:

$$P_x = [P_x(x_1), P_x(x_2),..., P_x(x_{2M})] \ , \ Q_x = [Q_x(x_1), Q_x(x_2),..., Q_x(x_{2M})]$$

Let us define two-dimensional Haar function

$$h_{kl}(x, y) = h_k(x)h_l(y) \qquad k,l = 1,2,...,2M$$

where expression of $h_k(x)$ and $h_l(y)$ is similar to the Eq. (1). And it is clear that these functions are orthogonal in $[0,1] \times [0,1]$ .Adapting the collocation points as follows

$$(x_i, y_j) = ((i - 0.5)/2M, (j - 0.5)/2M) \qquad i, j = 1,2,...,2M$$

and the discrete two-dimensional Haar wavelet function $h_{kl}(x, y)$, thus we get the two-dimensional Haar coefficient matrix, that is

$$H(l + (2M) \times (k - 1), j + (2M) \times (i - 1)) = h_{kl}(x_i, y_j) \qquad i, j, k, l = 1,2,...,2M$$

According to the theory of the two-dimensional multi-resolution analysis, the two-dimensional Haar coefficient matrix can be calculated by the kronecker product of one-dimensional Haar coefficient matrix, that is

$$H_{(2M)^2 \times (2M)^2} = H_x \otimes H_y$$

Correspondingly, the two-dimensional first-order integral operator matrix and second-order integral operator matrix can also be calculated directly by kronecker product, those are $P_{(2M)^2 \times (2M)^2} = P_x \otimes P_y$, $Q_{(2M)^2 \times (2M)^2} = Q_x \otimes Q_y$

3   Multi-scale Approximation of the Function

According to the theory of two-dimensional multi-resolution analysis, any function $f(x, y) \in L^2([0,1] \times [0,1])$, $f(x, y)$ can be expanded into two-dimensional Haar series, that is

$$f(x, y) = \sum_{k,l=1}^{+\infty} c_{kl} h_{kl}(x, y) \tag{4}$$

where the coefficients c_{kl} are determined by

$$c_{kl} = \langle f(x, y), h_{kl}(x, y) \rangle$$

Clearly, the expansion of $f(x, y)$ is infinite series. However, if it is approximated as piecewise constant in each sub-area, then it will be terminated at finite terms, namely

$$f(x,y) \approx f_J(x,y) = \sum_{k,l=1}^{(2M)^2} c_{kl} h_{kl}(x,y) = C_{(2M)^2}^T H_x(x) \otimes H_y(y) \tag{5}$$

where the coefficients vector and Haar function vector are defined as follows:

$$C_{(2M)^2}^T = [c_{11}, c_{12}, c_{1,2M}, ..., c_{2M,1}, c_{2M,2}, ..., c_{2M,2M}]$$

$$H_x(x) = [h_1(x), h_2(x), ..., h_{2M}(x)]^T, \quad H_y(y) = [h_1(y), h_2(y), ..., h_{2M}(y)]^T$$

4 Haar Wavelet Discretization Scheme of Burgers' Equation

Consider the two-dimensional Burgers' equation with initial-boundary value conditions

$$\begin{cases} \dot{u} + uu_x + uu_y = \upsilon(u_{xx} + u_{yy}) & ,(x,y) \in D, t \in (0,T] \\ u(x,y,0) = u_0(x,y), & (x,y) \in D (x,y) \in D \\ u(x,y,t) = a(x,y,t), & (x,y) \in \partial D, t \in (0,T] \end{cases} \tag{6}$$

where υ is viscosity coefficients, and $D = \{(x,y) : 0 < x, y < 1\}$.

∂D is boundary of area D.

Now divide the interval $(0,T]$ into equal parts of length $\Delta t = T/N$, and denote $t_s = (s-1)\Delta t, s = 1,2,...,N$.

$\dot{u}_{xxyy}(x,y,t)$ is viewed as constant function on t in sub-interval $[t_s, t_{s+1}]$, so it is assumed that $\dot{u}_{xxyy}(x,y,t)$ can be expanded into terms of Haar wavelets, that is

$$\dot{u}_{xxyy}(x,y,t) \approx C_{(2M)^2}^T H_x(x) \otimes H_y(y) \tag{7}$$

Integrating formula (7) with respect to t from t_s to t, we obtain

$$u_{xxyy}(x,y,t) = (t-t_s)C_{(2M)^2}^T H_x(x) \otimes H_y(y) + u_{xxyy}(x,y,t_s) \tag{8}$$

Integrating Eq. (8) twice with respect to x from 0 to x, with regard to y from 0 to y, combined with boundary conditions, we obtain

$$\begin{aligned} u_{xx}(x,y,t) &= (t-t_s)C_{(2M)^2}^T (H_x(x) \otimes Q_y H_y(y) - H_x(x) \otimes (P_y fy)) \\ &+ (1-y)(a_{xx}(x,0,t) - a_{xx}(x,0,t_s)) + y(a_{xx}(x,1,t) \\ &- a_{xx}(x,1,t_s)) + u_{xx}(x,y,t_s) \end{aligned} \tag{9}$$

$$\begin{aligned} u_{yy}(x,y,t) &= (t-t_s)C_{(2M)^2}^T (Q_x H_x(x) \otimes H_y(y) - (P_x fx) \otimes H_y(y)) \\ &+ (1-x)(a_{yy}(0,y,t) - a_{yy}(0,y,t_s)) + x(a_{yy}(1,y,t) \\ &- a_{yy}(1,y,t_s)) + u_{yy}(x,y,t_s) \end{aligned} \tag{10}$$

$$\begin{aligned} u_x(x,y,t) &= (t-t_s)C_{(2M)^2}^T (P_x H_x(x) \otimes Q_y H_y(y) - P_x H_x(x) \otimes (P_y fy)) \\ &+ (1-y)(a_x(x,0,t) - a_x(x,0,t_s) - a_x(0,0,t) + a_x(0,0,t_s)) \\ &+ y(a_x(x,1,t) - a_x(x,1,t_s) - a_x(0,1,t) + a_x(0,1,t_s)) + u_x(x,y,t_s) \end{aligned} \tag{11}$$

$$u_y(x,y,t) = (t-t_s)C^T_{(2M)^2}(Q_x H_x(x) \otimes P_y H_y(y) - (P_x fx) \otimes P_y H_y(y))$$
$$+ (1-x)(a_y(0,y,t) - a_y(0,y,t_s) - a_y(0,0,t) + a_y(0,0,t_s))$$
$$+ x(a_y(1,y,t) - a_y(1,y,t_s) - a_y(1,0,t) + a_y(1,0,t_s) + u_y(x,y,t_s)) \tag{12}$$

$$u(x,y,t) = (t-t_s)C^T_{(2M)^2}(Q_x H_x(x) \otimes Q_y H_y(y) - (P_x fx) \otimes Q_y H_y(y)$$
$$- Q_x H_x(x) \otimes (P_y fy) + (P_x fx) \otimes (P_y fy)) + (1-y)(a(x,0,t)$$
$$- a(x,0,t_s)) + (1-x)(a(0,y,t) + a(0,y,t_s)) + (x+1)(y-1)$$
$$\cdot(a(0,0,t) - a(0,0,t_s)) + x(a(1,y,t) + a(1,y,t_s)) - xy(a(1,1,t)$$
$$- a(1,1,t_s)) + x(y-1)(a(1,0,t) - a(1,0,t_s)) + y(a(x,1,t) - a(x,1,t_s))$$
$$+ (x-1)y(a(0,1,t) - a(0,1,t_s)) + u(x,y,t_s) \tag{13}$$

Derivation Eq. (18) with respect to t, we gain

$$\dot{u}(x,y,t) = C^T_{(2M)^2}(Q_x H_x(x) \otimes Q_y H_y(y) - (P_x fx) \otimes Q_y H_y(y) - Q_x H_x(x)$$
$$\otimes (P_y fy) + (P_x fx) \otimes (P_y fy)) + (1-y)\dot{a}(x,0,t) + (1-x)\dot{a}(0,y,t)$$
$$+ (x+1)(y-1)\dot{a}(0,0,t) + (x-1)y\dot{a}(0,1,t) + x\dot{a}(1,y,t) - xy\dot{a}(1,1,t)$$
$$+ x(y-1)xy\dot{a}(1,0,t) + y\dot{a}(x,1,t) \tag{14}$$

Where $f = [1,0,...,0]^T$
$\underbrace{\qquad}_{2\bar{M}-1}$

In Eqs. (7)~(14), obtained discrete results by assuming $x \to x_l, y \to y_l, t \to t_{s+1}$, and substituting these Eqs. into Eq. (6), we obtain nonlinear equations on wavelet coefficients $C^T_{(2M)^2}$. Here we use the Matlab function fsolve to solve it. Now letting $t \to t_{s+1}$, and substituting $C^T_{(2M)^2}$ into Eq.(13), So we can gain numerical solution of Burgers' equation through calculation layer by layer.

5   Numerical Result and Analysis

Consider the two-dimensional Burgers' equation

$$\dot{u} + uu_x + uu_y = \upsilon(u_{xx} + u_{yy}) \qquad 0 < x, y < 1, t > 0$$

with the periodic boundary conditions and initial condition

$$u(x,y,0) = \sin(2\pi x)\cos(2\pi y)$$

where υ is viscosity coefficients.

Where let us take initial value as $u_0 = \sin(2\pi x)\cos(2\pi y)$, and on boundary value, which firstly it is converted into general boundary, and is substituted into Eqs. (7)~(14), we see that all boundary values are offsetting each other.

From the above of calculation of Burgers' equation, adapting time step to $\Delta t = 0.004$, we calculate numerical results by using Haar wavelet method. The numerical results are shown in Table 1, Fig. 1 and Fig. 2.

Table 1 Numerical solution at different collocation points at T=0.125,0.25for v=0.01,0.001,J=3

− x	− y	− t=0.125 − v=0.01	− t=0.125 − v=0.001	− t=0.25 − v=0.01	− t=0.25 − v=0.001
− 1/32	− 1/32	− 0.0363	− 0.0474	− -0.1138	− -0.0921
− 1/32	− 15/32	− -0.3215	− -0.3411	− -0.4476	− -0.4862
− 1/32	− 31/32	− 0.0210	− 0.0330	− -0.1441	− -0.1204
− 15/32	− 1/32	− 0.3262	− 0.3502	− 1.2484	− 0.5041
− 15/32	− 15/32	− -0.0394	− -0.0547	− 0.1114	− 0.0777
− 15/32	− 31/32	− 0.3161	− 0.3416	− 1.0929	− 0.4871
− 31/32	− 1/32	− -0.0317	− -0.0427	− 0.1230	− 0.1012
− 31/32	− 15/32	− 0.3203	− 0.3391	− 0.4452	− 0.4822
− 31/32	− 31/32	− -0.0467	− -0.0569	− 0.0934	− 0.0734

Table 1 shows that the viscosity coefficients is selected v=0.01 and v=0.001, the time is used to t=0.125 and t=0.25. Then the use of Haar wavelets method obtained numerical solutions at different collocation points respectively. From the above of experimental results, it can be seen that in Table 1, the numerical solution at different collocation points of $(0,1) \times (0,1)$ are obtained.

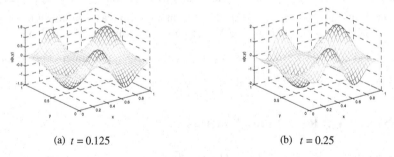

(a) $t = 0.125$ (b) $t = 0.25$

Fig. 1 Numerical solution for $t = 0.125$ and $t = 0.25$

Correspondingly, Fig. 1 indicates a behavior of the numerical solution for v=0.01 and J=4. Next, with the aid of Haar wavelet method, the numerical solutions of the equation for J=4, y=0.5 at t=0.125 and v=0.01, v=0.001 are derived, which shows in Fig. 2.

In Fig. 2 , the numerical results at different viscosity coefficients v=0.01 and v=0.001 for y=0.5, t=0.125 are shown. In Fig. 2(a), it can be concluded that the numerical solution of using Haar wavelet method , Compared with the Modified Local Crank-Nicolson method[8] , are much more smooth, in particular at the interval [0.2,0.8]. And in Fig. 2(b), when v=0.001, numerical results can also be gained. Therefor the Haar wavelet method for coping with the two-dimmentional Burger's equation is much berrer effectively.

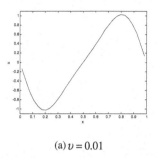

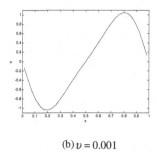

(a) $v = 0.01$ (b) $v = 0.001$

Fig. 2 The numerical solution of the equation for $y = 0.5$ $t = 0.125$.(a) $v = 0.01$ and (b) $v = 0.001$

6 Conclusion

The two-dimentional Haar wavelet numerical method, which based on the theory of two-dimentional multi-resolution analysis and two-tensor product wavelet analysis, is constructed for solving Burgers' equation. And this numerical method successfully convert partial differential equation into algebraic equation to solve. Numerical results demonstrate that the method is with reliable theory,algorithm is simple, and easy to program. In addition, dealing with boundary condition is not complicated. In another word, while discreting the equation, the boundary conditions are copied with. Thereby, algorithm complexity is reduced.

Acknowledgments. The authors sincerely thank the reviewers for their valuable suggestions and constructive comments which lead to improvement of our manuscript very much. This work was supported by Natural Science Foundation of Shaanxi Province of China(No.2011JM1013) and Education Department Foundation of Shaanxi Province of China(No. 11JK0524).

References

1. Hsiao, C.H.: Numerical solution of stiff differential equation via Haar wavelets. International Journal of Computer Mathematics 82(9), 1117–1123 (2005)
2. Lepik, U.: Numerical solution of differential equation using Haar wavelets. Mathematics and Computers in Simulation 68(2), 127–143 (2005)
3. Lepik, U.: Haar wavelet method for nonlinear intergro-differential equations. Applied Mathematics and Computation 176(1), 324–333 (2006)
4. Lepik, U.: Numerical solution of evolution equations by Haar wavelet method. Applied Mathematics and Computation 185(1), 695–704 (2007)
5. Lepik, U.: Application of the Haar wavelet transform to solving integral and differential equations. Proc. Estonian Acad. Sci. Phys. Math. 56(1), 28–46 (2007)
6. Shi, Z., Deng, L.: Haar wavelet method for solving the convection-diffusion Equation. Mathematics Application 21(1), 98–104 (2008)
7. Hariharan, G., Kannan, K., Sharma, K.P.: Haar Wavelet method for solving Fisher's equation. Applied Mathematics and Computation 211(1), 284–292 (2009)
8. Huang, P., Abduwali, A.: The Modified Local Crank-Nicolson method for one and two-dimentional Burgers'equations. Computers and Mathematics with Applications 59, 2452–2463 (2010)

A Summarization Strategy of Chinese News Discourse

Deliang Wang

Abstract. Due to the problem of information overloading, automatic text summarization is becoming more and more necessary. This paper proposes a strategy for Chinese news discourse summarization based on veins theory. This method can produce a summary of an original text without requiring its full semantic interpretation, but instead relying on the discourse structure.

1 Introduction

Summarization is the process of condensing a source text into a shorter version preserving its information content. Here, it refers specifically to the automatic text summarization in natural language processing, which can serve several goals — from survey analysis of a scientific field to quick indicative notes on the general topic of a text. Producing a quality informative summary of an arbitrary text remains a challenge which requires full understanding of the text [1].

Summarization has attracted many researchers and is studied from different perspectives. Generally speaking, there are two major approaches: deep semantic analysis and shallow linguistic analysis. For the semantic analysis, for example, McKeown and Radev [12] investigate ways to produce a coherent summary of several texts describing the same event, when a full semantic representation of the source texts is available. This type of source abstraction is the most expressive, but very domain dependent and hard to compute.

On the other hand, summaries can be built from a shallow linguistic analysis of the text. Former researchers attempt to extract summaries based on the word distribution [8], cue phrase [5], or location [5] [7]. These techniques are easily computed and rely on shallow formal clues found in the text.

The present study intends to adopt the latter approach. We propose a strategy for Chinese news discourse summarization based on Veins Theory [2]. This method can produce a summary of an original text without requiring its full semantic interpretation, but instead relying on the discourse structure.

Deliang Wang
School of Foreign Languages and Literatures, Beijing Normal University, Beijing, China,
e-mail: bright7883@126.com

F.L. Gaol et al. (Eds.): Proc. of the 2011 2nd International Congress CACS, AISC 145, pp. 389–394.
springerlink.com

Chinese is different from western languages typologically. Chinese has no morphological changes in lexis, which increases the difficulty of automatic segmentation. Chinese has no syntactic and paradigmatic markers in sentences, which increases the difficulty of parsing. Chinese has implicit constituents and special constructions, such as pivotal sentence, serial verbal sentence, etc. All these lead to the ineffectiveness of Chinese processing.

Veins Theory is proposed based on western languages. We also want to test whether it is applicable to Chinese and search for a more effective summarization method. News discourses will be adopted as a specific domain.

2   Veins Theory

Cristea et al. [2] proposed the Veins Theory which extends the applicability of centering rules [6] from local to global discourse. A key facet of the theory involves the identification of "veins" over discourse structure trees such as those defined in Rhetorical Structure Theory (RST [9]), which delimit domains of referential accessibility for each unit in a discourse.

Cristea et al. [2] represent discourse structures as binary trees, where terminal nodes represent discourse units and non-terminal nodes represent discourse relations. A polarity is established among the children of a relation, which identifies at least one node, the nucleus, considered essential for the writer's purpose; non-nuclear nodes, which include spans of text that increase understanding but are not essential to the writer's purpose are called satellites.

The following notations are used [2]:

- each terminal node (leaf node, discourse unit) has an attached label;
- $mark(x)$ is a function that takes a string of symbols x and returns each symbol in x marked in some way (e.g., with parentheses);
- $simpl(x)$ is a function that eliminates all marked symbols from its argument, if they exist; e.g. $simpl(a(bc)d(e))=ad$;
- $seq(x, y)$ is a sequencing function that takes as input two non-intersecting strings of terminal node labels, x and y, and returns that permutation of x/y (x concatenated with y) that is given by the left to right reading of the sequence of labels in x and y on the terminal frontier of the tree. The function maintains the parentheses, if they exist, and $seq(nil, y)=y$.

Heads
  1. The head of a terminal node is its label.
  2. The head of a non-terminal node is the concatenation of the heads of its nuclear children.

Vein expressions
  1. The vein expression of the root is its head.
  2. For each nuclear node whose parent node has vein v, the vein expression is:
    - if the node has a left non-nuclear sibling with head h, then $seq(mark(h), v)$;
    - otherwise, v.

3. For each non-nuclear node of head h whose parent node has vein v the vein expression is:
- if the node is the left child of its parent, then seq(h, v);
- otherwise, seq(h, simpl(v)).

Note that the computation of heads is bottom-up, while that of veins is top-down.

3 Veins-Based Summarization

Actually, previous researchers have noticed that the nuclei from RST trees may produce the summaries of the text [9][10]. After the introduction of veins theory in the last section, we may see that the veins of a text may reflect the main line of development. It surely covers the main idea of the text. Therefore, summaries may be generated based on the veins of the text. In addition to summarizing entire texts, Veins Theory can be used to summarize a given unit or sub-tree of that text. In effect, we reverse the problem addressed by text summarization efforts so far: instead of attempting to summarize an entire discourse at a given level of detail, we select a single span of text and abstract the minimal text required to understand *this span alone* when considered in the context of the entire discourse. This provides a kind of *focused abstraction,* enabling the extraction of sub-texts from larger documents [2].

The summarization strategy based on Veins Theory can be expressed as follows:

Step 1. POS-tag the text
Step 2. segment the text
Step 3. build the RS trees of each segment
Step 4. compute the heads and veins for all nodes in the tree structure
Step 5. extract summaries according to the depth required

In this strategy, step 2 and step 3 are very critical and need to be given further elaboration. The accuracy of their outputs may strongly influence the analysis of the next steps. On the other hand, there are no consistent and reliable analysis tools for them. For the second step, we may adopt the simplest method, that is, segment the whole text based on paragraph boundaries. Each paragraph will be taken as an independent segment. This is feasible because there are no overlong paragraphs in news discourse. Within a segment, label the units based on punctuations, eliminating those separate constituents. The basic rule for the unit labeling is that there must be at least one verb phrase in it.

For Step 3, we may borrow the method mentioned in Cristea et al. [3].Segment the sentences into *elementary discourse unit*s *(edus)* and then construct *elementary discourse trees (edts)* of each sentence. An *edt* is a discourse tree whose leaf-nodes are the *edus* of one sentence. Sentence-internal cue-words/phrases trigger the constituency of syntactically *edts* from each sentence [11][4]. Since usually, from a given sentence, more than one such tree can be drawn, for each sentence in the original text a set of *edts* is obtained. At this point a process that simulates the human power of incremental discourse processing is started. At any moment in the developing process, say after *n* steps corresponding to the first *n* sentences, a forest of trees is kept, representing the most promising structures built by combining in all possible ways all *edts* of all *n* sentences. Each such tree corresponds to one possible interpretation of the text processed so far. Then, at step *n+1* of the incremental discourse parsing, the following operations are undertaken: first, all *edts* corresponding to the next sentence are integrated in all possible ways onto all the trees of the existing forest; then the resulted trees are scored according to four independent criteria, sorted and filtered so that only a fraction of them is retained (again the most promising after *n+1* steps). From the final wave of trees, obtained after the last step, the highly scored is considered to be the discourse structure [3].

4   Example Analysis

In order to better elaborate how the summarization strategy works, we choose an example to make some analysis. The example is chosen from People's Daily, which is one of the most influential newspapers in mainland China. One segment is taken as analysis data. Firstly, the segment is labeled sentence by sentence as follows [13].

1) 受纽约股市回升的影响，
Influenced by the turning upward of the New York stock market,

2) 伦敦股市后市出现反弹。
The afternoon session of London stock market rebounded.

3) 《金融时报》指数收盘时为5068.8点，
The index of *Financial Times* was 5068.8 points when the market was closed,

4) 较前一交易日低69.5点，
which is 69.5 points lower than the previous exchange day

5) 降幅为1.35%。
The decline rate is 1.35%

6) 12日，法兰克福DAX30种股票指数收在4087.28点，
On 12th, the index of Frankfort DAX30 was 4087.28 points,

7) 全天下跌149.66点，
falling 149.66 point.

8) 跌幅为3.53%。
The falling rate is 3.53%.

9) 巴黎CAC30种股票指数上周连续4个交易日下跌，
The index of Paris CAC30 was keeping falling for 4 days last week.

10) 12日再跌57.27点，
It decreased 57.27 more points on 12th,

11) 收盘时为2862.54点。
closing at 2862.54 points.

(《人民日报》1998年1月14日第7版)
(Page 7, Jan 14, 1998, *People's Daily*)

Then the RS tree structure of the segment is built. Based on the veins theory, heads and veins of each node are computed as the following figure shows.

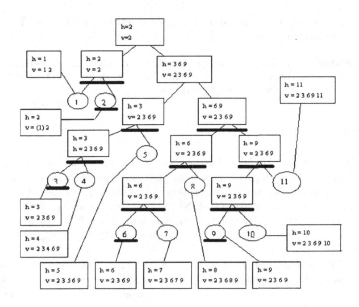

The highest node of the tree shows that "h=2, v=2". Therefore, we can extract "h=2" as the summary of this segment, which is "伦敦股市后市出现反弹 (The afternoon session of London stock market rebounded)". If we want a more specific summary, we may extract the heads from a lower level, that is "h=2" and "h=3 6 9". After putting them together, we can get the following summary: " 伦敦股市后市出现反弹，《金融时报》指数收盘时为5068.8点，12日，法兰克福 DAX30种股票指数收在4087.28点，巴黎CAC30种股票指数上周连续4个交易日下跌 (The afternoon session of London stock market rebounded. The index of *Financial Times* was 5068.8 points when the market was closed. On 12th, the index of Frankfort DAX30 was 4087.28 points. The index of Paris CAC30 was keeping falling for 4 days last week)".

Veins Theory can lead us to the focused abstraction. For example, if we want to get a summary of a certain part of the segment, such as <6 7 8>. As long as we compute its head, which is "h=6", we can achieve our goal.

5 Discussion and Conclusion

The veins-based summarization strategy proposed here has some strong points and weak points. For the strong points, it is a kind of shallow processing, easier to realize and faster to compute. It doesn't rely on semantic analysis. However, it's still not easy to segment discourse and build the RS tree structure. More efforts are needed in these aspects.

In this paper, we are focusing on proposing a summarization strategy. We dedicate our main attention to the theoretical basis and algorithm procedures. Although we demonstrate how the algorithm works with real examples chosen from mainstream newspapers, more empirical work should be done to test its effectiveness. This is our weak point and will be taken as the future work.

Acknowledgments. This study is funded by the Research Foundation of Humanities and Social Sciences from Ministry of Education, China under grant 08JC740001 and the Fundamental Research Funds for the Central Universities.

References

1. Barzilay, R., Elhadad, M.: Using Lexical Chains for Text Summarization. In: Proceedings of the ACL Workshop on Intelligent Scalable Text Summarization (1997)
2. Cristea, D., Ide, N., Romary, L.: Veins theory: A model of global discourse cohesion and coherence. In: Proceedings of COLING/ACL, Montreal, Canada (1998)
3. Cristea, D., Postolache, O., Pistol, I.: Summarisation Through Discourse Structure. In: Gelbukh, A. (ed.) CICLing 2005. LNCS, vol. 3406, pp. 632–644. Springer, Heidelberg (2005)
4. Cristea, D., Postolache, O., Puscasu, G., Ghetu, L.: Local and global information exploited in producing summaries. In: Proceedings of the International Symposium on Reference Resolution and Its Applications to Question Answering and Summarisation, Venice, Italy (2003)
5. Edmunson, H.: New methods in automatic extracting. Journal of the ACM 16(2), 264–285 (1969)
6. Grosz, B.J., Joshi, A.K., Weinstein, S.: Centering: A framework for modelling the local coherence of discourse. Computational Linguistics (1995)
7. Hovy, E., Lin, C.: Automated text summarization in summarist. In: ACL/EACL 1997 Workshop on Intelligent Scalable Text Summarization, Association for Computational Linguistics and the European Chapter of the Association for Computational Linguistics, Madrid, pp. 18–24 (1997)
8. Luhn, H.: The automatic creation of literature abstracts. IBM Journal of Research and Development 2(2) (1958)
9. Mann, W.C., Thompson, S.A.: Rhetorical structure theory: A theory of text organization. Text 8(3), 243–281 (1988)
10. Marcu, D.: The rhetorical parsing, summarisation and generation of natural language texts, Ph.D. thesis, Dept. of Computer Science, University of Toronto (1997)
11. Marcu, D.: The Theory and Practice of Discourse Parsing and Summarization. The MIT Press (2000)
12. McKeown, K., Radev, D.: Generating summaries of multiple news articles. In: Proceedings of 18th Annual International ACM SIGIR Conference on Research and Development in Information Retrieval, pp. 74–82. Special Interest Group on Information Retrieval, Washington (1995)
13. Wang, D.: A review of the veins theory. Modern Foreign Languages (3), 309–316 (2006) (in Chinese)

A Neuron Model Based on Hamilton Principle and Energy Coding

Yan Chuankui

Abstract. The studies on neural network and dynamics analysis are done by lots of researchers while there is few about single neuron. We get a dynamical model based on Hamilton principle from neural physical circuit. The discharge of neuron can be simulated successfully. Furthermore, we discuss the system generalized energy consumption when the neuron is firing. The variety patterns of energy maybe contain some coding about information transmission between neurons.

Keywords: model, Hamilton principle, neuron coding, energy function.

1 Introduction

Nowadays, the research on Computational Neuroscience is mainly about the neural network. The concrete neural network can form some specific behaviors and functions of humans, especially visual, olfactory, auditory and hippocampus network[1-4]. The common method is to extend the single neuron model to a network with coupling of synaptic connection model. As the basis of network research, neuron model shows even greater importance.

McCulloch and Pitts released the first relatively mature computational model of neuron in 1943[5]. Symbolic logic was used to describe neurons, which has brought durative effect on the study and application of neural network model later. A series of neural models have emerged as required. Hopfield's neural model made it possible to provide abundant impetus for the development of artificial neural network[6]. The ion channel model presented by Hodgkin and Huxley showed the properties of giant squid axon membrane[7]. The neural model by Fitzhugh and Nagumo reasonably simulated the physical chaotic phenomena in

Yan Chuankui
Institute for Cognitive Neurodynamics, School of Science,
East China University of Science and Technology, Shanghai, China
Department of Mathematics, School of Science,
Hang Zhou Normal University Hangzhou, China

F.L. Gaol et al. (Eds.): Proc. of the 2011 2nd International Congress CACS, AISC 145, pp. 395–401.
springerlink.com   © Springer-Verlag Berlin Heidelberg 2012

neura potentials and experiments[8,9]. Moreover, a simplified form of biophysical excitable equation was given by Chay[10]. Based on these models above, a majority of studies on network and dynamical analysis[11,12] were made.

The understanding of neural information transmission lies in encoding and decoding. Encoding is the foundation of understanding information. Nowadays, all kinds of neural models basically start from biological ion channels or aim at simulating the accuracy of action potentials. Based on these neural models, traditional coding methods include average frequency coding, population coding, timing coding, pattern coding, etc. However, what we know about brain is very limited. Thus, to develop new coding direction from a new perspective is to be indeed needed.

Since the generation of action potentials is also a mechanical process, we consider the problem from the point of analytical mechanics. A neural model is set up based on Hamilton principle. This may provide a new area for the simulations of neural network and dynamics analysis, even a new idea for neural encoding.

2 Model

Wherever Times is specified, Times Roman or Times New Roman may be used. If neither is available on your word processor, please use the font closest in appearance to Times. Avoid using bit-mapped fonts if possible. True-Type 1 or Open Type fonts are preferred. Please embed symbol fonts, as well, for math, etc.

A system with n particles is subject to whole ideal constraint. Particles position is determined by generalized coordinate $q_1, q_2, \cdots, q_k$ when the system degree of freedom is k. The initial position and terminal position are marked by $A(q_{kA}, t_1), B(q_{kB}, t_2)$. We name a real moving curve AMB as actual path, while we name any moving curve $AM'B$ as possible path subject to whole ideal constraint. The functional $S[q_\alpha] = \int_{t_1}^{t_2} L(q_\alpha, \dot{q}_\alpha t) dt$ is defined as Hamilton action, where $L(q_\alpha, \dot{q}_\alpha t)$ is Lagrange function of system.

Hamilton principle points out that in conservative system, compared with all possible paths with the same time interval, initial position and terminal position, the actual path's Hamilton action has an extreme value, that is, $\delta S[q_\alpha] = 0$, while in a non-conservative system, $\delta S[q_\alpha] = -\int_{t_1}^{t_2} \sum_\alpha (Q_\alpha \delta q_\alpha) dt$; here Q_α represents the non-conservative generalized force of the system.

Generalized coordinates can be general coordinates or other physical quantities to determine the system. Therefore, Hamilton principle is not only adequate for mechanical systems, but also applicable to systems in some other fields. Now the principle is applied to the physical circuit model of neuron.

A neural physical circuit is showed in Fig.1, where U_c is capacity C's electric potential difference representing membrane potential, $q_C, q_{Na}, q_K, q_L, q_S$ are branch electric quantities, U_{Na}, U_K, U_L, U_S are branch electromotive forces, and R_{Na}, R_K, R_L are resistances of sodium, potassium, leakage ion channel.

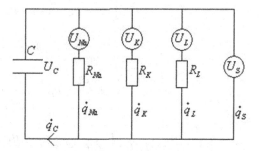

Fig. 1 Neural physical circuit model.

Assuming that q_α is generalized displacement, and that current $I = \dot{q}_\alpha$ is generalized velocity of circuit system, generalized displacement's change is driven by U_α. Therefore, U_α can be equated to generalized active force of system, where their dimension is volt. The resistance R_α is dissipation element. They always consume energy of system when currents pass them. The consumption strength $\dot{q}_\alpha R_\alpha$'s dimension is also volt. Hence, $\dot{q}_\alpha R_\alpha$ are equated to generalized dissipative force or generalized drag. They are non-conservative generalized forces.

From the system, what we know is as follows[13]:

Electric field energy: $W_C = \dfrac{q_C^2}{2C}$

Electromotive force energy:

$$W_U = -q_{Na}U_{Na} - q_K U_K - q_L U_L - q_S U_S$$

Thus, Lagrange function is:

$$L(q_\alpha, \dot{q}_\alpha, t) = -(W_C + W_U)$$

$$= -\frac{q_C^2}{2C} + q_{Na}U_{Na} + q_K U_K + q_L U_L + q_S U_S$$

According to Hamilton principle, the non-conservative generalized force makes the Hamilton action's functional variation of the actual path as follows: $\delta S[q_\alpha] = -\int_{t_1}^{t_2} \sum_\alpha (Q_\alpha \delta q_\alpha) dt$. The problem is to found a generalized displacement $q_\alpha(t)$ to satisfy this condition.

It was assumed that the stimulus current $\dot{q}_S(t)$ and the system constraint $q_C(t) = q_{Na}(t) + q_K(t) + q_L(t) + q_S(t)$ was known, so the system degree of freedom is three, and q_{Na}, q_K, q_L are the free generalized displacements of system.

Then,

$$\delta S[q_\alpha] = \delta \int_{t_1}^{t_2} L(q_{Na}, q_K, q_L, \dot{q}_{Na}, \dot{q}_K, \dot{q}_L, t) dt$$

$$= \int_{t_1}^{t_2} \delta L(q_{Na}, q_K, q_L, \dot{q}_{Na}, \dot{q}_K, \dot{q}_L, t) dt$$

$$= \int_{t_1}^{t_2} \begin{pmatrix} \dfrac{\partial L}{\partial q_{Na}} \delta q_{Na} + \dfrac{\partial L}{\partial q_K} \delta q_K + \dfrac{\partial L}{\partial q_L} \delta q_L \\[2mm] + \dfrac{\partial L}{\partial \dot{q}_{Na}} \delta \dot{q}_{Na} + \dfrac{\partial L}{\partial \dot{q}_K} \delta \dot{q}_K + \dfrac{\partial L}{\partial \dot{q}_L} \delta \dot{q}_L \end{pmatrix} dt$$

And

$$\frac{\partial L}{\partial q_\alpha} \delta q_\alpha + \frac{\partial L}{\partial \dot{q}_\alpha} \delta \dot{q}_\alpha = \left(-\frac{q_{Na} + q_K + q_L + q_S}{C} + U_\alpha \right) \delta q_\alpha$$

So

$$\delta S = \int_{t_1}^{t_2} \left(\sum_{\alpha = Na, K, L} \left(-\frac{q_{Na} + q_K + q_L + q_S}{C} + U_\alpha \right) \delta q_\alpha \right) dt$$

$$= -\int_{t_1}^{t_2} \left(\sum_{\alpha = Na, K, L} Q_\alpha \delta q_\alpha \right) dt$$

Where $Q_\alpha = -R_\alpha \dot{q}_\alpha$

Therefore, $\int_{t_1}^{t_2} \left(\sum_{\alpha = Na, K, L} \left(-\dfrac{q_{Na} + q_K + q_L + q_S}{C} + U_\alpha - R_\alpha \dot{q}_\alpha \right) \delta q_\alpha \right) dt = 0 \cdot$

$-\dfrac{q_{Na} + q_K + q_L + q_S}{C} + U_\alpha - R_\alpha \dot{q}_\alpha = 0$, because the variation δq_α is arbitrary,

We get the following results by derivation of t with the equation.

$$\frac{\dot{q}_{Na} + \dot{q}_K + \dot{q}_L + \dot{q}_S}{C} - \dot{U}_\alpha + R_\alpha \ddot{q}_\alpha = 0$$

where $\alpha = Na, K, L$.

The system is subject to the following constraint conditions

$$s.t. \begin{cases} \dot{U}_{Na} = I\{\dfrac{dU_C}{dt} > 0\} * A_1 \dfrac{1}{\sqrt{2\pi}\sigma_1} \exp(\dfrac{-(U_C - \mu_1)^2}{\sigma_1^2}) + I\{\dfrac{dU_C}{dt} \le 0\} * \sum_i a_i U_C^{\ i} \\[4mm] \dot{U}_K = I\{\dfrac{dU_C}{dt} > 0\} * A_2 \dfrac{1}{\sqrt{2\pi}\sigma_2} \exp(\dfrac{-(U_C - \mu_2)^2}{\sigma_2^2}) + I\{\dfrac{dU_C}{dt} \le 0\} * \sum_i b_i U_C^{\ i} \\[4mm] \dot{U}_L = I\{\dfrac{dU_C}{dt} > 0\} * \sum_i c_{1i} U_C^{\ i} + I\{\dfrac{dU_C}{dt} \le 0\} * \sum_i b_{2i} U_C^{\ i} \\[4mm] U_C = \dfrac{q_C}{C} \\[4mm] q_C = q_{Na} + q_K + q_L + q_S \end{cases}$$

3   Result

These equations of motion can not only simulate experiment results, but also define generalized energy, which H-H model cannot do. Then a new method of Hamilton function can be employed to discuss the information coding contained in potential action. It's the main purpose of the work.

We can get the membrane potential U_c and system Hamilton function

$$H = \sum_{\alpha} \frac{\partial L}{\partial q_{\alpha}} \dot{q}_{\alpha} - L(q_{\alpha}, \dot{q}_{\alpha}, t)$$ by solving the model. As we know, Hamilton func-

tion in mechanics system is mechanical energy. In our dynamical system, it is a generalized energy.

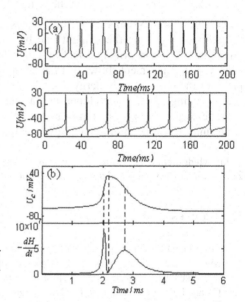

Fig. 2 (a) Discharge patterns under $a_0 = 252.549; 452.549$ (b) A potential action compared with energy variety.

Fig.2(a) reveals that neurons present a great diversity of discharge patterns under different stimulus. It is shown that spiking discharge and bursting discharge of experiments are successfully reproduced with the help of Hamilton principle. By solving the mechanical systems, the similar results to H-H equation and Chay model are achieved, so this method is relatively available. Furthermore, as is illustrated in Fig.2(b), we study the system generalized energy variety when an action potential goes. It is seen that system energy is consuming ceaselessly all the time because the derivative of H is always negative. The energy consumption goes high degree with high potential over threshold when it is low degree with resting potential or under threshold. Actually, although it doesn't consume energy when the ions diffuse by differential concentration, a mass of energy is consumed when the ions pump carry these ions diffusing via cell membrane back to cells inside or outside in order to revert to initial ionic conductance gradient. The whole process is equivalent to these ions diffusing via cell membrane consume energy indirectly. The more ions diffuse via cell membrane and ion channels are opened, the more energy consumed. At the ascend phase of action potential, energy consumption velocity reaches peak value rapidly while sodium ionic channels open ceaselessly and lots of sodium ions enter cell to lead to depolarization; At the peak of action potential, energy consumption velocity goes to a low degree because sodium ionic channels are closed on the whole and potassium ionic channels are not yet open; At the decline phase of action potential, energy consumption velocity gets to another peak slowly while a lot of potassium ionic channels are opened, and a large number of potassium ions depart from cell inside to induce repolarization.

According to the numerical calculation, we can easily get the result $q_\alpha(t) = \lambda q_\alpha(t_A) + (1-\lambda) q_\alpha(t_B)$ by check the data $q_\alpha(t)$. Therefore, the discharge process of neural systems from initial states to terminal states or generalized displacement goes along the straight line between two points. The path best illustrates the minimum principle and system's Hamilton function is its generalized energy. The experiment shows that the brain tries to decrease the size and range of parts for the reduction of energy losses[14]. Human brain has been going towards energy saving, and has developed into a relatively good state[15]. Energy saving neural enlarges the information of neural coding, so neural discharge should be in conformity with the minimum energy principle.

Acknowledgments. The work is Supported by the Youth Cultivating Foundation of Hang Zhou Normal University under Grant No. 2010QN02.

References

1. Lewicki, M.S.: Efficient coding of natural sounds. Nature Neurosci. 5, 356–363 (2002)
2. Mickey, B.J., Middlebrooks, J.C.: Representation of auditory space by cortical neurons in awake cats. J. Neurosci. 23, 8649–8663 (2003)
3. Galan, R.F., Fourcaud-Trocme, N., Ermentrout, G., Urban, N.: Correlation-induced synchronization of oscillations in olfactory bulb neurons. J. Neurosci. 26, 3646–3655 (2006)
4. Yan, C.K., Liu, S.Q.: The transitional function of DG to CA3 on hippocampus. Prog. Nat. Sci. 17, 1436–1444 (2007)
5. McCullock, W.S., Pitts, W.: A logical calculus of ideas immanent in nervous activity. Bulletin of Mathematical Biophysics 133, 115–133 (1943)
6. Hopfield, J.J.: Neural networks and physical systems with emergent collective computational abilities. Proceedings of the National Academy of Science 79, 2554–2558 (1982)
7. Hodgkin, A.L., Huxley, A.F.: A quentitative description of membrane current and its application to conduction and excitation in nerve. J. Physiol. 117, 500–544 (1952)
8. Nagumo, J., Arimoto, S., Yoshizawa, S.: An active pulse transmission line simulating nerve axon. In: Proc. IRE, vol. 50, pp. 2061–2070 (1962)
9. FitzHugh, R.: Mathematical models of excitation and propagation in nerve. In: Schwan, H.P. (ed.) Biological Engineering, pp. 1–85. McGraw-Hill, New York (1969)

10. Chay, T.R.: Chaos in three-variable model of an excitable cell. Physica. D. 16, 233–242 (1985)
11. Wang, Q.Y., Duan, Z.S., Lu, Q.S.: Average synchronzation and temporal order in a noisy neuronal network with couping delay. Chin. Phys. Lett. 24, 2759–2761 (2007)
12. Izhikevich, E.M.: Int. J. Bifurcat. Chaos 10, 117 (2000)
13. Zhao, H.X., Ma, S.J., Shi, Y.: Higher-Order Lagrangian Equations of Higher-Order Motive Mechanical System. Theor. Phys. 49, 479–481 (2008)
14. Attwell, D., Laughlin, S.B.: J. Cerebr. Blood F Met. 21, 1133 (2001); Sarpeshkar, R.: Neural Comput. 10, 1601 (1998)
15. Levy, W.B., Baxter, R.A.: Energy-efficient neural codes. Neural Comp. 8, 531–543 (1996)

Formation of Bonded Exciplex in the Excited States of Dicyanoanthracene-Pyridine System: Time Dependent Density Functional Theory Study

Dani Setiawan, Daniel Sethio, Muhamad Abdulkadir Martoprawiro, and Michael Filatov

Abstract. Strong quenching of fluorescence was recently observed in pyridine solutions of 9,10-dicyanoanthracene chromophore. It was hypothesized that quenching may be attributed to the formation of bound charge transfer complexes in the excited states of the molecules. In this work, using time-dependent density functional calculations, we investigate the possibility of formation of bonded exciplex states between DCA and pyridine molecules. On the basis of theoretical calculations, it is proposed that a partial electron transfer occurs in the lowest excited state of the dicyanoanthracene-pyridine system which leads to the formation of bonded exciplex species and to quenching of fluorescence from dicyanoanthracene.

1 Introduction

Recent studies by Wang and co-workers [19] show a number of interesting phenomena in charge transfer complexes between 9,10-dicyanoanthracene (DCA) with pyridine (Pyr). DCA which is a chromophore widely used in electron transfer photochemical and photophysical studies [2, 7, 9, 12, 13, 17], has long-lived singlet excited state (11-15 ns), high fluorescence quantum yield in many solvents (>0.9) and well-characterized radical anion. It is reported that fluorescence quenching of DCA has been observed due to the presence of pyridine. A careful investigation of the reaction kinetics enabled one to rule out a possible mechanism whereby the

Dani Setiawan · Michael Filatov
Zernike Institute of Advanced Material, University of Groningen, Nijenborgh 4,
Groningen 9741PV, Netherlands
e-mail: D.Setiawan@rug.nl

Daniel Sethio · Muhamad Abdulkadir Martoprawiro
Department of Chemistry, Bandung Institute of Technology, Jalan Ganesha 10,
Bandung 40132, Indonesia

F.L. Gaol et al. (Eds.): Proc. of the 2011 2nd International Congress CACS, AISC 145, pp. 403–409.
springerlink.com   © Springer-Verlag Berlin Heidelberg 2012

quenching occurs due to strong Coulombic stabilization of n-donor radical ion [8]. Indeed the results of the experimental observations [19] could not be explained in terms of simple charge-transfer model [18].

The absorption spectra of DCA in neat pyridine give no indication of formation of a sufficiently stable charge transfer complex in the ground state of the species. Indeed, formation of such a complex would be an endothermic reaction with an enthalpy effect of 0.5-0.8 eV. Because the rate of quenching is sufficiently rapid, it was hypothesized that the reaction proceeds via the formation of bonded exciplex species, that is transient intermediates due to solvent-solute electron transfer in the excited states. This conjecture was indirectly confirmed by the experimental study of the steric and Coulombic effect on the quenching rate constants. Quantum chemical investigation of the ground state potential energy curve of the DCA-Pyr complex using a popular density functional confirmed that its formation is energetically unfavorable. However, these calculations did not provide direct evidence of the formation of a bonded exciplex species. It is therefore the purpose of the present work to study the excited states of DCA-Pyr and to investigate whether the formation of bonded exciplex is possible.

This study is carried out with the use of time-dependent DFT (TD-DFT) which is a computationally inexpensive, though sufficiently accurate method to investigate the excited states. Although TD-DFT is sufficiently accurate for localized valence excitations, it may yield poor results for Rydberg and charge transfer excitations [1, 4, 5, 6, 11]. This is a consequence of the self-interaction error (SIE) which results in the incorrect asymptotic behavior of approximate functionals [3]. SIE is an incomplete compensation of the electron self-repulsion by the approximate exchange-correlation functional. As a result of SIE, the energy of charge-transfer excitation (CT) density may be strongly underestimated and may even fall below the energies of intramolecular excitations (IE) [10, 16].

In the present work, a series of meta-GGA M06 density functionals developed recently by Zhao and Truhlar [20, 21, 22] are applied to the investigation of the excited states of DCA-Pyr complex. Indeed, the M06-HF, M06-2X, M06 and M06-L density functionals have been extensively parametrized to improve the accuracy of description of the ground as well as the excited states of organic and inorganic compounds. It is noteworthy that with the use of meta-GGA functionals TD-DFT is capable of yielding excitation energies in good agreement with the high-level *ab initio* calculations.

2   Method of Calculation

M06-HF, M06-2X, M06 and M06-L meta-GGA density functionals are used as implemented in the Jaguar package version 7.5 [14]. The geometry optimization used the 6-31G* basis set and the TD-DFT calculations employed the 6-31+G** basis set. A relaxed scan of the potential energy surfaces of the ground and the lowest excited states was carried out by optimizing all internal coordinates of DCA and pyridine while keeping the CX-N distance fixed.

To evaluate the accuracy of TD-DFT, the asymptotic value of the CT excitation energy was estimated using the ΔSCF approach:

$$E_{\Delta SCF} = (E_{D^+} - E_{D^0}) - (E_{A^0} - E_{A^-}) \qquad (1)$$

where D and A are the donor and the acceptor molecules respectively.

The geometry of DCA-Pyr complex was set up as shown in Figure 1. The geometry was optimized for the ground state of the DCA-Pyr species using M06/6-31G* method. The vertical excitation energies were calculated using the so obtained geometries and the M06, M06-2X, or M06-HF functionals in connection with 6-31+G** basis set.

3   Results and Discussion

The numeric results obtained with the M06, M06-2X and M06-HF functionals for the C9, C2 and C1 DCA-Pyr complexes are collected in Table 1. The data in Table 1 show that there is only inessential dependence of the obtained CT excitation energies on the position at which pyridine approaches DCA (C9, C2, or C1) and, especially at short distances, on the choice of the density functional. Although at longer distances the M06-HF functional yields excitation energies in better agreement with the ΔSCF excitation energies, the difference between the excitation energies obtained with different functionals reduces at shorter CX-N separations.

The potential energy curves of the ground and 10 lowest excited states of the C9 DCA-Pyr complex obtained with the use of M06 functional are shown in Figure 3. Analysis of the orbital excitations contributing to the S_1 state reveals that at longer distances it has a character of local intramolecular excitation (IE) of DCA. The energy of this excitation gradually increases as the pyridine molecule approaches DCA and at a CX-N distance of approximately 2.5 Å, it crosses a state with CT

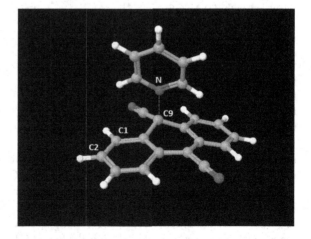

Fig. 1 Complex between DCA and Pyr: nitrogen atoms are shown in blue, carbons in grey and hydrogens in white. During the relaxed scan, the distance between CX (X=9, 2, or 1) and N atoms was kept fixed while optimizing all other geometric parameters. The following discrete values of the CX-N distance were used in the scan: 20.0 Å, 4.2 Å, 3.9 Å, 3.6 Å, 3.3 Å, 3.0 Å, 2.7 Å, 2.3 Å, 2.0 Å, 1.7 Å, 1.6 Å, 1.5 Å and 1.4 Å.

Table 1 CT excitation energies (eV) of different conformation of DCA-Pyr, calculated with 6-31+G** basis sets. The blue numbers indicate energies of reversed CT states $\pi_{DCA} \to \pi^*_{Pyr}$.

Methods	Conf.	1.4Å	1.6Å	1.7Å	2.0Å	2.3Å	2.7Å	3.0Å	3.3Å	3.6Å	4.2Å	20Å
	C1	1.91	2.06	2.14	2.62	3.91	4.72	4.80	7.15	7.11	7.27	8.13
M06-HF	C2	1.81	2.27	2.46	3.57	5.65	6.89	6.79	7.53	7.70	7.77	8.17
	C9	2.06	2.46	2.64	2.84	3.06	4.82	4.85	4.86	4.85	7.12	8.14
	C1	1.04	1.52	1.89	2.50	4.07	4.28	4.46	4.98	4.93	4.97	5.88
M06-2X	C2	0.93	1.62	2.03	2.90	3.79	4.53	4.77	5.05	5.39	5.39	5.90
	C9	1.44	1.72	2.01	2.99	3.24	4.50	4.53	4.54	4.55	4.95	5.88
	C1	0.54	0.97	1.46	2.23	3.62	3.56	3.64	3.74	3.68	3.62	4.06
M06	C2	0.39	1.07	1.49	2.68	3.53	3.66	3.60	3.78	4.00	3.95	4.07
	C9	1.31	1.46	1.68	2.75	3.09	3.75	3.69	3.69	3.71	3.71	4.06

Fig. 2 Occupied molecular orbitals of DCA-Pyr: a) π_{Pyr}, b) π_{DCA}, c) n_{Pyr} and unoccupied molecular orbitals: d) n^*_{Pyr}, e) π^*_{Pyr}, f) π^*_{DCA}. There are four most probable CT excitations: 1) CT from $n_{Pyr} \to \pi^*_{DCA}$, 2) CT from $\pi_{Pyr} \to \pi^*_{DCA}$, 3) Reversed CT from $\pi_{DCA} \to \pi^*_{Pyr}$ and 4) Reversed CT from $\pi_{DCA} \to n^*_{Pyr}$.

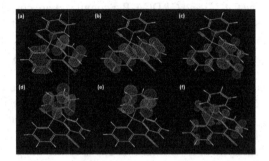

Fig. 3 Potential energy surface of ground and excited states in DCA-Pyr system at C9, along R<4.5 Å, calculated with TD-M06/6-31+G**. A local minimum is present in the lowest excited state at R~1.5 Å. The steric repulsion between DCA and Pyr increases at R<1.5 Å which is reflected in the potential energy curve of the S_1 state.

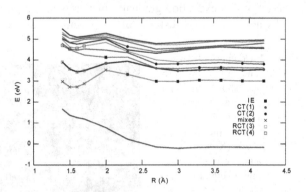

character. Due to avoided crossing of these states, a low potential barrier is formed on the S_1 curve and the S_1 state acquires a substantial CT character.

The further analysis of the lowest excited states from TD-M06/6-31+G** and TD-M06-HF/6-31+G** using amplitudes of orbital excitations revealed that near $R_{CX-N} \sim 2.5$ Å the intramolecular $\pi_{DCA} \to \pi^*_{DCA}$ state experiences avoided crossing with the $n_{Pyr} \to \pi^*_{DCA}$ state, which is the lowest CT state at longer distances. At distances shorter than 2.0 Å the reversed CT state formed by the $\pi_{DCA} \rightarrow \pi^*_{Pyr}$ orbital transition approaches the lowest excited state and forms another avoided crossing with this state. As a result of the mixing between these excited states, the lowest excited state near the minimum can be best described as a superposition of the intramolecular excitations on DCA and on pyridine and of the CT excited states, as shown in Eq. 2.

$$\Psi_{Bond.Exc.} = c_1 \Psi_{(DCA^--Pyr^+)} + c_2 \Psi_{(DCA^*\cdots Pyr)} + c_3 \Psi_{(DCA\cdots Pyr^*)} + c_4 \Psi_{(DCA\cdots Pyr)} \quad (2)$$

Thus the character of the lowest excited state of DCA-Pyr complex changes from pure intramolecular excited state at long distances (>2.5 Å) to CT excited state at intermediate distances ($\sim$2.0-1.5 Å) to reversed CT state at short distances ($\sim$1.5 Å). Besides confirming the existence of bonded exciplex in the DCA-Pyr system, this result also suggests that a transfer of excitation from DCA to pyridine may occur as a consequence of sequential double electron transfer, as shown in Figure 4.

It is also noteworthy that, at CX-N distances shorter than 2.5 Å, a substantial electron transfer occurs in the ground state of DCA-Pyr complex. This is illustrated in Figure 5 where the results of NBO analysis of the ground state density matrix of DCA-Pyr are shown. Thus, because the ground and the lowest excited states of DCA-Pyr both have a charge transfer character, the oscillator strength of the $S_0 \leftarrow S_1$ transitions decreases considerably (from $f \sim 0.17$-0.25 at $R_{CX-N} = 3.0$ Å to $f \sim 0.00$-0.09 at $R_{CX-N} \leq 1.7$ Å). Taken together with the possibility of excitation transfer from DCA to pyridine, this decrease of the oscillator strength enables one to explain the experimentally observed quenching of fluorescence in DCA-Pyr system.

Fig. 4 Possible schema of electronic transition on the excited states of DCA-Pyr. Intramolecular excitation in DCA (1) might be followed by double electron transfer: from DCA* to Pyr* (2) and from Pyr to DCA (3).

Fig. 5 Natural bond orbital analysis (NBO) of DCA-Pyr system at C9 of DCA, calculated with 6-31+G** basis sets. At distance which correspond to the minimum on the lowest excited states curve, more than half of electron is transferred from pyridine to DCA.

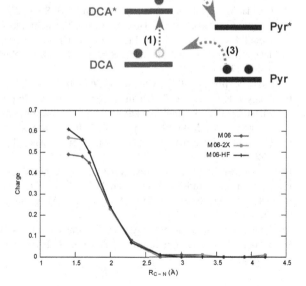

4 Conclusions

Theoretical investigation of the excited states of DCA-Pyr complex was carried out with the use of TDDFT formalism in connection with the M06, M06-2X and M06-HF density functionals suggests that formation of a bonded exciplex occurs in the lowest excited state of this species. The character of the lowest excited state of DCA-Pyr changes from an intramolecular $\pi_{DCA} \rightarrow \pi^*_{DCA}$ excitation at larger distances (>2.5 Å) to CT $n_{Pyr} \rightarrow \pi^*_{DCA}$ at intermediate distances (~ 2.0-1.5 Å) to reversed CT $\pi_{DCA} \rightarrow \pi^*_{Pyr}$ at short distances (~ 1.5 Å). Because near the minimum on the lowest excited state potential energy curve, the radiative transition probability is considerably reduced, as compared to longer distance the formation of bonded exciplex should result in substantial quenching of fluorescence from DCA. Furthermore, as a result of varying character of the lowest excited state of DCA-Pyr, a transfer of excitation from DCA to pyridine may take place in the lowest excited state which should lead to further reduction of fluorescence. Thus, the results of theoretical investigation obtained in the present work provide additional evidence to the hypothesis of formation of bonded in DCA-Pyr system proposed in Ref. [18].

References

1. Appel, F., Gross, E.K.U., Burke, K.: Phys. Rev. Lett. 90, 43005 (2003)
2. Baciocchi, E., Crescenzib, C., Lanzalunga, O.: Tetrahedron 53, 4469 (1997)
3. Casida, M.E.: Time-dependent density functional response theory for molecules. In: Chong, D.P. (ed.) Recent Advances in Density Functional Methods, vol. 1, p. 155. World Scientific, Singapore (1995)
4. Casida, M.E., Gutierrez, F., Guan, J., Gadea, F.X., Salahub, D.R., Daudley, J.P.: Phys. Chem. 113, 7062 (2000)
5. Casida, M.E., Salahub, D.R.: Phys. Chem. 113, 8918 (2000)
6. Dreuw, A., Weisman, J.L., Head-Gordon, M.: Chem. Phys. 119, 2943 (2003)
7. Gould, I.R., Ege, D., Moser, J.E., Farid, S.: Chem. Soc. 112, 4290 (1990)
8. Jacques, P., Haselbach, E., Henseler, A., Pilloud, D., Suppan, P.: Chem. Soc. Faraday Trans. 87, 3811 (1991)
9. Kizu, N., Itoh, M.: Phys. Chem. 96, 5796 (1992)
10. Magyar, R.J., Tretiak, S.: Chem. Theory. Comp. 3, 976 (2008)
11. Matsuzawa, N., Ishitani, A., Dixon, D.A., Uda, T.: Phys. Chem. 105, 4953 (2001)
12. Nakamura, M., Miki, M., Majima, T.: Bull. Chem. Soc. Jp. 72, 2103 (1999)
13. Okamoto, M., Yamada, K., Nagashima, H., Tanaka, F.: Chem. Phys. Lett. 342, 578 (2001)
14. Jaguar, ver 7.5, p. 5. Schrödinger, LLC, New York, NY (2008)
15. Sension, R.J., Hudson, B.S.: Chem. Phys. 90, 1377 (1989)
16. Setiawan, D., Kazaryan, A., Martoprawiro, M.A., Filatov, M.: Phys. Chem. Chem. Phys. 12, 11238 (2010)
17. Vauthey, E., Högemann, C., Allonas, X.: Phys. Chem. A 102, 7362 (1998)
18. Wang, Y., Haze, O., Dinnocenzo, J.P., Farid, S., Farid, R.S., Gould, I.R.: Org. Chem. 72, 6970 (2007)

19. Wang, Y., Haze, O., Dinnocenzo, J.P., Farid, S., Farid, R.S., Gould, I.R.: Phys. Chem. A 112, 13088 (2008)
20. Zhao, Y., Truhlar, D.G., Phys, J.: Phys. Chem. A 110, 13126 (2006)
21. Zhao, Y., Truhlar, D.G.: Chem. Phys. 125, 194101 (2006)
22. Zhao, Y., Truhlar, D.G.: Theor. Chem. Acc. 120, 215 (2008)

Simulation of a Cockpit Display System under the Sunlight

Wei Heng-yang, Zhuang Da-min, and Wan-yan Xiao-ru

Abstract. More and more visual information is being displayed by the aircraft Cockpit Display System (CDS) nowadays. In order to keep flight safety and avoid tiredness, the design of CDS must meet every light condition during the flight to ensure the readability of the information. As the main factor of ambient light in the cockpit, sunlight may heavily affect the visual ergonomy of CDS. Therefore, simulating the CDS under the sunlight within the design phase is required. As an example, depending on different sunlight incident angle and front sunshade layout, varied affections were simulated by the SPEOS simulation software from OPTIS. The results directly showed the effects of sunlight on the CDS and tested the design of front sunshade, which could be of value in evaluating the CDS.

1 Introduction

Aircraft cockpit display system (CDS) is the most important man-machine interface in the cockpit, and it is crucial to make the design of CDS be capable of displaying information clearly and effectively. In order to meet the requirements of increasing flight safety and decreasing flying fatigue, the design of such complicated man-machine system has to satisfy every light condition during the flight [1].However, in the traditional design cycle of aircraft, this has to be done with experience and evaluated up to the prototype test. Once the design does not pass the test, the designers have to go back to the design phase and test it again, costing lots of time and money. For the above reason, the technology of test and evaluation of the CDS in the design phase (test-on-design) is quite necessary.

Wei Heng-yang · Zhuang Da-min · Wan-yan Xiao-ru
School of Aeronautics Science and Engineering, Beihang University, Beijing, China
e-mail: keanu131@163.com, dmzhuang@buaa.edu.cn
    wwwchuan2003@yahoo.com.cn

F.L. Gaol et al. (Eds.): Proc. of the 2011 2nd International Congress CACS, AISC 145, pp. 411–416.
springerlink.com   © Springer-Verlag Berlin Heidelberg 2012

Some researches have been carried out on this topic by scholars. One idea is to build an ambient light simulator to test the cockpit interior and display system. [2] shows that the NASA had built a flexible cockpit sunlight simulator in 1986, and this method has been further developed nowadays, as in [3] and [4]. However, these simulators cost a lot and cannot completely meet the needs of test-on-design. The simulator built above can be used to test the real aircraft or the prototype easily; however, it can do nothing to the design on the blueprint. Another idea was introduced by [5], where the whole cockpit ambient light was simulated with computer software. Because of its combination with the computer aided design (CAD) software (e.g. CATIA from Dassault), test-on-design became possible and the cost was significantly decreased. Furthermore, the simulation results can be quantized by the parameters of luminance and illumination, resulting in convenient test and evaluation. In this way, such method gave a new solution to the design cycle of aircraft and was used in the present study for investigating the impact of ambient light on CDS.

2   Key Factor Definition

While evaluating the ambient light of the cockpit, the sunlight has always been the most important factor. This could be inferred from the luminance and illuminance levels. The brightness of common-used display in the commercial airplane is a little more than 500 cd/m^2 at its highest level. The GJB5187-2003 suggested a luminance of 6,852 cd/m^2 for cockpit sunlight evaluation, which is still at the middle level. The maximum horizontal diffuse illuminance produced by the West-European sky is in the order of 56,000 lx (Hunt, 1979), and the maximum illuminance produced by the direct sunlight can be as high as 110,000 lx (Hunt, 1979) [5].

The sunlight may come from every direction during the flight, and the direction may change from time to time. To reduce the disadvantages of sunlight, relative technologies have been developed and applied into engineering practice. For example, special windshield glass can keep out 30% of the sunlight, proper designed front sunshade and displays can greatly diminish incident and reflect rays, and reasonable layout of the cockpit can also be of help. Considering all the above factors, the analysis will be very complicated. However, using the simulation method, all the factors could be considered and simulated together in an integrated system. In our study, a CDS under the sunlight with respect to the factors of sunlight direction and front sunshade layout was simulated as an example.

3   Simulation Setup

The simulation was preceded with the help of CATIA from the Dassault System and SPEOS CAA from the OPTIS.

3.1   Cockpit Model Setup

Based on a newly developed feeder
liner of China, its digit model of the
cockpit was built up with CATIA by
reverse engineering. The model had
been calibrated to control the error below 1mm, as shown in figure 1.

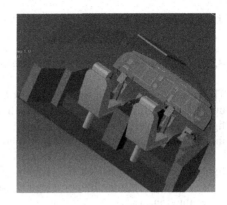

3.2   Ambient Light Setup

According to 3.8.3.1 of the General
specification for airborne active matrix
liquid crystal displays (GJB 5187-

Fig. 1 Digital cockpit model

2003), the ambient light of the cockpit was set to 21,528 lx (2,000 fc) as the diffused
illuminance and 6,852 cd/m^2 (2,000 fl) as the sunlight luminance. The directions of
the sunlight included four conditions, i.e. vertex, right side, ahead and no sunlight,
which are common senses during the flight. In addition, the brightness of the display
was set to 400 cd/m^2, which is at the common level of commercial airplanes.

3.3   Simulation Algorithm

The algorithm can be represented by figure 2. The rays emitted by the light
sources (e.g. ambient light, displays, and lighting systems) are propagated in the
built 3D model. With geometrical optics and ray-tracing, the part of the light
which enters observer's view generates a spectral luminance map. Then the map is
transformed through a hu-
man eye model, which can
model the eye response
(spectral/spatial) to the energetic image. After that, a detection box analyzes and
compares the results with
standards and physiological
parameters of the human vision, and the final results are
displayed as hyper-realistic
energetic image, informing
the actual color seen by the
pilot, and the seen/non seen
information. [5]

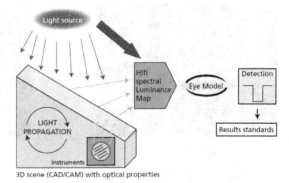

Fig. 2 Simulation algorithm

4　Result and Discussion

4.1　Different Sunlight Directions

The simulation results of the sunlight from different directions can be seen in figure 3. Under different incident angles, the effects were of great differences: sunlight from the vertex and ahead decreased the contrast of the display and made the information unreadable. However, if there was no sunlight, the normal diffused environment would make the CDS perfectly perceived. It is clearly that the sunlight has a great impact on the readability of the CDS. Therefore, in the design process of CDS, it may be not adequate to take tests under certain circumstances for evaluation.

(a) Sunlight from vertex

(b) Sunlight from right side

(c) Sunlight from ahead

(d) No sunlight

Fig. 3 Sunlight from different directions

4.2　Different length of Front Sunshade

The simulation results of front sunshade with different lengths can be seen in figure 4. If no front sunshade was installed, the display would expose to the sunlight, resulting in the unacceptable visual ergonomy. When the sunshade became longer, it offered a better shield of sunshine, which was positive for the display of the information. That's why front sunshade was found in nearly all of modern airplanes. However, as a long sunshade may obscure pilot's view field, such design should be weighted carefully.

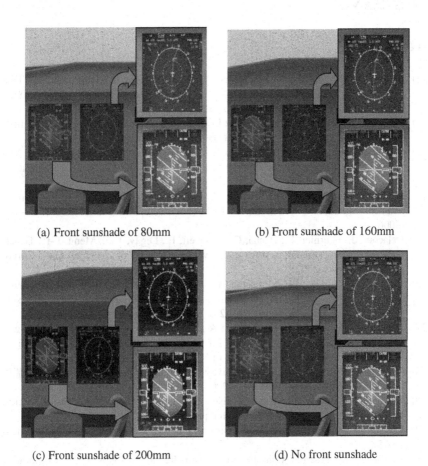

(a) Front sunshade of 80mm   (b) Front sunshade of 160mm

(c) Front sunshade of 200mm   (d) No front sunshade

Fig. 4 Front sunshade of different length

5 Conclusion

The approach adopted in the study offers a new way of achieving visual ergonomy in the cockpit, which is of value of in the test and evaluation the design of CDS under ambient light. Though the factors that affect the readability of CDS are complicated, the simulation method can take them into consideration and display the finally results. In the present example, the application of the approach was il-

luminated, and the results obtained are consistent with the fact. Other objects of
the CDS may join to the test as well, which together would be the evaluation of
the whole system.

The research is supported by National Basic Research Program of China (Program Grant No. 2010CB734104). Thanks to the OPTIS for the evaluation copy of
SPEOS CAA V4.0.

References

[1] DOT/FAA/AM-01/17. Human Factors Design Guidelines for Multifunctional Displays   (2001),
    http://www.hf.faa.gov/docs/508/docs/cami/0117.pdf
[2] Tarrant, A.: A Feasibility Study of the Flexible Simulation of Sunlight (cockpit simulators) (Final Report) NTIS N8722693 (1986)
[3] Duranti, Pierluigi, Barneschi, Angiolo: System for On-Ground Simulation of the Ambient Light Conditions in the Cockpit of an Aircraft during Flight. European Patent,
    EP2125520 (2010)
[4] Ma, R., Ma, C.-B., Song, D., Zhang, J.-S.: Design and Implementation of Electron Instrument Simulation System for Civil Aircraft. Computer Simulation 12, 62– 64
    (2006)
[5] Delacour, J., Fournier, L., Hasna, G., Humbert, E., Legay, J.-F., Menu, J.-P.: Development of a new simulation software for visual ergonomy of a cockpit. Optis Corporation (November 2002)
[6] Delacour, J., Cuinier, J.-L.: Presentation of the first PLM integrated Optical Simulation Software for the Design and Engineering of Optical Systems. In: Optical Design
    and Engineering, Proc. of SPIE, vol. 5249, pp. 42–53
[7] Delacour, J., Fournier, L., Menu, J.-P.: A global simulation approach to optics, lighting, rendering, and human perception for the improvement of safety in automobiles.
    In: Photonics in the Automobile, Proc. of SPIE, vol. 5663, pp. 65–75
[8] General specification for airborne active matrix liquid crystal displays (机载有源矩阵液晶显示器通用规范) GJB 5187–2003 (2003)
[9] Zhang, C., Zhang, W.: Research of Visual Simulation For Decoration Inside Aircraft
    Cockpit. Aeronautical Computing Technique 1, 89 – 92 (2009)
[10] Federal Aviation Administration. Instrument Flying Handbook. FAA-H-8083-15A
     (2008)
[11] Visual ergonomics and visibility analysis in digital mock-ups. Optis Corporation
     (2002)
[12] Light Modeling - Reference Manual - SPEOS CAA V5 Based - V4.0 - v.01. Optis
     Corporation (2002)

Dynamic Analysis of Nonlinear Elasticity Microbeam with Electromechanical Couping[*]

Yang Liu, Peng Jian-she, Xie Gang, and Luo Guang-bing

Abstract. The material nonlinearity is one of the many nonlinear factors in MEMS(Micro Electro Mechanical System). The microbeam is a basic component in MEMS. In this paper, the influence of nonlinear elasticity factor has been considered in dynamic analysis of the microbeam. A nonlinear modal is set up that as a clamped-clamped microbeam subjected to a transverse electrostatic force. The nonlinear governing equation is transformed into a linear differential equation system through the use of Linstedt-Poincaré perturbation method, which are then solved by using the Galerkin method. Numerical results show that, the nonlinear factor can't be ignored when the number of the nonlinear material constant, B is big. The amplitude and the period of the microbeam will increase when the number of B is increasing.

Keywords:-MEMS, microbeam, nonlinear elasticity, dynamic analysis, Linstedt-Poincaré method, Galerkin method.

1 Introduction

In recent years, various nonlinear behaviors in MEMS have been researched by people. The response of the microbeam subjected to nonlinear electrostatic force

Yang Liu
Department of Physics and Electronic Information, China West Normal University
Nan Chong, China
e-mail: 184312526@qq.com

Peng Jian-she
Department of Industrial Manufacturing, Cheng Du University
Cheng Du, China,
e-mail: pengjianshe2005@163.com

Luo Guang-bing
Traction Power State Key Laboratory of Southwest Jiao tong University

* This work was supported by the applied basic research foundation of Science and Technology Department of Sichuan Province under contract 2011JY0076.

have been researched through the use of theory methods and experiment methods by Zook, Ahn, Hung et al.[1,2,3]. Chu et al.[4] analysed the dynamic response of the micro actuator with nonlinear electrostatic force and nonlinear squeeze film damping by using the modal of equivalent single-degree-of-freedom. Xu, L et al.[5] researched the nonlinear dynamic response of the microbeam with Electromechanical Coupling. Y.M.Fu et al.[6] researched the nonlinear static and dynamic response of a viscoelastic microbeam. But, there is no paper for materials nonlinearity. However, with developing of the MEMS, the choice of materials will more and more diverse, and the nonlinear elasticity factor exist in many materials. So, It is essential to research the influence of the nonlinear elasticity material on the MEMS.

In this paper, a nonlinear dynamic response of little deflection modal, based on the classic elasticity theory, considering the nonlinear elasticity material, is set up that as a clamped-clamped microbeam subjected to electrostatic force. Using the Linstedt-Poincaré method and Galerkin method solved the nonlinear governing equations. The conclusions can provide useful references for related personnel.

2   The Governing Equation of the Microbeam

For an isotropic, homogeneous slender microbeam with considering transverse vibration only. Suppose $\overline{w}(\overline{x},\overline{t})$ be the transverse displacement in the $\overline{x}$ position and in the $\overline{t}$ moment; $p(\overline{t})$ be the load per unit length on the microbeam; Considering the nonlinear elasticity material, the unit stress-strain relationship is[7]

$$\sigma = E(\chi - B\chi^3) \tag{1}$$

where σ is the stress; E is the Young's modulus; χ is the strain; B is the nonlinear material constant (the value is positive).

The dimensionless dynamics equation of the microbeam is given as follows when the edge effect is neglected

$$\frac{\partial^4 w}{\partial x^4} - \beta\left[\left(\frac{\partial^2 w}{\partial x^2}\right)^2\left(\frac{\partial^4 w}{\partial x^4}\right) + 2\left(\frac{\partial^2 w}{\partial x^2}\right)\left(\frac{\partial^3 w}{\partial x^3}\right)^2\right] + \Omega^2\frac{\partial^2 w}{\partial t^2} = \frac{\alpha}{(1-w)^2} \tag{2}$$

where $x = \dfrac{\overline{x}}{L}, t = \omega\overline{t}, \Omega = \dfrac{\omega}{\omega_L}, \omega_L = \dfrac{1}{L^2}\sqrt{\dfrac{EI}{\rho A}}, w = \dfrac{\overline{w}}{g}, \beta = \dfrac{3\overline{B}g^2}{L^4}, \overline{B} = \dfrac{3h^2}{20}B, \alpha = \varepsilon V^2, \varepsilon = \dfrac{6\varepsilon_0 L^4}{Eh^3 g^3}$, L is the length; I is the second moment of inertia; ρ is the mass density; A is the cross-sectional area; g is the Initial gap between the substrate and microbeam; h is the thickness; b is the width; V is the voltage; ε_0 is the vacuum dielectric constant.

The excitation force is a function of the lateral deflection of the microbeam, and continues to act on it whenever the voltage applies, so the system is self-excited since the motion itself produces the excitation force. Note that $w \ll 1$, we can approximate the electrostatic force in Taylor expansion: $1/(1-w)^2 \approx 1 + 2w + 3w^2 \cdot$

So, (2) can be rewritten as

$$\frac{\partial^4 w}{\partial x^4} - \beta\left[\left(\frac{\partial^2 w}{\partial x^2}\right)^2\left(\frac{\partial^4 w}{\partial x^4}\right) + 2\left(\frac{\partial^2 w}{\partial x^2}\right)\left(\frac{\partial^3 w}{\partial x^3}\right)^2\right] + \Omega^2\frac{\partial^2 w}{\partial t^2} = \varepsilon V^2(1 + 2w + 3w^2). \tag{3}$$

3 The Perturbation Equations of the Microbeam

Using Linz Ted-Poincaré method, both the dimensionless deflection w and the dimensionless frequency parameter Ω^2 can be expanded into the series forms in terms of the small perturbation parameter ε.[8]

$$w = \varepsilon w_1 + \varepsilon^3 w_3 + \cdots \tag{4}$$

$$\Omega^2 = 1 + \varepsilon^2\sigma_2 + \cdots \tag{5}$$

Substituting (4) and (5) into (3), so (3) becomes

$$\varepsilon^1 : \frac{\partial^4 w_1}{\partial x^4} + \frac{\partial^2 w_1}{\partial t^2} = V^2 \tag{6}$$

$$\varepsilon^2 : 2V^2 w_1 = 0 \tag{7}$$

$$\varepsilon^3 : \frac{\partial^4 w_3}{\partial x^4} + \frac{\partial^2 w_3}{\partial t^2} = \beta\left[\left(\frac{\partial^2 w_1}{\partial x^2}\right)^2\left(\frac{\partial^4 w_1}{\partial x^4}\right) + 2\left(\frac{\partial^2 w_1}{\partial x^2}\right)\left(\frac{\partial^3 w_1}{\partial x^3}\right)^2\right] - \sigma_2\frac{\partial^2 w_1}{\partial t^2} + 3w_1^2 V^2. \tag{8}$$

According to the requirements of precision, we can intercepted any number order equations. Now, we intercepted front three order equations. In these equations, there are two results to w_1. But the value of w_1 is zero all the time in the second-order equation, and it should be eliminated. So we consider the first-order and third-order equations only.

4 Solving Equations by Galerkin Method

The first-order perturbation equation is

$$\frac{\partial^4 w_1}{\partial x^4} + \frac{\partial^2 w_1}{\partial t^2} = V^2 \tag{9}$$

For the clamped-clamped microbeam, the trial function is

$$w_{1m} = T_{1m}(t)w_{1m}^* , w_{1m}^* = 1 - \cos 2m\pi x, \tag{10}$$

Substituting (10) into (9) and using Galerkin method, (9) can be rewritten as

$$\int_0^1\left(\frac{\partial^4 w_{1m}}{\partial x^4} + \frac{\partial^2 w_{1m}}{\partial t^2} - V^2\right)w_{1m}^* dx = 0 \tag{11}$$

The simplify equation of (11) is

$$\frac{d^2 T_{1m}}{dt^2} + \omega_m^2 T_{1m} = f_1 \tag{12}$$

where $\omega_m^2 = \frac{(2m\pi)^4}{3}, f_1 = \frac{2V^2}{3}$.

The initial conditions of (12) are

$$T_{1m}(0) = 0 , T_{1m}'(0) = 0 . \tag{13}$$

Take Laplace transform to (12), there has

$$T_{1m}(t) = \frac{f_1}{\omega_m^2}(1 - \cos \omega_m t) \tag{14}$$

Substituting (14) into (10), there has

$$w_{1m} = \left[\frac{f_1}{\omega_m^2}(1 - \cos \omega_m t)\right](1 - \cos 2m\pi x) \tag{15}$$

The third-order perturbation equation is

$$\frac{\partial^4 w_3}{\partial x^4} + \frac{\partial^2 w_3}{\partial t^2} = f_3(x,t) \tag{16}$$

Where
$$f_3(x,t) = \beta\left[\left(\frac{\partial^2 w_1}{\partial x^2}\right)^2\left(\frac{\partial^4 w_1}{\partial x^4}\right) + 2\left(\frac{\partial^2 w_1}{\partial x^2}\right)\left(\frac{\partial^3 w_1}{\partial x^3}\right)^2\right] - \sigma_2 \frac{\partial^2 w_1}{\partial t^2} + 3w_1^2 V^2.$$

For the clamped-clamped microbeam, the trial function is

$$w_{3m} = T_{3m}(t)w_{3m}^* , w_{3m}^* = w_{1m}^* = 1 - \cos 2m\pi x . \tag{17}$$

Substituting (10) and (17) into (16), and using Galerkin method, (16) can be rewritten as

$$\int_0^1 \left\{\frac{\partial^4 w_{3m}}{\partial x^4} + \frac{\partial^2 w_{3m}}{\partial t^2} - \beta\left[\left(\frac{\partial^2 w_{1m}}{\partial x^2}\right)^2\left(\frac{\partial^4 w_{1m}}{\partial x^4}\right) + 2\left(\frac{\partial^2 w_{1m}}{\partial x^2}\right)\left(\frac{\partial^3 w_{1m}}{\partial x^3}\right)^2\right] + \sigma_{2m}\frac{\partial^2 w_{1m}}{\partial t^2} - 3w_{1m}^2 V^2\right\}w_{3m}^* dx = 0. \tag{18}$$

The simplify equation of (18) is

$$\frac{d^2 T_{3m}}{dt^2} + \omega_m^2 T_{3m} = s_m T_{1m}^3 - \sigma_{2m}\frac{d^2 T_{1m}}{dt^2} + 5V^2 T_{1m}^2 \tag{19}$$

$$= s_m \left(\frac{f_1}{\omega_m^2}\right)^3\left(-\frac{1}{4}\cos 3\omega_m t + \frac{3}{2}\cos 2\omega_m t - \frac{15}{4}\cos \omega_m t + \frac{5}{2}\right) - f_1 \sigma_{2m}\cos \omega_m t + 5V^2\left(\frac{f_1}{\omega_m^2}\right)^2\left(\frac{1}{2}\cos 2\omega_m t - 2\cos \omega_m t + \frac{3}{2}\right)$$

where $\omega_m^2 = \frac{(2m\pi)^4}{3}, s_m = \frac{(2m\pi)^8 \beta}{12}$.

To remove the secular terms in (19), the sum of the coefficients related to $\cos \omega_m t$ must be zero, which gives the expressions of σ_{2m} as below

$$\sigma_{2m} = -\frac{15 s_m f_1^2}{4\omega_m^6} - \frac{10V^2 f_1}{\omega_m^4} \tag{20}$$

So, without secular terms, (19) can be rewritten as

$$\frac{d^2 T_{3m}}{dt^2} + \omega_m^2 T_{3m} = s_m \left(\frac{f_1}{\omega_m^2}\right)^3 \left(-\frac{1}{4}\cos 3\omega_m t + \frac{3}{2}\cos 2\omega_m t + \frac{5}{2}\right) + 5V^2 \left(\frac{f_1}{\omega_m^2}\right)^2 \left(\frac{1}{2}\cos 2\omega_m t + \frac{3}{2}\right) \qquad (21)$$

The initial conditions of (21) are

$$T_{3m}(0) = 0 , T_{3m}{}'(0) = 0 , \qquad (22)$$

Take Laplace transform to (21), there has

$$T_{3m}(t) = \frac{s_m f_1^3}{\omega_m^8}\left(\frac{1}{32}\cos 3\omega_m t - \frac{1}{2}\cos 2\omega_m t - \frac{65}{32}\cos \omega_m t + \frac{5}{2}\right) + \frac{5V^2 f_1^2}{\omega_m^6}\left(-\frac{1}{6}\cos 2\omega_m t - \frac{4}{3}\cos \omega_m t + \frac{3}{2}\right) \qquad (23)$$

So, the deflection of the microbeam in the third-order equation is

$$w_{3m} = T_{3m}(t) \cdot (1 - \cos 2m\pi x) \qquad (24)$$

The deflections response of the microbeam is

$$w = \sum_{m=1}^{\infty} \left(\varepsilon w_{1m} + \varepsilon^3 w_{3m}\right) = \sum_{m=1}^{\infty} \begin{bmatrix} \varepsilon \frac{f_1}{\omega_m^2}(1 - \cos \omega_m t)(1 - \cos 2m\pi x) \\ + \varepsilon^3 [\frac{s_m f_1^3}{\omega_m^8}(\frac{1}{32}\cos 3\omega_m t - \frac{1}{2}\cos 2\omega_m t \\ - \frac{65}{32}\cos \omega_m t + \frac{5}{2}) + \frac{5V^2 f_1^2}{\omega_m^6}(-\frac{1}{6}\cos 2\omega_m t \\ - \frac{4}{3}\cos \omega_m t + \frac{3}{2})](1 - \cos 2m\pi x) \end{bmatrix} \qquad (25)$$

The nonlinear vibration dimensionless frequency can be obtained from

$$\Omega^2 = \sum_{m=1}^{\infty}\left[1 - \varepsilon^2 \left(\frac{15 s_m f_1^2}{4\omega_m^6} + \frac{10V^2 f_1}{\omega_m^4}\right)\right] \qquad (26)$$

5   Numerical Results

The physical and material properties of the microbeam are: Length $L = 610\mu m$, Width $b = 40\mu m$, Thickness $h = 2.2\mu m$, Initial gap $g = 2.3\mu m$, Young's modulus $E = 149GPa$, Mass density $\rho = 2330 kg/m^2$, $\varepsilon_0 = 8.854 \times 10^{-12} F/m$.

The linear displacement of the midpoint of the microbeam with ignoring the nonlinear of the material have been listed in Table 1 and illustrated in Fig. 1. The value of the current method is close to [9] method already when we take m=1.

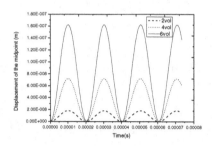

Fig. 1 Displacement of the midpoint

Table 1 The Comparison of Different Methods

Voltage	Current method	Literature [9] method
2	0.0180	0.021588
4	0.0720	0.073430
6	0.1626	0.164503
7	0.2222	0.227081
8	0.2920	0.303027
9	0.3727	0.398113

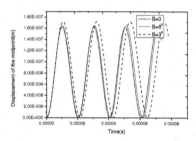

Fig. 2 Displacement of the midpoint

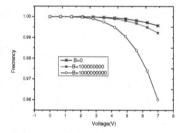

Fig. 3 Frequency response of the microbeam

Fig. 2 illustrates the response of the microbeam under different nonlinear material constants, B when load voltage is 6(V). The nonlinear factor can be ignored when the number of the nonlinear material constant, B is small. But, when the number of B is big, the amplitude and the period of the microbeam will increase obviously with the number of B increasing, and we must consider the influence of the nonlinear elasticity material in dynamic response of the microbeam.

Fig. 3 shows the frequency response, Ω of the microbeam under different step voltages and nonlinear material constants, B. The frequency, Ω will decrease when step voltages is increasing. When the number of B is big, the frequency, Ω will decrease obviously with the number of B increasing.

6 Conclusion

In this paper, a nonlinear dynamic response of little deflection modal, considering the nonlinear elasticity material, is set up that as a clamped-clamped microbeam subjected to electrostatic force. The nonlinear governing equation has been deduced. Approximate the electrostatic force using the Taylor expansion. The deflection response and frequency response are solved by perturbation Galerkin method. From the numerical results, we know the nonlinear factor can be ignored when the number of the nonlinear material constant, B is small. When the number of B is big, the amplitude and the period of the microbeam will increase obviously with the number of B increasing. We can predict that the frequency and the pull-in voltage of the microbeam with nonlinear elasticity materials are more lower than the microbeam with linear elasticity materials. For research and development of MEMS, we can do reasonable selection and design by making full use of the non-linear elasticity factor in materials.

References

1. Zook, J.D., Bums, D.W.: Characteristics of polysilicon resonant microbeams. Sensors and Actuators 35, 51–59 (1992)
2. Ahn, Y., Guckel, H., Zook, J.D.: Capacitive microbeam resonator desings. J. Micro-mech. Microeng. 11, 70–80 (2001)
3. Hung, E.S., Yang, Y.-J., Senturia, S.D.: Low-order models for fast dynamical simula-tion of MEME microstructures, transducers. In: Proceedings of the 1997 International Conference on Solid-State Sensors and Actuators, pp. 1101–1104. IEEE, New York (1997)
4. Chu, P.B., Lo, N.R., et al.: Optical communication using micro corner cube reflectors. In: Proceedings of IEEE MEMS 1997, Nagoya, Japan, January 26, vol. 30, pp. 350–355 (1997)
5. Xu, L., Jia, X.: Electromechanical coupled nonlinear dynamics for microbeams. Arch. Appl. Mech. 77, 485–502 (2007)
6. Fu, Y.M., Zhang, J.: Nonlinear static and dynamic responses of an electrically actuated viscoelastic microbeam. Research Paper. Acta Mech. Sin. 25, 211–218 (2009)
7. Qiang, H.: Nonlinear Vibration of a Rectangular Plate with Physical Nonlinearity. Jour-nal of South China University of Technology (Natural Science Edition) 28(7), 78–82 (2000)
8. Chia, C.-Y.: Nonlinear Analysis of Plates. Science Press (1989)
9. Xie, W.C., Lee, H.P., Lim, S.P.: Nonlinear Dynamic Analysis of MEMS Switches by Nonlinear Modal Analysis. Nonlinear Dynamics 31, 243–256 (2003)

Continuous Analysis Based on Profile Information for Vectorization Optimization

Yuan Yao and Rong-cai Zhao

Abstract. Aiming at the non-continuous reference problem inside the loop, a method which can implement the continuous analysis based on feedback-directed compiling optimization technique is proposed. It can make up for the traditional vector-based static analysis identified inadequate for the program appears contains a lot of pointers, array of structures and other complex data structures, and to provide more accurate identification and optimization of the vector.

I Introduction

The continuous of data reference is an important prerequisite for the loop vectorization. Currently, many programs have a lot of pointers, array of structures and other complex data structures, the discontinuous of data references are always caused by which. This makes the loop vectorization difficult to identify. Even that can be identified in this case, the vectorization has to bring the inefficient code. The traditional vector identification based on static analysis for large-scale use pointers or array reference subscript too complex and so difficult to achieve good results. This paper gives new analysis algorithm based profile information to solve the data references continuous in the loop.

2 Access to the Profile Information

In this paper, FDO is used to obtain the profile information of the data reference continuity in the loop that needed by the analysis algorithm. Feedback-directed optimization improves the performance of programs by applying optimizations

Yuan Yao · Rong-cai Zhao
ZhengZhou Information Science and Technology Institute,
Zhengzhou Henan, China
e-mail: {yaoyuan3070,rczhao126}@126.com

F.L. Gaol et al. (Eds.): Proc. of the 2011 2nd International Congress CACS, AISC 145, pp. 425–430.
springerlink.com   © Springer-Verlag Berlin Heidelberg 2012

which use the information gathered during the profile run. This allows the compiler to be more precise when applying optimizations.

FDO involves three stages. First is the generation of the instrumented binary. Second is running the instrumented binary with a representative data set to generate the profile for the data set which captures the hot path information. In the third step this profile information is used by the compiler to apply precise optimizations, given the knowledge of the hot paths in the code from the profile data. This targeted optimization is what gives the FDO compilation the performance advantage over a single stage compilation. The process is shown in Figure 1.

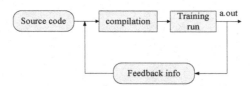

Fig. 1 Feedback directed compilation

The above method is implemented based on the Open64 [1] compiler, as an open source compiler, which is a good platform for researching and developing, and achieve a complete set of compiler optimization technology.

Open64 compiler has a clear framework, which is a well-written Modularized, robust, state-of-the-art compiler with support for C/C++ and Fortran 77/90. The major modules of Open64 are the multiple language frontends, the inter-procedural analyzer (IPA) and the middle end/back end, which is further subdivided into the loop nest optimizer (LNO), global optimizer (WOPT), and code generator (CG). Five levels of a tree-based intermediate representations (IR) called WHIRL exist in Open64 to facilitate the implementation of different analysis and optimization phases. They are classified as being Very High, High, Mid, Low, and Very Low levels, respectively. Most compiler optimizations are implemented on a specific level of WHIRL [3]. In the Open64 framework, the front-ends generate the highest level of WHIRL. Optimization proceeds together with the process of continuous lowering, in which a WHIRL lowered is called to translate WHIRL from the current level to the next lower level. At the end, the code generator translates the lowest level of WHIRL to its own internal representation that matches the target machine instructions. WHIRL thus serves as the common IR interface among all the back-end components. Because lowering is done gradually, a secondary benefit is that each lowering step is simpler and easier.

The main steps are shown as follow:

2.1 The Loop Analysis and Instrumentation Stage

Profile information needs to include data references address, the element size of arrays or pointer variable [2], the reference type which is read or write. The algorithm pseudo-code description is shown in Figure 2.

```
Input: PUs' WHIRL tree
Output: New PUs trees
Description:
SELECT-INSTRUMENT(PUs)
      foreach PU in PUs
            CREATE-MAP(PU_entry)
            DFS-LOOP(PU_entry)

DFS-LOOP(wn)
      foreach kid in wn_kids
            DFS-LOOP(kid)
      if wn ∈ {DO_LOOP, WHILE_DO, DO_WHILE}
            if false = CHECK-DFS(wn)
                  instr_count++
                  loop_count++
                  DFS-ARRAY (wn)
                  GEN-CALL-BEFORELOOP(wn)
                  GEN-CALL-LOOPEND(wn, linenum, instr_count, loop_count)


CHECK-DFS(wn)
      foreach kid in wn_kids
                  if CHECK-DFS(kid) is true
                        return true
            if  wn ∈ {DO_LOOP, WHILE_DO, DO_WHILE}
                  return true
            return false

DFS-ARRAY (wn)
      foreach kid in wn_kids
                  DFS-ARRAY (kid)
            if wn ∈ {STID、 ISTORE、 LDID、 ILOAD}
                  linenum ←GET-LINENUM(wn)
                  instr_count++
                  addr_count++
                  while parent_wn is not a statement
                        parent_wn ← GET-PARENT(parent_wn)
                  GEN_ADDR_BEFORE_WN(wn, parent_wn, instr_count, addr_count)
```

Fig. 2 The instrumentation algorithm pseudo-code

In this algorithm, CREATE-MAP is used to construct the corresponding WHIRL Parent Map with WHIRL tree. This is because the instrumenting before the data access operation needs to find the inserted pile where the statement quoted in the statement before the insert into the corresponding block pile, which requires data access node can be traced back to the parent node. DFS-LOOP using depth-first traversal algorithm traverses each node in PU and completes the instrumentation.

When determining the location of the data reference Instrumentation, the algorithm traverses and finds the WHIRL node in loop which contains OPR_STID, OPR_LDID, OPR_ILOAD, OPR_ISTORE operation code. The algorithm also gets element size, name of array or pointer variable from the corresponding symbol table, and reference type from operation code.

2.2   The Training Runs and Obtains the Profile Information

Through the training run, we can get real-time data references address in the loop. The continuous analysis algorithm based profile information analyzes the real-time address. The result was saved by some form into the feedback files for the compiler optimization.

2.3   The Feedback Stage

The second compilation put the profile information used to optimization into the corresponding node in the WHIRL tree. The prerequisite for success is that the position of instrumenting and feedback must be consistent. The instrumentation stage and the feedback stage must be completed in the same modules to ensure consistent.

The feedback algorithm description is shown as follows:

Input:Profile information, WHIRL tree
Output:WHIRL with profile information

1) read profile information.

2) traverses and finds the loop.

3) get the loop information in 2) . That traverses the WHIRL node in the loop, if the loop is the inner loop, incrementing the total instrumentation number, incrementing the loop instrumentation number and use the number as the ID of the loop.

4) get the data reference instrumenting position. Traversing the loop body, get the data reference node. Incrementing the total instrumentation number, and incrementing the data reference instrumentation number and use the number as the ID of the data reference.

5) check instrumentation number.

Comparing the total number and other number with the CHKSUM computed in the instrumentation stage. If it is same, that proves the feedback stage is right, else it is wrong.

6) read the profile information from the feedback file according the ID of loop and the ID of data reference.

---

3   The Continuous Analysis Algorithm Based on the Profile Information and Optimization Methods

Through the training run, the algorithm analyzes the profile information to get the continuous information of corresponding data reference, which used to guide the vectorization optimization.

The algorithm can identify the structure and continuity of the array. The algorithm designs as follows:

Firstly, initialize the profile information. Secondly, comparing the address of the current iteration with the last iteration, step_size can be obtained. The algorithm pseudo-code description is shown in Figure 3.

According to the CONTIN_TYPE and ADDR_TYPE, the data reference is divided into the following categories:

- ✓ PROFILE_STRUC_REGUNCONTINUITY. Such as: structure array a[i*2].k.
- ✓ PROFILE_STRUC_CONTINUITY. Such as structure array a[i].k.
- ✓ PROFILE_ARRAY_CONTINUITY. Such as array a[i].
- ✓ PROFILE_ARRAY_REGUNCONTINUITY. Such as a[i*2].

According to the data reference type, we will take different optimization methods to achieve the continuous requirement of vectorization.

```
Input: the real-time address of data reference , elem_szie
Output: CONTIN_TYPE, step_size etc.
CONTINUITY-ANALYSIS()
    foreach addr in one iter of loop
            if cur_iter = 0
                    step_size  ←  -1
                        CONTIN_TYPE  ←  PROFILE_ADDR_NONE
            if CONTIN_TYPE > 4,
                    return

            if step_size = -1
                    step_size  ←  |current address – last address| / elem_size
            else if |current address – last address| / elem_size = 1
                        SET-CONTIN_TYPE(PROFILE_ADDR_CONTINUITY)
            else
                        SET-CONTIN_TYPE(PROFILE_ADDR_REGUNCONTINUITY)

            if |current address – last address| = 0
                        SET-CONTIN_TYPE(PROFILE_ADDR_CONST)
            else if |current address – last address| < elem_size
                        SET-CONTIN_TYPE(PROFILE_INSTRUC_CONTINUITY)

    SET-CONTIN_TYPE(current CONTIN_TYPE)
        if last CONTIN_TYPE = PROFILE_ADDR_NONE
                CONTIN_TYPE  ←  current CONTIN_TYPE
        else if last CONTIN_TYPE =current CONTIN_TYPE
                return
        else
                CONTIN_TYPE  ←  PROFILE_ADDR_NONE
```

Fig. 3 The analysis algorithm pseudo-code

4   Test and Analysis

In order to verify the correctness and accuracy of the program, we tested this application using spec2000 [4]. We analyze the 179.art, 181.mcf, 300.twolf, 197.parser as follow. The test is completed in a dedicated test platform with 256-bit SIMD Extensions.

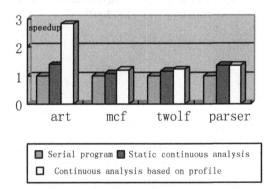

Fig. 4 The continuous analysis and optimization test

According to the test data, the maximum speedup of the algorithm and optimization is about 2.8(art), and the minimum is 1.2(mcf). It is higher than the static effect [5].

In the 179.art program, the structure variables in the hot loop are global variables, the fields are simple types such as float. These structure variables were fully optimized by the algorithm and structure peeling method.

In the 300.mcf program, the main data structure in the hot loop is node_t and arc_t two structure array, which arc_t domain contains a number of node_t structures for this complex structure and its associated call. Intel's automatic vectorization optimization would bring greater overhead, which offsets the quantifiable benefits. But compiled by the continuity of the feedback analysis and optimization, we can get a good speedup, which also verified the correctness of the algorithm and optimize performance.

5   Conclusion

This paper implements a new method of continuous analysis based on the profile information. According to the test, the method has been improved. It can improve the capability of the traditional compiler's vectorization identification effectively.

References

[1] Overview of the open64 Compiler Infrastructure,
    http://open64.sourceforge.net
[2] Nuzman, D., Rosen, I., Zaks, A.: Auto-Vectorization of Interleaved Data for SIMD. In: PLDI 2006, Ottawa, Canada, pp. 132–143 (2006)
[3] WHIRL Symbol Table Specification,
    http://open64.sourceforge.net/symtab_Pro64_SGI.pdf
[4] SPEC2000 overview, http://www.spec.org/cpu2000/
[5] Slingerland, N., Smith, A.J.: Design and characterization of the Berkeley multimedia workload. Multimedia Systems 8(4), 315–327 (2002)

A New Three-Dimensional Spherical Terrain Rendering Method Based on Network Environment

X.F. Dai, H.J. Xiong, and X.W. Zheng

Abstract. A new three-dimensional spherical terrain rendering method based on network environment is proposed in this study. Delay of fetching data or failure of data transmission could happen on network environment, and this method can deal with the situation without affecting real-time rending. In this method, a set of nest regular grids (NRG) are used to cache the terrain, Graphics hardware GPU is applied to accelerate the terrain rendering, and cracks elimination is discussed.

1 Introduction

Building 3D real time digital elevation model from multi resolution imagery and elevation data has been one of the most important components in the practical applications, the terrain geometry can be used in the virtual reality, games, cartography and so on, in this paper, real-time terrain height-field rendering is concerned.

Related work has been done by the scholars. Progress meshes is introduced by Hugues hoppe [1], which can keep the appearance consistent during the simplify progress and transit smoothly among different terrain levels; in his study, pre-processing [2] and storage[3] technology of massive terrain data is proposed; Real-time Optimally Adapting Meshes (ROAM) [4] is one of the most widely used algorithms in triangular mesh construction, the recursive bisection of right triangles is used to reduce memory use and traversal; Lindstrom-Koller[5] algorithm is proposed by David Koller and Peter Lindstrom and the vertexes are organized by bin-tree; Renato Pajarola[6] do the farther based on Lindstrom-Koller algorithm, and a

X.F. Dai · H.J. Xiong
The State Key laboratory of information Engineering,
Mapping and Remote Sensing
Wuhan, China
e-mail: daixuefeng203@gmail.com, xionghanjiang@163.com

F.L. Gaol et al. (Eds.): Proc. of the 2011 2nd International Congress CACS, AISC 145, pp. 431–436.
springerlink.com   © Springer-Verlag Berlin Heidelberg 2012

top-down spatial storage structures are designed; P. Cignoni proposes the BDAM[7] algorithm which combine the irregular and regular meshes, and P-BDAM[8] is used to render spherical surface, and wavelet compression is used for data transmission in C-BDAM[9]; Geometry clipmaps [10] organize the terrain data a pyramidal multi resolution scheme and the residual between levels is compressed using an advanced image coder that supports fast access to image regions [11].

2   Nest Regular Grids

2.1   Terrain Tile

Terrain tile is the basic unit of the nest regular grids (NRG). It can represent the different terrain level by changing the length of the side. It is a square mesh consists of four equal parts. Each part can be quartered turning to the next terrain level. The relation between terrain level and side length can be described by the formula 1, s(n) means side length of level n, L means the side length of the first level which equals to 18 degrees in this study and n means level number.

$$s(n) = \frac{L}{2^n} \tag{1}$$

2.2   Terrain Center

Terrain center is square and consists of 16 terrain tiles, which is used to display the highest resolution terrain data at the current view point that required. Terrain center lies in the center of the NRG, and the position is decided by the height of the view point, which can cover most part of the area that viewer interested.

2.3   Terrain Ring

One Terrain ring (figure 1) consists of 12 terrain tiles. Each ring represents a level of terrain data, and the difference between two adjacent rings' level is 1. If the terrain center is at terrain level N, then the level of the first ring is N-1, the level of the second ring is N-2, and so on.

And the whole NRG will be like figure 2.

3   Construction of the Terrain Grid

3.1   Construction of the Terrain Center Objects

The terrain center level can be calculated by the formula 2 taking the height of the view point as the parameter, and the side length of the terrain tiles that make up the terrain center can be calculated by formula 1. From the coordinate of the view point, the coordinate center of the terrain center can be deducted, so does the coordinate of each terrain tile center, and then the row and column of the terrain tile in this terrain level can be calculated by formula 2 and formula 3, lon and la means longitude and latitude the coordinate of the terrain tile.

$$x = \frac{\left\| \left| -90 - la - 180 * \left| \frac{-90 - la}{180} \right| \right| \right\|}{s(n)} \tag{2}$$

$$y = \frac{\left\| \left| -90 - lon - 360 * \left| \frac{-180 - lon}{360} \right| \right| \right\|}{s(n)} \tag{3}$$

Since the level, row and column of the terrain tile are known, the terrain center object can be created, and the real mesh can be constructed when the terrain data is fetched from the data server on the net.

3.2 Construction of the Terrain Ring Objects

The terrain rings are constructed from inside to outside. In figure 3, VR means the view range; S(n) means the side length of the terrain tiles which make up the terrain ring n; L(n) means the terrain level of the terrain ring n, the relationship between terrain ring n and n-1 is: $S(n-1) = 2*S(n)$ & $L(n) = L(n-1)+1$. The first ring is outside of the terrain center, if the terrain center level is n, the level of first terrain ring $L(1) = n-1$, and $L(m) = n-m(n-m \geq 0)$. When the terrain ring object x is constructed, if $S(x) < VR/2$, then a new terrain ring object should be constructed, otherwise, no more terrain ring object should be created. The row and column of the terrain tiles can be calculated the same as the terrain center tiles mentioned above.Only terrain tile objects which record the information used for fetching data from the data server on the net are created. The real terrain mesh has not been created in the computer memory.

3.3 Improvement on Network Environment

Delay of fetching data or failure of data transmission could happen on network environment, under this situation, part of the NRG cannot be constructed without the terrain data, which could affect real-time rendering, and any application based on the terrain data could cause abnormal results. The virtual terrain tile (VTT) is introduced to deal with this kind of situation.

In figure 4, VTT(N) means the four virtual terrain tiles which are added to the terrain ring N. The VTT(N) becomes the parent of terrain ring N-1 and VTT(N-1), if the terrain data missing in terrain ring N-1, VTT(N) could be used to represent the missing part after simple interpolation.

The difference between the VTT and terrain tiles is that the real mesh for VTT is only created when the terrain data cannot be acquired in time, but the VTT object is always created like the other normal terrain tiles to record the necessary information used to fetching the terrain data from the data server on the net.

The terrain tile objects are constructed from inside to outside, and the data download request are sent in the same order, but real terrain meshes are created oppositely order in the computer memory, in order to determine whether real meshes for the VTT should be created or not.

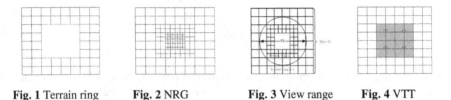

Fig. 1 Terrain ring   **Fig. 2** NRG   **Fig. 3** View range   **Fig. 4** VTT

4   Gap Elimination

Gaps will appear between the terrain rings at different levels. To eliminate the gaps, two methods are can be used during the construction of the NRG. The first one is to add a "skirt" to every terrain tile, this method is simple and effective, but in some applications, the objects under the terrain are the targets that the viewer interested in, the "skirt" will affect the visualization; the other one is to morph the outer boundary of each terrain ring tile, so that the mismatch point transitions to the point at the coarser level, then the terrain tile can be divided to four classes (figure 5).

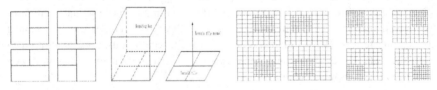

Fig. 5 Four classes   **Fig. 6** Bounding box   **Fig. 7** Offset A   **Fig. 8** Offset B

5   Optimization

5.1   View-Frustum Culling

For the view-frustum culling, each terrain tile is extended by the terrain bounds to form an axis-aligned bounding box (figure 6), here the axis has the same direction with the normal direction of this terrain tile. The view frustum culling can reduce the rendering load, actually, the culling operation is executed right after the terrain tile objects are constructed, so there is no chance for the invisible terrain tiles to create the real mesh in computer memory.

5.2 Acceleration Based on GPU

If the gap elimination used the "skirt" method, all the terrain tiles have the same structure, if the other method is used, the terrain tiles can be divided into four classes, whichever method is applied, the terrain tiles have fixed kinds of stricture, and the Geometry Instancing technique can be adopted, the terrain can be created and rendered in batch based on Graphics hardware GPU; the terrain data is stored in the height map, then the vertex textures (e.g. as in DirectX9 Vertex Shader 3.0) can be used to get the terrain data and do the interpolation in GPU.

5.3 Grid Reconstruction

When the view point position is changed, the terrain center should move, and the NRG has to be reconstructed, little change of the viewer can cause the reconstruction which is not efficient. In fact, NRG is enabled to shift terrain center or part of the terrain rings to reduce the reconstruction, which profit from the structure of the terrain tile (figure 7), because a terrain tile consists of four similar parts, each part has chance to change the visibility or replaced by a finer level terrain tile, but the situation like figure 8 shows must avoid because the difference between adjacent rings has to be 1.

6 Implementation and Results

An experimental a rendering application supporting the NRG method has been implemented on Windows platforms using C++ with Direct 3D and HLSL. The application shows that the terrain mesh can be constructed quickly and terrain level changes smoothly while the view point is moving (figure 9 and 10).

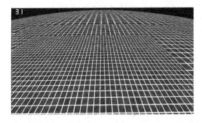

Fig. 9 Terrain mesh without height data   **Fig. 10** Terrain mesh with Texture

Acknowledgments. This research is partially supported by the national "863" Project (2009AA12Z229).

References

1. Hoppe, H.: Progressive meshes (1996)
2. Hoppe, H.: Smooth view-dependent level-of-detail control and its applications to terrain rendering. IEEE Visualization, 35–42 (1998)
3. Koller, D., Lindstrom, P.: Level-of-Detail Management for Real-Time Rendering of Photo—textured Terrains[EB/OL],
   `ftp://ftp.gvu.gatech.edu/pub/gvu/tr/1995/95-06.pdf.1995`
4. Duchaineau, M.A., Wolinsky, M., Sigeti, D.E., Miller, M.C., Aldrich, C., Mineev-Weinstein, M.B.: ROAMing terrain:Real-time optimally adapting meshes. In: Proceedings IEEE Visualization, pp. 81–88 (1997)
5. Koller, D., Lindstrorm, P.: Real-Time Continuous LOD Rendering of Height Fields. In: Proceedings of the ACM SIGGRAPH, pp. 109–118 (1996)
6. Pajarola, R.: Large scale terrain visualization using the restricted quadtree triangulation. In: Rushmeier, H., Elbert, D., Hagen, H. (eds.) Proceedings of Visualization, pp. 19–26 (1998)
7. Cignoni, P., Gannovelli, F., Gobbetti, E., Ponchio, F., Scopigno, R.: BDAM: Batched dynamic adaptive meshes for high performance terrain visualization. Computer Graphics Forum 22(3), 505–514 (2003)
8. Cignoni, P., Gannovelli, F., Gobbetti, E., Marton, F., Ponchio, F., Scopigno, R.: Planet-sized batched dynamic adaptive meshes(P-BDAM). In: IEEE Visualization, pp. 147–154 (2003)
9. Gobbetti, E., Moarton, F., Cignoni, P., Di Benedetto, M., Ganovelli, F.: C-BDAM-Compressed Batched Dynamic Adaptive Meshes for Terrain Rendering. In: EUROGRAPHICS, pp. 333–342 (2006)
10. Losasso, F., Hoppe, H.: Geometry clipmaps: terrain rendering using nested regular grids (2004)
11. Bettio, F., Gobbetti, E., Marton, F., Pintore, G.: High-quality networked terrain rendering from compressed bitstreams. In: Proc. ACM Web3D International Symposium, pp. 37–44 (April 2007)

Combining Probabilistic Dependency Models and Particle Swarm Optimization for Parameter Inference in Stochastic Biological Systems

Michele Forlin, Debora Slanzi, and Irene Poli

Abstract. In this work we present an efficient method to tackle the problem of parameter inference for stochastic biological models. We develop a variant of the Particle Swarm Optimization algorithm by including Probabilistic Dependency statistical models to detect the parameter dependencies. This results in a more efficient parameter inference of the biological model. We test the Probabilistic Dependency-PSO on a well-known benchmark problem: the thermal isomerization of α-pinene.

1 Introduction

Biology is the science of life and living organisms. This discipline has seen a tremendous growth during the last 20 years, empowered by the deployment of auto-mated experimental frameworks. The increase in computational power pushed the research community to investigate more accurately, at an unprecedented finer level of granularity, how living systems behave.

Modeling is an essential step for fully understanding the dynamics of biological systems. Although models can be used to give a static picture of the whole system, they may also easily include the dynamical information required to study its evolution over time. In order to build a valid model we need to identity the components constituting the biological system as well as their interactions and dynamical behavior. Once the current knowledge of the system is incorporated into the model, it can be used to provide new insights and predictions for conditions of the systems that have not previously been explored. The available knowledge about a system to be

Michele Forlin
European Centre for Living Technology, S. Marco 2940, Venice, IT
e-mail: michele.forlin@ecltech.org

Debora Slanzi · Irene Poli
Department of Environmental Sciences, Informatics and Statistics,
University Ca'Foscari of Venice, Cannaregio 873, Venice, IT
e-mail: {debora.slanzi,irenpoli}@unive.it

F.L. Gaol et al. (Eds.): Proc. of the 2011 2nd International Congress CACS, AISC 145, pp. 437–443.
springerlink.com   © Springer-Verlag Berlin Heidelberg 2012

modeled is only partial. One may know which are the components of the systems under study and how they interact, but not the parameters that govern their dynamical behavior. This problem is known as parameter inference or model calibration. Given a set of experimental data, the goal is to calibrate model parameters in order to reproduce the experimentally observed behavior at the best.

Most of the available methods and tools have been created and used for deterministic models of biological systems ([15], [1]), while only recently works have concentrated on stochastic settings ([14], [2]). In particular, in [7] the author developed an inference scheme using evolutionary computation techniques based on Particle Swarm Optimization iterating the modeling and simulation steps in an intelligent manner. Following this paradigm we propose a novel approach which combines the evolutionary strategy proposed in [7] with the information achieved by a Probabilistic Dependency Model [11] estimated through the iterative evolutionary process. This information let us to disentangle the dependency relations among the model parameters and thus improving the efficiency of the inference procedure.

This work is organized as follows: in Section 2 we introduce the methodology based on the combination of Particle Swarm Optimization and Probabilistic Dependency Models and Section 3 contains the results of the approach on a well-known test case and some conclusion.

2 Methods

Particle Swarm Optimization (PSO) is a stochastic optimization algorithm first developed and introduced by Eberhart and Kennedy [10]. PSO is a population-based algorithm, in which the population is called *swarm* while the search points are called particles. Each particle moves in the search space with an adaptable velocity recording the best position it has ever visited , i.e. the position with the lowest function value when dealing with minimizing objective problems. Moreover, each particle has a neighborhood that consists of some pre-specified particles and the best position ever attained by any member of its neighborhood is communicated to the particle itself and influences its movement.

Technically, a swarm $S = (X_1, X_2, \ldots, X_M)$ consists of M particles. Every particle $X_i = (x_{i1}, x_{i2}, \ldots, x_{in})^T$ is composed by n components. The velocity, V_i, of the ith particle, as well as its best position, P_i, at each iteration step t, are n-dimensional vectors, $V_i = (v_{i1}, v_{i2}, \ldots, v_{in})^T$ and $P_i = (p_{i1}, p_{i2}, \ldots, p_{in})^T$. The best position of neighbourhood particles is $P_{gi} = (p_{g1}, p_{g2}, \ldots, p_{gn})^T$.

Let t be the iteration counter. The velocity and position of X_i are updated according to the equations

$$V_i(t+1) = V_i(t) + c_1 r_1 (P_i(t) - X_i(t)) + c_2 r_2 (P_{gi}(t) - X_i(t)) \tag{1}$$

and

$$X_i(t+1) = X_i(t) + V_i(t+1) \tag{2}$$

where ω is the inertia factor, c_1, c_2 are positive acceleration parameters called cognitive and social parameter respectively, and r_1, r_2 are vectors with components uniformly distributed in the range $[0, 1]$.

Starting from the standard version of the PSO algorithm we introduce the Adaptive Inertia Weight Factor (AIWF) presented in [12] to efficiently control the global search and convergence to the global best solution. In particular, AIWF is determined as follows:

$$\omega = \begin{cases} \omega_{min} + \frac{(\omega_{max}-\omega_{min})\cdot(f-f_{min})}{f_{avg}-f_{min}} & f \leq f_{avg} \\ \omega_{max} & f > f_{avg} \end{cases}$$

where ω_{max} and ω_{min} denote the maximum and minimum of ω respectively, f is the current objective value of the particle, f_{avg} and f_{min} are the average and minimum objective values of all particles, respectively. All vectors operations in equations 1 and 2 are performed componentwise. In the context of parameter estimation for stochastic models of biological systems, particles are composed by the set of parameters to be estimated, generating thus a real-valued vector. The iterative evolutionary process of PSO depends on some stochastic components represented by vectors r_1, r_2. In this work, we claim that a more intelligent sampling of this stochastic vectors could be of help in the overall optimization task. Each component of these vectors is related to a model parameter: if there exist some conditional relations among parameters leading to optimal system behaviors, then a linked dependency sampling could be able to enhance the optimization process.

To detect the relations among the model parameters, we propose to use a class of statistical models, namely the class of Probabilistic Dependency Models [11]. These models are able to assess joint probability distributions in terms of direct graphs with probabilistic semantics. We focus on a particular class of models, that is the class of Conditional Gaussian Bayesian Networks (CGBN) [8, 13], where nodes in the graph correspond to gaussian random variables and arcs between nodes describe the dependence structure that may characterize the set of variables on which we then develop statistical inference. In this context, each model parameter of the biological system composing the particle in the PSO, $X_i(t+1) = (x_{i1}, x_{i2}, \ldots, x_{in})^T$, is considered as a random variable and it is represented by a node N_k, $k = 1, \ldots, n$, in the graph. The relations (directed dependencies) among the random variables are then represented by directed arcs. Any pair of unconnected nodes of the graph indicates (conditional) independence between the variables represented by these nodes. In this way the graph can give insights on the (conditional) dependence and independence relations associated with the variables characterizing the system.

Because of the role taken by the variables in the dependence relation, if there is an arc pointing from node N_h to node N_k, then N_h is usually named parent of N_k and $Pa(N_k)$ as the set of N_k parents.

Several methods for automatic induction of Conditional Gaussian Bayesian Networks have been derived [13, 4]. In this work we adopt the Max-Min Hill-Climbing [16]: an hybrid algorithm which is able to discover the interaction network among the variables in the system in an effective way.

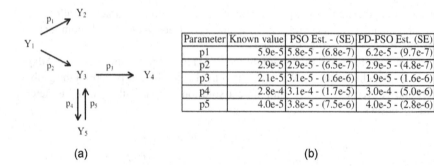

Parameter	Known value	PSO Est. - (SE)	PD-PSO Est. (SE)
p1	5.9e-5	5.8e-5 - (6.8e-7)	6.2e-5 - (9.7e-7)
p2	2.9e-5	2.9e-5 - (6.5e-7)	2.9e-5 - (4.8e-7)
p3	2.1e-5	3.1e-5 - (1.6e-6)	1.9e-5 - (1.6e-6)
p4	2.8e-4	3.1e-4 - (1.7e-5)	3.0e-4 - (5.0e-6)
p5	4.0e-5	3.8e-5 - (7.5e-6)	4.0e-5 - (2.8e-6)

(a)   (b)

Fig. 1 (a) Mechanism for thermal isomerization of α-pinene (b) Estimated parameters with standard errors and best known value

We combine the principles of the PSO algorithm with the information on the dependence relations among the model parameter to build what we call the *Probabilistic Dependency model-based Particle Swarm Optimization* (PD-PSO). In particular, at each generation t of the algorithm, the $CGBN_t$ is estimated from the 30 best fitting particles $X_i(t) = (x_{i1}, \cdots, x_{in})$, $i = (1), \ldots (30)$. The $CGBN_t$ model contains the information regarding the dependence relations among the particle components, i.e. the biological model kinetic parameters. To determine the new position $X_i(t+1)$ of the particle i at generation $t+1$, the PD-PSO approach is in general driven by the equations described in 1 and 2 combined with the AIWF. Afterwords, when the condition $f_i \leq f_{arg}$ in AIWF is satisfied, then the position $X_i(t+1) = (x_{i1}, \cdots, x_{in})$ is updated with the information achieved by the $CGBN_t$. In particular, for each k-th component x_{ik}, $k = 1, \ldots, n$, with set of parent $Pa(x_{ik}) \neq \emptyset$ in the graph, the value of x_{ik} is updated with the $CGBN_t$ model prediction conditioned on the values assumed by the components defining $Pa(x_{ik})$.

3   Results

To fully understand how the evolutionary inference framework works we develop a test example. The system represents the thermal isomerization of α-pinene in which we want to estimate 5 rate constants $(p_1,...,p_5)$ of a complex biochemical reaction system composed by 5 species $(Y_1,...,Y_5)$. This pathway has been originally studied by Box and coworkers [3], and it is also part of COPS (Collection of large-scale Constrained Optimization ProblemS) [6]. The system is depicted in figure 1(a).

α-pinene (Y_1) is converted into dipentene (Y_2) and alloocimen (Y_3) which in turn yields α- and β-pyronene (Y_4) and a dimer (Y_5). Experiments on this process have been conducted by Hunt and Hawkins [9], reporting the concentrations of the reactant and the four products at eight time intervals.

We build a stochastic model of the pathway using the BlenX language [5]. We apply the PD-PSO approach within the general evolutionary inference framework combining stochastic modeling and simulation presented in [7]. We consider as

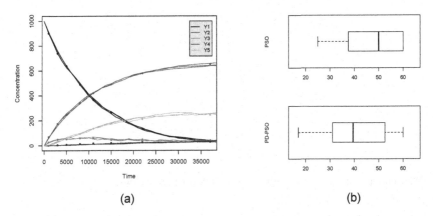

Fig. 2 (a) Experimental data and predictions with inferred parameters. (b) Number of generations required to convergence

fitness function the sum of square errors between model simulation and experimental data. Halting criteria are twofold. The first one, based on number of iterations, has been arbitrarily fixed to 60, while the second, based on a fitness function threshold, has been fixed to an average 2.5% distance from experimental data, which in this system is equal to 100. We repeat the evolutionary inference procedure for 100 times, letting us to properly check the convergence properties of the method and to compare it with a standard PSO algorithm. Optimal estimated parameters and the corresponding standard errors for both PSO and PD-PSO are reported in table 1(b) together with the known parameters in literature. As can be seen by glancing at table 1(b), the parameter estimates are very close to the best known values. To confirm the goodness of fit we report in figure 2 the behavior of the best solutions identified together with experimental points. Traces on the plot represent a single stochastic simulation which has shown a good fitting. The procedure identify 3 solutions with a fitness function value lower than the predefined threshold.

To compare the convergence performance of the PD-PSO with respect to PSO, we represent the distribution of generations required to convergence in 100 replicas. Figure 2 (b) clearly shows how PD-PSO requires less generations of the iterative evolutionary process to detect optimal parameter values. While standard PSO requires an average of 48 generations to converge, PD-PSO requires only an average of 41 generations. A statistical t-test confirms that PD-PSO average generation convergence is lower than the standard PSO one at a significance level greater than 99%. The information from the Probabilistic Dependency Models estimated during the evolutionary process is thus able to drive in a more efficient way the inference of biological model stochastic parameters. Figure 2 (a) shows the probability of occurrence of each possible arc in the CGBN in 100 replication of the evolutionary inference procedure through the 60 generations. We can notice that the direct dependency of F from p_1 (arc n. 31) is captured only at the beginning of the

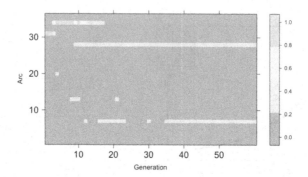

Fig. 3 Evolution of arc presence in estimated CGBN over generations

procedure. The dependency of F from p_4 (arc n. 34) is detected slightly later, while the dependency relations between p_4 and p_5 (arc n. 28) and between p_1 and p_2 (arc n.7) clearly arise converging towards the optimal solution. This confirms the quality of the information derived by the CGBN and its active role in enhancing the overall performance of the evolutionary estimation process.

References

1. Balsa-Canto, E., Peifer, M., Banga, J.R., Timmer, J., Fleck, C.: Hybrid optimization method with general switching strategy for parameter estimation. BMC Systems Biology 2(1), 26 (2008)
2. Boys, R.J., Wilkinson, D.J., Kirkwood, T.B.L.: Bayesian inference for a discretely observed stochastic kinetic model. Statistics and Computing 18(2), 125–135 (2008)
3. Box, G.E.P., Hunter, W.G., MacGregor, J.F., Erjavec, J.: Some problems associated with the analysis of multiresponse data. Technometrics 15(1), 33–51 (1973)
4. Darwiche, A.: Modeling and reasoning with Bayesian networks. Ebooks Corporation (2009)
5. Dematté, L., Priami, C., Romanel, A.: Modelling and simulation of biological processes in BlenX. ACM SIGMETRICS Performance Evaluation Review 35(4), 32–39 (2008)
6. Dolan, E.D., Moré, J.J., Munson, T.S.: Benchmarking optimization software with COPS 3.0. Argonne National Laboratory Research Report (2004)
7. Forlin, M.: Knowledge discovery for stochastic models of biological systems. University of Trento, PhD Thesis (2010)
8. Geiger, D., Heckerman, D.: Learning gaussian networks (1994)
9. Hunt, H.G., Hawkins, J.E.: The rate of thermal isomerization of α-pinene and βpinene in the liquid phase. Journal of the American Chemical Society 72, 5618–5620 (1950)
10. Kennedy, J., Eberhart, R.C.: Particle swarm optimization. In: Proceedings of IEEE International Conference on Neural Networks, vol. 4, pp. 1942–1948 (2005)
11. Koller, D., Friedman, N.: Probabilistic graphical models: Principles and techniques. MIT Press (2009)
12. Liu, B., Wang, L., Jin, Y.H., Tang, F., Huang, D.X.: Improved particle swarm optimization combined with chaos. Chaos, Solitons & Fractals 25(5), 1261–1271 (2005)

13. Neapolitan, R.E.: Learning bayesian networks. Pearson Prentice Hall, Upper Saddle River (2004)
14. Reinker, S., Altman, R.M., Timmer, J.: Parameter estimation in stochastic biochemical reactions. IEE Proc. -Syst. Biol. 153(4), 168 (2006)
15. Rodriguez-Fernandez, M., Egea, J.A., Banga, J.R.: Novel metaheuristic for parameter estimation in nonlinear dynamic biological systems. BMC Bioinformatics 7(1), 483 (2006)
16. Tsamardinos, I., Brown, L.E., Aliferis, C.F.: The max-min hill-climbing Bayesian network structure learning algorithm. Machine learning 65(1), 31–78 (2006)

Neuro-aided H₂ Controller Design for Aircraft under Actuator Failure

Zhifeng Wang and Guangcai Xiong

Abstract. To advocate the development of new robust and reliable controllers, it has been defined a benchmark problem, where robust controllers are required for controlling a 6-Degree of Freedom (6-DOF) nonlinear F16 aircraft in auto-landing phase undergoing actuator failures. This paper attempt to provide a solution by developing a robust Neural Network (Neuro) aided H₂ controller. Simulation results show that the fault tolerant performance of the proposed Neuro-aided H₂ controller is better than H₂ controller under actuator failure condition, both the number of hidden neurons and the time of on-line learning are small enough to be used in engineer problems.

1 Introduction

Currently, nearly all the airplanes are equipped with automatic control systems, which can be coupled with Instrument Landing Systems (ILS) or Global Positioning Systems (GPS) for portions of landing. These automatic landing systems, with controllers mostly being PID, give a fairly satisfactory control performance on a wide variety of airplanes [1][2][3][4]. However, under the severe wind and/or loss of control surfaces, safe auto landing is difficult to achieve.

Because of their powerful ability to approximate linear or nonlinear functions, neural network aided controllers can adapt to changes in system dynamics quickly and still provide good performance. A detailed survey about using neural networks in nonlinear flight control was presented by Calise [5] and Steiberg [6]. They outlined a number of potential areas in flight control for neural networks. The advantage of neural network is also quite useful for aircraft auto landing control, as during the descent if there is severe wind or partial loss in the control surfaces or actuator effectiveness, safe landing is generally not possible using the PID

Zhifeng Wang · Guangcai Xiong
School of Automation, Northwestern Polytechnical University, Xian, China
email: {wangzhifeng,xionggc}@nwpu.edu.cn

F.L. Gaol et al. (Eds.): Proc. of the 2011 2nd International Congress CACS, AISC 145, pp. 445–450.
springerlink.com   © Springer-Verlag Berlin Heidelberg 2012

controller. In [6], a neural network is trained off-line to generate the desired trajectories for auto-landing under wind disturbances. Although the auto-landing controller is still PID, this paper explores the use of neural network in auto-landing control system by incorporating it into the inner loop to generate the desired pitch angle command. Further, in [7], a neural network is used as an auto-landing controller, where a feed-forward network is trained off-line to replace the original PID controller. In [8], Napolitano and Kincheloe present the investigation of on-line learning neural controllers in the context of the aircraft autopilot functions and of stability augmentation systems for both longitudinal and lateral directional dynamics. The extended Back-Propagation (BP) algorithm was used in the on-line learning neural controllers, and its accuracy and learning speed were studied.

However, in all the above applications using Neuro-schemes, a prior training is needed and a feed-forward neural network with back propagation learning algorithm has been used as the main paradigm. Alternatively, since early nineties, Radial Basis Function Networks (RBF) with Gaussian function has been widely used as the basic structure for neural networks due to their good local interpolation and global generalization ability. For sequential learning of RBF, Lu *et. al.* had developed an algorithm known as Minimal Resource Allocation Network (MRAN), in which a pruning strategy was introduced to remove the neurons that consistently made little contribution to the network output [9].

In the past two decades, robust multi-variable feedback controllers such as H_2 and H_{inf} controllers have been designed for a variety of systems [10][11], and good performance had been observed. In [12], a comprehensive control law is developed which combined the gain scheduling and H_{inf} techniques, to control a VSTOL aircraft. The proposed schemes have been tested on simulation platform and good results are observed.

This paper is organized as follows. Section 2 describes the nonlinear F16 aircraft model. In section 3, a Neuro- H_2 controller is designed for wing level flight phase of auto landing to demonstrate the effectiveness of MRAN network controlling aircraft under actuator stuck. The performance of the controller is evaluated through a comparison study with the traditional H_2 controller approach. Section 4 summarizes the conclusions from the simulation results.

2   Aircraft Model

The candidate aircraft used here is a high performance fighter aircraft similar to F16 as reported in [13]. Under the rigid body assumption and referenced to a body-fixed axis system, the mathematical equations describing the motions of the aircraft are,

$$\dot{u} = rv - qw - g \sin \theta + \frac{\bar{q}}{m} SC_{x,t} + \frac{T}{m}$$

$$\dot{v} = pw - ru + g \cos \theta \sin \phi + \frac{\bar{q}}{m} SC_{y,t}$$

$$w = qu - pv + g\cos\theta\cos\phi + \frac{\bar{q}}{m}SC_{z,t}$$

$$\dot{p} = (c_2 p + c_1 r + c_4 H_e)q + \bar{q}\,Sb(c_3 C_{l,t} + c_4 C_{n,t})$$

$$\dot{q} = (c_5 p - c_7 H_e)r + c_6(r^2 - p^2) + \bar{q}\,S\,\bar{c}\,c_7 C_{m,t}$$

$$\dot{r} = (c_8 p - c_2 r + c_9 H_e)q + \bar{q}\,Sb(c_4 C_{l,t} + c_9 C_{n,t}) \tag{1}$$

The total aerodynamic coefficients $C_{x,t}$, $C_{y,t}$, $C_{z,t}$, used in the simulation were derived from low-speed static and dynamic wind tunnel. Total coefficient equations were used to sum the various aerodynamic contributions to a particular force or moment.

Because most analysis and control theory require a linear representation of a system, the above highly nonlinear aircraft equations should be linearized at the steady-state conditions. The linearized equations of motion can be derived from small perturbation theory and in Matlab library. The linearized F16 model for auto-landing phase can be described using the following equations:

$$\dot{x} = A\,x + B\,u$$
$$y = C\,x + D\,u \tag{2}$$

X=[h V$_t$ dlat βαθφψ]$'$ are states of this model, where, α、 β are attack angle and sideslip angle (deg); φ、 θ、 ψ are roll, pitch, yaw angles (deg); V_t is Total air speed (m/s); h is altitude (m); dlat is the lateral error (m).

The control inputs to the aircraft are u=[δ_{hl} δ_{hr} δ_{al} δ_{ar} δ_r δ_T]$'$, means deflection angle of left/right elevator, left/right aileron, rudder respectively (deg) and throttle.

3   Neuro- H$_2$ Controller Design

In this section, the H$_2$ controller is designed firstly. To improve the auto-landing quality, MRAN Network aided H$_2$ control strategy is designed and implemented here.

Fig. 1 present the strategy used in this paper. The H$_2$ controller in the inner loop plays an important role in the proposed strategy. It is not only used to stabilize the overall system, but also provide error signals to train the network. Under normal operating condition, the robust controller will provide better performance, and when the environment change suddenly the neural network controller can quickly learn these changes and provide appropriate control signals to maintain satisfactory performance.

Fig. 1 This control architecture uses a conven-
tional H₂ controller in
the inner loop to stabilize
the system dynamics,
and the Neuro-controller
acts as an aid to the H₂
controller through on-
line learning.

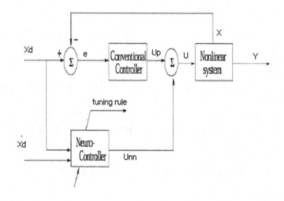

Fig. 2 The H₂ controller
used here is an LMI-
based H₂ design method
which can be done with
the help of MATLAB
[14] [15] [16]. MRAN
controller learns the
needed control inputs
quickly through on-line
learning and generates a
larger control signal,
driving the aircraft to
follow the desired
outputs.

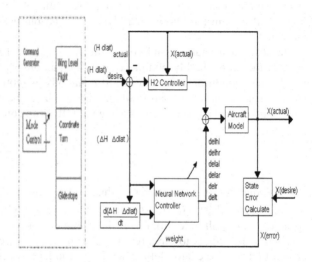

Fig. 2 shows MRAN aided H₂ control scheme for auto-landing. MRAN is a sequential learning algorithm for implementing "minimum" RBF neural network,
which is developed by Lu *et. al.* [9][14][17][18]. As in [19], the error for updating
the neural network is based on the conventional controller's output signals as it reflects the error between the commands and actual outputs. And in [5], the error for
updating the neural network is the model state's error. Compared with the RBFN
scheme where only tuning the weights is done, the MRAN algorithm adapts all the
parameters of the network (including widths and centers) via an Extended Kalman
Filter, and hence improves the tracking accuracy.

The typical landing scenario can be divided into wing level flight、 coordinated
turn、 glide slope and landing flare. The simulation result of wing level flight
phase with left elevator stuck at 7deg can be seen in Fig. 3. All the state variables 、 hidden neuron-history and control signals are showed here. Specification of variables and control signals can be seen in Equation 2.

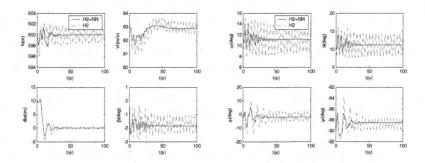

Fig. 3 Simulation result of wing level flight under left elevator stuck 7deg

From the Fig. 3, it can be seen that Neuro-aided H$_2$ controller decreases the "dlat" and smooth other variables than the H$_2$ controller under actuator stuck. And the number of hidden neurons can be added or pruned automatically.

4 Concluding Remarks

From the simulation results, it can be seen that the H$_2$ controller designed creates large errors under the aileron (or elevator) stuck condition, but the Neuro-aided H$_2$ controller is able to remedy the big error and track the desired landing trajectories better. The Neuro-aided H$_2$ control system can enlarge the stability zones than H$_2$ control system.

References

[1] Pallett, E.H.J., Coyle, S.: Automatic Flight Control. Blackwell Scientific (1993)
[2] Donald, M.: Automatic Flight Control Systems. Prentice-Hall (1990)
[3] Bole, M., Svoboda, J.: Design of a Reconfigurable Automated Landing System for VTOL Unmanned Air Vehicles. In: Proceedings of the IEEE 2000 National Aerospace and Electronics Conference, NAECON 2000, pp. 801–806 (2000)
[4] Jorgensen, C.C., Scheley, C.: Neural Network Baseline Problem for Control of Aircraft Flare and Touchdown. In: Neural Networks for Control, pp. 402–425. MIT Press, Cambridge (1990)
[5] Calise, A.J.: Neural Networks in Non-Linear Aircraft Flight Control. IEEE Aerospace and Electronic Systems Magazine, 5–19 (July 1996)
[6] Steinberg, M., Di Giralamo, R.: Applying Neural Network Technology to Future Generation Military Flight Control Systems. In: International Joint Conference on Neural Networks, Seattle, WA (1991)
[7] Iiguni, Y., Akiyoshi, H., Adachi, N.: An Intelligent Landing System Based On a Human Skill Model. IEEE Trans. on Aerospace and Electronic System 34(3), 877–882 (1998)

[8] Napolitano, M.R., Kincheloe, M.: On-Line Learning Neural-Network Controllers for Autopilot Systems. AIAA Journal of Guidance, Control and Dynamics 33(6), 1008–1015 (1995)

[9] Lu, Y., Sundararajan, N., Saratchandran, P.: Performance Evolution of a Sequential Minimal Radial Basis function(RBF) Neural Network Learning Algorithm. IEEE Tran. on Neural Network 9(2), 308–318 (1998)

[10] Cao, Y.Y., Lin, Z.L., Ward, D.G.: H_{inf} Antiwindup Design for Linear Systems Subject to Input Saturation. Journal of Guidance, Control, and Dynamics 25(3), 455–463 (2002)

[11] Naidu, D.S., Banda, S.S., Buffington, J.L.: Unified Approach To H_2 and H_{inf} Optimal Control of a Hypersonic Vehicle. In: Proc. of the American Control Conference, pp. 2737–2741, San Diego, California (June 1999)

[12] Hyde, R.A., Glover, K.: The Application of Scheduled H_{inf} Controllers to a VSTOL Aircraft. IEEE Trans. on Automatic Control 38(7), 402–425 (1993)

[13] Nguyen, L.T., Marilyn, E.O., William, P.G., Kemper, S.K., Philip, W.B., Deal, P.L.: Simulator study of stall/post stall characteristics of a fighter airplane with relaxed longitudinal static stability. NASA Technical Paper 1538, USA (1979)

[14] Li, Y., Wang, Z.F., Sundararajan, N.: Robust Neural Controller Design for Aircraft Auto Landing Under Failures., DSOCL01144, Final Report (December 2003)

[15] Che, J., Chen, D.G.: Automatic Landing Control using Hinf Control and Stable Inversion. In: Orlando, F.U. (ed.) Proceedings of the 40th IEEE Conference on Decision and Control (December 2001)

[16] Shue, S., Agarwal, R.K.: Design of Automatic Landing Systems using Mixed H2/Hinf Control. Journal of Guidance, Control and Dynamics 22(1), 103–114 (1999)

[17] Li, Y., Sundararajan, N., Saratchandran, P., Wang, Z.F.: Robust Neuro-Hinf Controller Design for Aircraft Auto-Landing. IEEE Transactions on Aerospace and Electronic Systems 40(1), 167–168 (2004)

[18] Li, Y., Wang, Z.F., Sundararajan, N., Saratchandran, P.: Robust and Adaptive Neuro-Controller for Aircraft Undergoing Microburst. In: Proceedings of the American Control Conference, Denver, Colorado (2003)

[19] Gomi, H., Kawato, M.: Neural network control for a closed-loop system using feedback-error-learning. Neural Networks 6(7), 933–946 (1993)

Transcriptome to Reactome Deterministic Modeling: Validation of *in Silico* Simulations of Transforming Growth Factor-β1 Signaling in MG63 Osteosarcoma Cells, TTR Deterministic Modeling

Clyde F. Phelix, Bethaney Watson, Richard G. LeBaron, Greg Villareal, and Dawnlee Roberson

Abstract. Integrated Systems Biology was used to study bone cancer via an iterative process of *in vitro* testing for validation of an *in silico* computer simulation where the transcriptome was used to derive the parameters of a kinetic model. A computer simulation model of the transforming growth factor-beta (TGF-β1) signaling pathway was obtained from Reactome®. The transcriptome of MG-63 cells was accessed from NCBI GEO GSE11414. With this method the model is not trained to match the biological system. The *in vitro* study on osteosarcoma (MG-63) cells was used to compare with the results from the computer simulation. MG-63 cells were grown in culture and exposed to TGF-β1 to identify differences in expression of a target-gene, TGF-β-Induced 68kDa protein (TGFBI), at serial time intervals. Real-time PCR was used to measure TGFBI mRNA levels and the temporal profile was identical with that predicted by the *in silico* model. A sensitivities test was performed through the *in silico* model and a candidate target for gene-knock-down in the TGF-β1 signaling pathway, Smad3, was identified. An 80% reduction of this reactant in the model attenuated TGFBI expression by 64%, an effect that matched such knockdown of Smad3, *in vitro*, for other target genes reported in the literature. The assumption that the transcriptome drives the reactome is validated and substantiates a novel method for deriving parameters for kinetic deterministic models of biological systems.

Clyde F. Phelix · Bethaney Watson · Richard G. LeBaron
University of Texas at San Antonio, San Antonio, TX, USA
e-mail: {clyde.phelix, Richard.LeBaron}@utsa.edu,
        bethaney.watson@gmail.com

Greg Villareal
CFPAL Biomedical Consultants, San Antonio, TX, USA
e-mail: gregvillarealphd@gmail.com

Dawnlee Roberson
AL Phahelix Biometrics Inc, San Antonio, TX, USA
e-mail: dawnlee.roberson@gmail.com

F.L. Gaol et al. (Eds.): Proc. of the 2011 2nd International Congress CACS, AISC 145, pp. 451–457.
springerlink.com   © Springer-Verlag Berlin Heidelberg 2012

Keywords: kinetic model, gene expression profile, parameter estimation, sensitivities analysis.

1 Introduction

The promise of *in silico* simulations in biological research was to reduce the number of actual studies by allowing researchers to control parameters in systems individually, which is normally not possible *in vivo*. This research integrates work by a significant number of researchers, publically available genome-wide microarray data sets, and validates one particular cell signaling collection of reactions from publically available pathway models. More importantly, a novel method of parameter determination from gene expression profiles is substantiated. This method generates kinetic models that simulate the biological systems without training and parameter estimations; all parameters are determined by derivation from the transcriptome. Phelix and coworkers [1] recently reported the utility of *in situ* hybridization as a source of gene expression profiles to model metabolic pathways in brain tissue; this report extends the method to the use of genome-wide microarray transcription profiles for determining model parameters and for signaling pathways.

Integrated Systems Biology includes *in silico, in vitro* and *in vivo* studies for the iterative process of validating model approaches. Such approaches for modeling the transforming growth factor-beta (TGF-β) signaling pathway have used results of the biological systems studies, including parameter estimations, to train the models to fit the biological data [2-5]. TGFβ-Induced Gene Human Clone 3 (BIGH3), also known in the literature as TGF-β Induced 68kDa protein (TGFBI) and keratoepithelin, is an integrin adhesion-class extracellular matrix protein. LeBaron, Phelix and coworkers have reported that TGFBI mediates the TGF-β1-dependent apoptosis in cultured osteosarcoma, MG-63, cells [6]. TGFBI is a widely recognized target gene of TGF-β1 signaling via the Smad protein pathway [7], but has not been included in any model system yet reported [2-5]. This study is the first to report kinetic modeling of the target gene expression in the TGF-β signaling pathway and to validate those expression dynamics by *in vitro* testing of temporal expression patterns and gene-knock-down studies [8,9].

The purposes of the *in vitro* study were to document that TGF-β1 evokes TGFBI gene expression in MG-63 osteosarcoma cells with a characteristic temporal pattern. Osteosarcoma cells were treated with TGFβ-1 to stimulate Smad signaling leading to TGFBI expression. After TGF-β1 was added to cells, TGFBI mRNA expression was measured. The result was compared to the *in silico* TGFBI mRNA levels in a kinetic model where the parameters were determined by the genome-wide transcription profile of MG-63 osteosarcoma [10] cells to determine if the computer model could be validated. The overall goal is to develop successful biomarker identification as novel cancer therapies for the individual patient through individualized personalized medicine. Sensitivities analyses were performed to identify a single knock-down candidate and thus Smad3 was chosen for its strongest sensitivity value. Knock-down, *in silico*, of Smad3 protein expression attenuated target gene expression comparable to that reported for *in vitro* studies by others [8,9]. We have established the Transcriptome-To-Reactome™Method for computational modeling of biological systems networks.

2   Methods

Transcriptome-To-Reactome™Method *in silico* model: The TGF-β signaling pathway model was obtained from Reactome®[11], downloaded as an SBML file that was imported into COmplex PAthway SImulator (COPASI®) for computational modeling of biochemical networks [12]. Manual curation was required to adapt the model for the TGFBI gene as a target for Smad transcription factor effector functions [2-5].

The model incorporated 36 ordinary differential equations for 29 reactions and 63 reactants or species, using initial conditions determined from the genome-wide microarray data, GSE11414 [10] accessed from NCBI GEO [13]. There were 8 compartments. The model included 43 parameters derived by globalization [14] from expression levels of 35 genes for initial species levels and k-values of the mass action reactions. The time course simulation as deterministic (LSODA) method was run to generate the species level and reaction flux value reports. The relative tolerance was 1^{-06}, absolute tolerance was 1^{-12}, and maximum internal steps were 10^3. The simulation duration was 6,000 model minutes. Additionally a sensitivities analysis was run on the time series simulation with the function as all variables of the model and variable as all parameter values. The parameter delta factor was 1^{-06} and delta minimum was 1^{-12}. Microsoft Excel 2007 was used to assess and graph all data.

Cells: MG-63 cells [ATCC CRL-1427] were grown as a monolayer in growth medium containing 10% fetal bovine serum [6]. For experiments, 10^3 cells were seeded in each well of a 24-well plate. Then 10 ng/mL of TGF-β1 was added to a subset of the wells. The TGF-β1 vehicle was added to other control wells. RNA extractions were performed in duplicates along with a control at 8 time points, 1, 5, 9, 17, 33, 41, 49, and 57 hours.

In vitro quantitative Real-Time PCR (qRT-PCR): MG-63 RNA was extracted using RNeasy Mini Kit protocol (Qiagen) and cDNA was synthesized with TaqMan reverse transcription reagents (Applied Biosystems). For TGFBI qRT-PCR analysis, the primer pairs were used 5 GTCTCCCTTCAGGACATCCA (forward) and 5 TGGACAGACCCTGGAAACTC (reverse). Results were normalized to 18s RNA. These *in vitro* results were then compared to the *in silico* results.

3   Results and Discussion

3.1   *In Silico Results*

The TGF-β1 was administered in a bolus amount in the simulation matching the exposure of cells for the *in vitro* study. All signaling events, i.e., species levels and reaction flux values, displayed an essentially bell-shaped response with a rapid rise phase to a peak value and a rapid fall phase back to baseline levels. Species levels, as concentration (Fig.1A), and reaction flux values (Fig.1B) were measured as peak values and time to peak values. For the Smad protein complexes that act as transcription factors, both the cytosolic and nucleoplasmic concentrations peaked at about 9 hours. The TGFBI mRNA levels peaked at this time also (Fig.2).

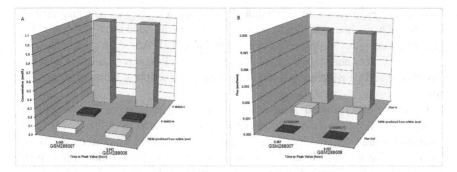

Fig. 1 MG-63 Osteosarcoma cells, 3-D graphs showing concentration or flux on the y-axis, time to peak value and sample identifier on the x-axis and dependent variables measured on the z-axis. Results of phospho-Smad in the cytoplasm (P-SMAD-C) and in the nucleus (P-SMAD-N) are shown in (A). The P-SMAD-N acts as a transcription factor to change gene expression; TGFBI is one of those target genes. Note uniformity of simulation results from two independent replicate microarray data sets. In (B) the flux of P-SMAD-C into the nucleus and of P-SMAD-N out of the nucleus are shown. These results show a consistent effect on the TGFBI levels as predicted from the mRNA level within each microarray data set. This TGFBI level is not the result of the simulation, but is a result from the original experiment on the MG-63 cells for which the microarray test was run.

Fig. 2 Depicts the temporal profile of TGFBI gene expression as mRNA levels for the *in vitro* (straight line curve) and *in silico* (smooth curve) results. This result is validation of the model. The values for relative expression on the y-axis were adjusted such that the values for both the simulation and quantitative RT-PCR can be seen.

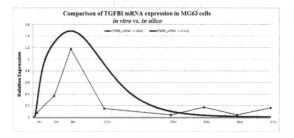

3.2   In Vitro Results

Post TGF-β1 addition, the *in vitro* results showed a roughly bell-shaped TGFBI mRNA expression pattern peaking at 9 hours. The *in silico* model revealed the TGFBI mRNA level peaked at exactly the same time as the *in vitro* results (Fig.3). The temporal profiles of target gene expression dynamics have similar appearances and provide strong evidence to validate the transcriptome to reactome model.

3.3   Model Sensitivities

The results of sensitivities analysis are shown in Figure 4. Essentially two reactions showed the greatest sensitivities to subsets of reactants. The TGFBI mRNA expression reaction showed strongest positive sensitivity to the Smad2 and Smad3 reactants; Smad3 was of interest as a candidate for *in silico* knock-down for model validation. A common *in vitro* method for selective knock-down of protein expression is via addition of small interfering ribonucleic acids (siRNAs) [8,9], in particular against Smad3 for investigating TGF-β1 signaling and target gene expression.

A strong knock-down effect of siRNAs can achieve greater than 80% reduction in protein expression; thus we chose that level of reduction in the level of Smad3 mRNA that was a reactant for cytosolic Smad3 protein production in the model. When the time course simulation was run after this knock-down, the peak value of

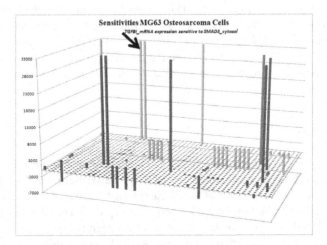

Fig. 3 Sensitivities analysis of the TGF- signaling for the MG-63 cells shows only two major reactions (z-axis) with sensitivity values in the range of 3300 (y-axis). These reactions (TGFBI mRNA expression in background arrow, and TGF-1 dimer binding to the TGFβ receptor-2 in foreground) are sensitive (positive values) and insensitive (negative values) to subsets of reactants (x-axis). At the arrow, the reactant is Smad-3. This was used as a biomarker for testing by simulating the use of siRNA to attenuate the expression of the candidate target, Smad-3 mRNA in the model, down to 80% below control.

Fig. 4 Note in the knock-down model (hatched line) that the target gene expression (TGFBI-mRNA) is suppressed by approximately 64% of control (solid line). This target gene attenuation with the simulation matches that reported by Kobayashi et al [8,9].

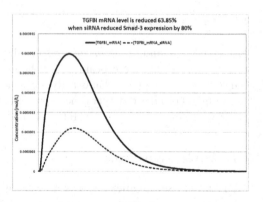

the TGFBI mRNA was reduced by 64% (Fig.4). Kobayashi et al. [8,9] used immunoblot and PCR methods to measure the effectiveness of Smad3 siRNAs on cultured cells and found similar magnitudes of attenuated Smad3 protein expression, as well as, target gene expression, respectively.

3.4 Model Validation

Nearly exact results were achieved for the simulation as compared to the *in vitro* studies in our hands for TGFBI mRNA expression and for additional TGF-β1 target genes as reported by others [8,9]. This accuracy of mimicking the actual biological systems is accomplished without training the model or parameter estimations that are commonly used by other modeling methods [2-5]. This present study is the second reported by our group where the derivation of model parameters from gene expression levels replicated biological systems, accurately, reproducibly, and reliably [1].

4 Summary

The Transcriptome-To-Reactome™(TTR™) Method for determining kinetic model parameters is validated by this study. The TTR approach determines model behavior that mimics and matches the actual biological system from which the cell sample was collected to generate the genome-wide microarray expression profile (transcriptome). This TTR™Method allows for the generation of an integrated system biological computational model for each individual sample for which the transcriptome is generated. We will use this method to develop individualized personalized medicine. The sensitivities analyses are useful for biomarker identification and for detection and testing of candidate targets for novel drug development [15]. The TTR™Method provides solutions for many of the obstacles identified by the pharmaceutical industry [16].

References

1. Valdez, C.M., Phelix, C.F., Smith, M.A., Perry, G., Santamaria, F.: Modeling cholesterol metabolism by gene expression profiling in the hippocampus. Molecular BioSystems 7, 1891–1901 (2011)
2. Zi, Z., Klipp, E.: Constraint-based modeling and kinetic analysis of the Smad dependent TGF- signaling pathway. PLoS One 2(9), e936 (2007)
3. Clarke, D.C., Liu, X.: Decoding the quantitative nature of TGF-β/Smad signaling. Trends Cell Biol. 18(9), 430–442 (2008)
4. Melke, P., Jonsson, H., Pardali, E., ten Dijke, P., Peterson, C.: A rate equation approach to elucidate the kinetics and robustness of the TGF-beta pathway. Biophysical Journal 91(12), 4368–4380 (2006)
5. Chung, S.W., Miles, F.L., Sikes, R.A., Cooper, C.R., Farach-Carson, M.C., Oqunnaike, B.: Quantitative modeling and analysis of the transforming growth factor beta signaling pathway. Biophysical Journal 95(5), 1733–1750 (2009)
6. Zamilpa, R., Rupaimoole, R., Phelix, C.F., Somaraki-Cormier, M., Haskins, W., Asmis, R., LeBaron, R.G.: C-terminal fragment of transforming growth factor beta-induced protein (TGFBIp) is required for apoptosis in human osteosarcoma cells. Matrix Bio. 28(6), 347–353 (2009)
7. Yuan, C., Yang, M.-C., Zins, E.J., Boehlke, C.S., Huang, A.: Identification of the promoter region of the human IGH3 gene. Molecular Vision 10, 351–360 (2004)
8. Kobayashi, T., Liu, X., Wen, F.Q., et al.: Smad3 mediates TGF-beta1 induction of VEGF production in lung fibroblasts. Biochem. Biophys. Res. Commun. 327(2), 393–398 (2005)
9. Kobayashi, T., Liu, X., Wen, F.Q., et al.: Smad3 mediates TGF-beta1-induced collagen gel contraction in human lung fibroblasts. Biochem. Biophys. Res. Commun. 339(1), 290–295 (2006)
10. Sadikovic, B., Yoshimoto, M., Al-Romaih, K., Maire, G., Zielenska, M., Squire, J.A.: In vitro analysis of integrated global high-resolution DNA methylation profiling with genomic imbalance and gene expression in osteosarcoma. PLoS One 3(7), e2834 (2008)
11. Croft, D., et al.: Reactome: a database of reactions, pathways and biological processes. Nucleic Acids Res. 39, D691–D697 (2011) (accessed April 15, 2010)
12. Hoops, S., et al.: COPASIa COmplex PAthway SImulator. Bioinformatics 22(24), 3067–3074 (2006)
13. Barrett, T., et al.: NCBI GEO: archive for high-throughput functional genomic data. Nucleic Acids Res. 37, D885–D890 (2009)
14. Fundel, K., Kuffner, R., Aigner, T., Zimmer, R.: Normalization and gene p-value estimation:issues in microarray data processing. Bioinform. Biol. Insights 2, 291–305 (2008)
15. van Riel, N.A.W.: Dynamic modeling analysis of biochemical networks: mechanism-based models and model-based experiments. Briefings in Bioinformatics 7(4), 364–374 (2006)
16. Gomes, G.: Practical applications of systems biology in the pharmaceutical industry. Int. Drug. Disc 5(2), 54–57 (2010)

Advanced Co-phase Traction Power Supply Simulation Based on Multilevel Converter*

Zeliang Shu, Xifeng Xie, and Yongzhi Jing

Abstract. Serious power quality and neutral sections disadvantage restrict the evolution of traditional traction power supply system. An advanced co-phase traction power supply system is proposed in this paper based on three-phase to single-phase converter as the key equipment in substation. This converter adopts multi-module diode clamped multi-level algorithm, connects between industrial grid and feeder line without using output transformer. It transfers active power from three-phase grid isolated to one-phase traction line, and provides the frequency, phase and amplitude controlled output voltage. Then, the feeder line between different substation can be connected directly, and constructs an ideal co-phase traction supply system in the same phase. The performance of the converter is validated by the discussion and simulation results in this paper.

1 Introduction

Feeding electric railway traction loads using transformer from public three-phase industrial grid may lead to critical problems of power quality, including harmonics, reactive currents, and power unbalance. The distortion degree depends on the movement, tractive profile and converter scheme of electric locomotives, and supply scheme of traction power [2]. These problems will present a potentially huge impact to the grid as the substations (SS) are rated up to a huge power.

Zeliang Shu · Yongzhi Jing
School of Electrical Engineering, Southwest Jiaotong University
e-mail: {mailtosunny, jyzbenben}@163.com

Xifeng Xie
Guangxi Hydraulic and Electric Polytechnic
e-mail: xiexifeng7901@163.com

* The research are supported by the National Natural Science Foundation of China, No.51007075.

F.L. Gaol et al. (Eds.): Proc. of the 2011 2nd International Congress CACS, AISC 145, pp. 459–465.
springerlink.com © Springer-Verlag Berlin Heidelberg 2012

Meanwhile, the traditional power supply system are always separated into many of substations as illustrated in Fig. 1. In this scheme, one substation commonly feds two-phase feeding wires from three-phase utility grid. The two different track wires are connected to the two secondary windings of transformer, which must be kept strictly isolated for different phase. The neutral sections (NS) are used at the joints between different feeding wires in order to prevent the risk of phase mixing. For example, there are five NSs between three SSs as shown in Fig. 1. But the neutral section will cause electrical separation from power supply continually. In other word, the pantograph can not transfer power in neutral section. Therefore, the excessive neutral sections will bring about losses of traction power and train speed. It becomes more seriously for high speed and heavy-loaded railway transportation [4].

Some static compensators using static synchronous compensator (STATCOM) and active power filter (APF) are usually adopted [1, 5]. An approach using railway static power conditioner (RPC) and Scott transformer is adopted to correct the unbalanced load current introduced in [6]. A co-phase traction power supply system is adopted using single-phase back-to-back converter in [7], which has only one phase feeding wire instead of two-phase feeding wires. The neutral section is required only at the end of the feeding wire in one SS. One half numbers of neutral sections in the whole traction line can be reduced than traditional two-phase traction networks. But this scheme can not connect all the SSs to each other and compose one feeder wire of the whole traction power supply system.

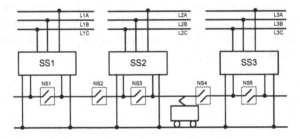

Fig. 1 Traditional traction power supply scheme.

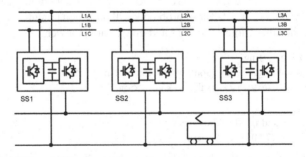

Fig. 2 Advanced co-phase traction power supply scheme.

In this paper, an advanced co-phase traction power supply system is proposed as illustrated in Fig. 2. It adopts three-phase to single-phase diode-clamped multilevel converters (DCMC) as the main equipments of the SSs, which transfer power from three-phase industrial power grid to single-phase co-phase traction power networks. Meanwhile, because the frequency, phase and amplitude of the output voltage are fully controlled, the feeder lines from different subsections can be directly connected into one network. It inherits the advantages of power electronics-based converters such as more flexibility, unity power factor, perfect steady-state and dynamic performances.

2   Model of the Converter-Based System

The three-phase to single-phase converter is the key equipment of the advanced co-phase traction power supply system, so it will be discussed in details in this paper. Figure 3 illustrates a $n + 1$ level diode-clamped three-phase to single-phase converter with dc balancer circuit. As indicated in this figure, it is composed of the following four submodules: 1. Three-phase rectifier: a converter that transfers active power between three-phase grid and dc capacitor; 2. Capacitor: serial capacitors store power to provide dc voltage that the back-to-back converter needs; 3. DC balancer: an auxiliary circuits to balance the voltage of each dc capacitor using capacitors and switches; 4. Single-phase inverter: a single-phase converter that transfers active power between feeder line and dc capacitor, meanwhile outputs the controlled traction supply voltage.

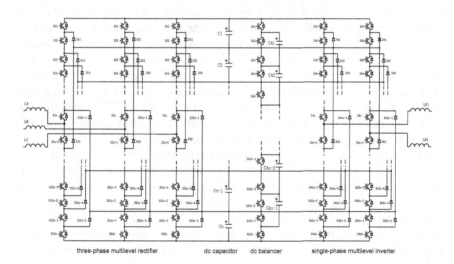

Fig. 3 Multi-Level diode-clamped three-phase to single-phase converter with dc balancer circuit.

The three-phase rectifier circuit with dc capacitor in this scheme operates as the common PWM rectifier. It controls the amplitude and direction of active power currents between three-phase grid and dc capacitor to maintain the dc voltage. In some rectifier control strategies such as synchronous reference frame based control. The three-phase input currents are transformed into dqz vector. Then, the fundamental components in abc frame becomes a dc term (i_d, i_q, i_z) in dqz frame. The fundamental components, active power i_d and reactive power i_q are relatively easy to be controlled through error-based loop control, and reference v_d, v_q can be transformed with PI controller. Then, the instantaneous waveforms (v_a, v_b, v_c) can be produced by applying the inverse synchronous reference transformation. They can be set as the modulation reference waveforms of the rectifier.

The serial dc capacitor module provide the rectifier and inverter a stable dc voltage. On the other hand, these capacitors are the temporary storage for active power absorbed from grid through the three-phase rectifier module. Meanwhile, they supply traction feeder with the equivalent active power through the single-phase inverter. In spite of their own power loss, their input and output power are equal, therefore, the stable dc voltage can be guaranteed with the dc voltage control strategy in the i_d loop of the three-phase rectifier.

Balanced dc capacitor voltage is the necessary precondition for stable operation of a DCMC. Due to the active power drawn and the transients or unbalanced components of the load, the dc capacitor voltage drift phenomenon will occur in diode-clamped multilevel converters [3]. In the proposed traction power converter, the primary target is transferring active power from three-phase to single-phase grid. Therefore, this converter needs a dc balancing control to avoid dc capacitor voltage unbalance. In our scheme, an auxiliary circuit-based equalizing circuit is adopt on the dc side and utilizing balancing control strategies as shown in Fig. 3. The most simple control scheme is alternatively turning on/off the adjacent series switches in two modulation sequences. According to this modulation, the switches are divided into two groups for each of the two adjacent devices. There are one active switch and diode are in conduction in each period, which construct two paths for the charge or discharge currents. Though it requires additional circuits and control hardware, which add the system cost and complexity, but it extends the DCMC operating stable region and generally is independent from PWM modulation scheme of main circuits, and is applicable for DCMC with any number of levels.

The single-phase inverter module outputs the traction voltage to feeder the locomotives active power current. The frequency, amplitude and phase of voltage should be well controlled same as the other substation. Under this condition, the converter can be combined to traction line in advanced co-phase application as illustrated in Fig. 2. In the inverter control strategy, a phase-locked loop can be used to obtain the phase and frequency of the feeder line voltage. Simultaneously, dq-based control for single-phase inverter will output the equivalent active power as rectifier absorbed from three-phase grid.

Based on the three-phase to single-phase converter, a output transformerless advanced co-phase power supply system can be config as illustrated in Fig. 4. This system cascades numbers of three-phase to single-phase converter module discussed before. Though one module output voltage are limited by present power electronic technology, the series modules can output voltage several times that of one module. Because the input voltage of three-phase grid can be in several grade, the input transformers are still used in this scheme. Meanwhile, the transformers provide isolated input voltages between different converters. Then the outputs of different converters can be serial connected, and combined to form one high voltage of traction supply.

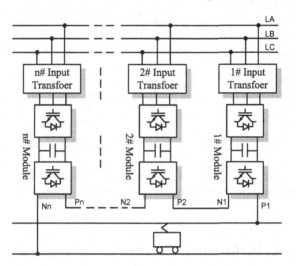

Fig. 4 Transformerless converter system configuration.

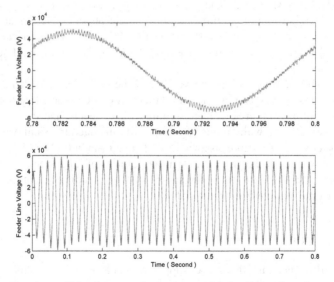

Fig. 5 Output voltage of the five cascade converters, (a) detail waveform in one cycle, (b) transient waveform when resistance load change from 200 Ω to 66.7 Ω.

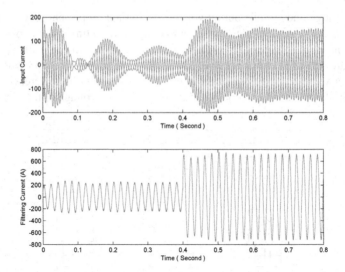

Fig. 6 Transient output and input current waveforms of the five cascade converters when resistance load change from 200 Ω to 66.7 Ω, (a) input three-phase currents, (b) output single-phase current.

3 Simulation Verification

In order verify the system operation proposed above, simulation has been conducted using MATLAB/SIMULINK. The simulation adopts discrete modeling algorithm with $2e-6$ sample time. Five three-phase to single-phase converters are used in our study, which are cascaded as shown in Fig. [?]. Each converter can provide maximum 6kV output voltage, then five modules are sufficient to supply a 27.5kV traction feeder line.

Fig. 5(a) shown the detail output voltage of the cascaded converters. Because the phase-to-phase voltage of one five-level converter output 9-level staircase wave-form, The resulting output ac voltage swings with 45 levels, and the waveform is nearly sinusoidal, even without filtering. Fig. 5(b) illustrated transient waveform of the cascade converters when resistance load change from 200 Ω to 66.7 Ω at 0.4s. Simultaneously, Fig. 6 shown the transient waveforms of input tree-phase currents and output single-phase current of the supply system. It indicates that the rectifiers needs more time than inverters to be stable.

4 Conclusion

An advanced co-phase traction power supply system is discussed in this paper. This system adopts cascaded three-phase to single-phase diode-clamped multilevel converters (DCMC) with dc voltage auxiliary balancer. This converter system transfer active power from three-phase industrial power grid to single-phase co-phase traction power networks. It inherits the advantages of power electronics-based converters such as light weight, small size, more flexibility, unity power factor, perfect steady-state and dynamic performances. The basic function and dynamic performance of the system is validated by the simulation results in this paper.

References

1. Cetin, A., Ermis, M.: Vsc-based d-statcom with selective harmonic elimination. IEEE Transactions on Industry Applications 45(3), 1000–1015 (2009)
2. Heising, C., Bartelt, R., Oettmeier, M., Staudt, V., Steimel, A.: Analysis of single-phase 50-kw 16.7-hz pi-controlled four-quadrant line-side converter under different grid characteristics. IEEE Transactions on Industrial Electronics 57(2), 523–531 (2010)
3. Marchesoni, M., Tenca, P.: Theoretical and practical limits in multilevel mpc inverters with passive front ends. In: Proc. Eur. Conf. Power Electronics and Applications, Graz, Austria (August 2001)
4. Pao-Hsiang, H., Shi-Lin, C.: Electric load estimation techniques for high-speed railway (hsr) traction power systems. IEEE Transactions on Vehicular Technology 50(5), 1260–1266 (2001)
5. Pee-Chin, T., Poh Chiang, L., Holmes, D.G.: A robust multilevel hybrid compensation system for 25-kv electrified railway applications. IEEE Transactions on Power Electronics 19(4), 1043–1052 (2004)
6. Uzuka, T., Ikedo, S.: Railway static power conditioner field test. Quarterly Report of RTRI 45(2), 64–67 (2004)
7. Zeliang, S., Shaofeng, X., Qunzhan, L.: Single-phase back-to-back converter for active power balancing, reactive power compensation, and harmonic filtering in traction power system. IEEE Transactions on Power Electronics 26(2), 334–343 (2011)

Exponential Convergence of a Randomized Kaczmarz Algorithm with Relaxation

Yong Cai, Yang Zhao, and Yuchao Tang*

Abstract. The Kaczmarz method is a well-known iterative algorithm for solving linear system of equations $Ax = b$. Recently, a randomized version of the algorithm has been introduced. It was proved that for the system $Ax = b$ or $Ax \approx b + r$, where r is an arbitrary error vector, the randomized Kaczmarz algorithm converges with expected exponential rate. In the present paper, we study the randomized Kaczmarz algorithm with relaxation and prove that it converges with expected exponential rate for the system of $Ax = b$ and $Ax \approx b + r$. The numerical experiments of the randomized Kaczmarz algorithm with relaxation are provided to demonstrate the convergence results.

1 Introduction and Preliminaries

Image reconstructive problems arising in computerized tomography (CT) often can be modelled by the equation

$$Ax = b, \tag{1}$$

where $A_{m \times n}$ is a big and sparse matrix; $b \in \mathbb{R}^m$ is the observed data and $x \in \mathbb{R}^n$ is the original image.

The Kaczmarz algorithm[8] is one of the most important methods for solving linear system (1) and is a projected method. It is also referred as the algebraic reconstruction technique (ART)[6] in image reconstruction problems. The Kaczmarz algorithm is presented as follows:

Yong Cai · Yang Zhao
Department of Mathematics, NanChang University, Nanchang, P.R. China 330031

Yuchao Tang
Institute for Information and System Science, Faculty of Science, Xi'an Jiaotong University
Xi'an, P.R. China 710049
e-mail: hhaaoo1331@yahoo.com.cn

* Corresponding author.

F.L. Gaol et al. (Eds.): Proc. of the 2011 2nd International Congress CACS, AISC 145, pp. 467–473.
springerlink.com

Algorithm 1.
Initialization: $x_0 \in \mathbb{R}^n$ arbitrary,
Iterative Step:

$$x_{k+1} = x_k + \frac{b_{i(k)} - \langle a_{i(k)}, x_k \rangle}{\|a_{i(k)}\|_2^2} a_{i(k)}, \qquad (2)$$

where $i(k) = (k \bmod m) + 1$ and $\| \cdot \|_2$ denotes the Euclidean norm.

The Kaczmarz's algorithm is sequential. At each iteration, the previous iteration is projected orthogonally onto the solution hyperplane $\langle a_i, x \rangle = b_i, i = 1, 2, \cdots, m$, where a_i denotes the row of A. There are also some other important iterative algorithms for solving linear system (1), such as Cimmino algorithm[4], symmetric Kaczmarz algorithm[1], Component Averaging algorithm[2], etc. Although these algorithms are popular, the rate of convergence for these methods have been difficult to obtain in theory. Since the Kaczmarz algorithm cycles through the rows of A sequentially, its convergent rate depends on the order of the rows. In order to accelerate the rate of convergence, one way is to use the rows of A in a random order instead of sequence. Some observations are made on the improvements of this randomized version[5, 7], but the ice has been broken recently. In [10, 11], Strohmer and Vershynin first proposed a randomized version of the Algorithm 1 as follows:

Algorithm 2. Randomized Kaczmarz algorithm
Initialization: $x_0 \in \mathbb{R}^n$ arbitrary,
Iterative Step:

$$x_{k+1} = x_k + \frac{b_{r(i)} - \langle a_{r(i)}, x_k \rangle}{\|a_{r(i)}\|_2^2} a_{r(i)}, \qquad (3)$$

where $r(i)$ is chosen from the set $\{1, 2, \cdots, m\}$ at random, with probability proportional to $\|a_{r(i)}\|_2^2$.

They proved the following exponential bound on the expected rate of convergence for the randomized Kaczmarz method,

$$\mathbb{E}\|x_k - x\|_2^2 \leq (1 - \kappa(A)^{-2})^k \|x_0 - x\|_2^2, \qquad (4)$$

where $\kappa(A) = \|A\|_F \|A^{-1}\|_2$, $\|A\|_F$ denotes the Frobenius norm of matrix A, and $\mathbb{E}$ denotes the expectation. Here and throughout, we will assume that A has full column rank so that $\|A^{-1}\|_2 = \inf\{M : M\|Ax\|_2 \geq \|x\|_2 \text{ for all } x\}$ is well defined.

Since the results of [10, 11], there has been some further discussion about the benefits of this randomized Kaczmarz algorithm (see e.g., [3, 12]). Strohmer and Vershynin[10, 11] first provided the proof on the rate of convergence of the randomized Kaczmarz algorithm. In terms of standard matrix properties, the rate is exponential in expectation. We are not aware of any other Kaczmarz method that provably achieves exponential convergence. Very recently, Needell[9] has considered the system of (1) is corrupted by noise r, i.e.,

$$Ax \approx b + r. \qquad (5)$$

He proved the randomized Kaczmarz algorithm reach an error threshold dependent on the matrix A with the same rate as in the error-free case.

Let x_k^* be the kth iteration of the noisy randomized Kaczmarz method running with $Ax \approx b + r$. He proved

$$\mathbb{E}\|x_k^* - x\|_2 \leq (1 - \kappa(A)^{-2})^{\frac{k}{2}}\|x_0 - x\|_2 + \kappa(A)\gamma, \tag{6}$$

where $\gamma = \max_i \frac{|r_i|}{\|a_i\|_2}$.

Remark 1. If the error term $r = 0$, the inequality (6) reduces to (4).

We consider the randomized Kaczmarz algorithm with relaxation in the present paper.

Algorithm 3. Randomized Kaczmarz algorithm with relaxation
Initialization: $x_0 \in \mathbb{R}^n$ arbitrary,
Iterative Step:

$$x_{k+1} = x_k + \lambda_{k,i}\frac{b_{r(i)} - \langle a_{r(i)}, x_k \rangle}{\|a_{r(i)}\|_2^2}a_{r(i)}, \tag{7}$$

where $r(i)$ is the same as Algorithm 2, $\lambda_{k,i}, i = 1, 2, \cdots, m$ are relaxation parameters and $\lambda_{k,i} \in (0, 2)$.

Remark 2. In (7), the parameters $\lambda_{k,i}$ vary with the iterative process. However, for the sake of convenience, we assume it to be constant here, then the iterative process (7) could be rewritten out

$$x_{k+1} = (1 - \lambda)x_k + \lambda P_{r(i)}x_k,$$

where $P_{r(i)}$ denotes the orthogonally projection of x_k onto the hyperplane $\langle a_{r(i)}, x \rangle = b_{r(i)}$, i.e., $P_{r(i)}x_k = x_k + \frac{b_{r(i)} - \langle a_{r(i)}, x_k \rangle}{\|a_{r(i)}\|_2^2}a_{r(i)}$. The Algorithm 2 is a special case of Algorithm 3, when the relaxation parameter $\lambda = 1$.

Strohmer and Vershynin[11] showed the randomized Kaczmarz algorithm computationally outperforming the conjugate gradient method (CGLS). Numerical experiments in [11] also demonstrate the randomized Kaczmarz algorithm is superior to the standard method in many cases. Using the above selection strategy, they are able to prove a proof for the expected rate of convergence that shows exponential convergence. No such convergent rate for any Kaczmarz method has been proved before. Needell[9] considered the noisy version of (1) and also proved the Kaczmarz method converges exponentially to the solution within a specified error bound. Motivated and inspired by the above works, in this paper we consider the randomized Kaczmarz algorithm with relaxation and prove that it converges with expected exponential rate for the system of (1) and (5), respectively. Our results generalize the results of Strohmer and Vershynin[11] and Needell[9].

2   Main Results

We shall use the following notations: Define a random vector Z whose values are the normals to all the equations of (1),

$$Z = \frac{a_j}{\|a_j\|_2} \quad \text{with probability} \quad \frac{\|a_j\|_2^2}{\|A\|_F^2}, j = 1, \cdots, m.$$

Theorem 1. *Let the system (1) is consistent and x be the solution of (1). Then Algorithm 3 converges to x in expectation, and*

$$\mathbb{E}\|x_k - x\|_2^2 \leq \left(1 - \lambda(2 - \lambda)\kappa(A)^{-2}\right)^k \|x_0 - x\|_2^2. \tag{8}$$

Proof. With the help of the well known inequality in Hilbert space, $\|\alpha x + (1 - \alpha)y\|^2 = \alpha\|x\|^2 + (1 - \alpha)\|y\|^2 - \alpha(1 - \alpha)\|x - y\|^2$, for any $\alpha \in (0, 1)$. We have

$$\begin{aligned}
\|x_k - x\|_2^2 &= \|(1 - \lambda)(x_{k-1} - x) + \lambda(P_{r(i)}x_{k-1} - x)\|_2^2 \\
&= (1 - \lambda)\|x_{k-1} - x\|_2^2 + \lambda\|P_{r(i)}x_{k-1} - x\|_2^2 \\
&\quad - \lambda(1 - \lambda)\|x_{k-1} - P_{r(i)}x_{k-1}\|_2^2.
\end{aligned} \tag{9}$$

Since the orthogonal projection P satisfying $\|Px - y\|_2^2 \leq \|x - y\|_2^2 - \|Px - x\|_2^2$, for any $x, y \in R^n$. Set $x := x_{k-1}, y := x$, then

$$\|P_{r(i)}x_{k-1} - x\|_2^2 \leq \|x_{k-1} - x\|_2^2 - \|P_{r(i)}x_{k-1} - x_{k-1}\|_2^2. \tag{10}$$

Substituting (10) into (9), we obtain

$$\begin{aligned}
\|x_k - x\|_2^2 &\leq (1 - \lambda)\|x_{k-1} - x\|_2^2 + \lambda\|x_{k-1} - x\|_2^2 \\
&\quad - \lambda\|P_{r(i)}x_{k-1} - x_{k-1}\|_2^2 - \lambda(1 - \lambda)\|x_{k-1} - P_{r(i)}x_{k-1}\|_2^2 \\
&= \|x_{k-1} - x\|_2^2 - \lambda(2 - \lambda)\|x_{k-1} - P_{r(i)}x_{k-1}\|_2^2.
\end{aligned} \tag{11}$$

By the definition of x_k, we have

$$\|x_{k-1} - x_k\|_2 = \lambda\|x_{k-1} - P_{r(i)}x_{k-1}\|_2 = \lambda\langle x_{k-1} - x, z_k \rangle,$$

where $z_1, z_2, \cdots$ are independent realizations of the random vector Z. Then

$$\begin{aligned}
\|x_k - x\|_2^2 &\leq \|x_{k-1} - x\|_2^2 - \lambda(2 - \lambda)|\langle x_{k-1} - x, z_k \rangle|^2 \\
&\leq \left(1 - \lambda(2 - \lambda)\left|\left\langle \frac{x_{k-1} - x}{\|x_{k-1} - x\|_2}, z_k \right\rangle\right|^2\right)\|x_{k-1} - x\|_2^2.
\end{aligned}$$

The following proof follows from Theorem 2 of [11] immediately. This completes the proof.

Theorem 2. *Let x_k^* be the kth iteration of the noisy random Kaczmarz algorithm with relaxation running with $Ax \approx b + r$, then we have*

$$\mathbb{E}\|x_k^* - x\|_2 \le \left(1 - \lambda(2-\lambda)\kappa(A)^{-2}\right)^{\frac{k}{2}}\|x_0 - x\| + \kappa(A)\gamma. \tag{12}$$

Proof. Due to the proof of Theorem 2.1 of Needell[9], we can get the above theorem immediately.

Remark 3. It is easy to see that the exponential convergence of Theorem 1 and Theorem 2 include the results of Strohmer and Vershynin[11] and Needell[9].

3 Numerical Experiments

In this section, we provide some numerical experiments for the randomized Kaczmarz algorithm with relaxation.

First, suppose the equation of (1) is consistent and the matrix A is a 500×200 whose entries are independent $N(0,1)$ random variables. There are three types of the randomized Kaczmarz algorithm with relaxation. If $\lambda = 1$, it reduces to the standard randomized Kaczmarz algorithm. $\lambda \in (0,1)$ is called underrelaxation and $\lambda \in (1,2)$ is called overrelaxation. We take $\lambda = 1, \lambda = 0.5$ and $\lambda = 1.8$ to represent the three cases, respectively. Specially, the relaxation parameter $\lambda = 1 + n/m$ is chosen from the Section 5 of [11]. The numerical results are in Figure 1.

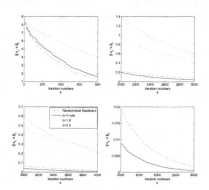

Fig. 1 The comparison of rate of convergence for the randomized Kaczmarz algorithm with and without relaxation parameter. The relaxation parameters take the value of $\lambda = 1 + n/m$, $\lambda = 1.8 \in (1,2)$, $\lambda = 0.5 \in (0,1)$.

From the upper left of Figure 1, we can see that the standard randomized Kaczmarz algorithm converges faster than others at the beginning of iteration. However, with the increasing of the iteration numbers, the randomized Kaczmarz algorithm with relaxation $\lambda = 1 + n/m$ performs better and better. The lower right of Figure 1 shows that the randomized Kaczmarz algorithm with relaxation $\lambda = 1 + n/m$ is much faster than the standard randomized Kaczmarz algorithm. The randomized

Kaczmarz algorithm with underrelaxation and overrelaxation are slower than the other two. Second, we consider the linear system equations is corrupted by noise, i.e., $Ax \approx b+r$. The experimental matrix A is randomly generated whose entries are i.i.d. Gaussian distribution with mean 0 and variance 1. The matrix A has 500 rows and 200 columns. The noise r is Gaussian distribution with mean 0 and variance 0.01. The results obtained are plotted in Figure 2.

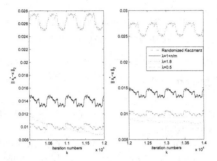

Fig. 2 x_k^* is generated by the noisy linear system of (5) and x is the true value of system (1). When the iteration numbers are increasing, we can see that the randomized Kaczmarz algorithm with underrelaxation performs the best.

Third, we consider the linear system equations (1) is inconsistent. It is aim to find the least square solution of it. The size of matrix A is 500×100, whose entries are i.i.d. Gaussian distribution with mean 0 and variance 1.

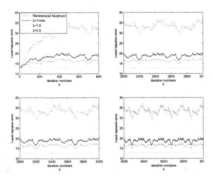

Fig. 3 Shows the least square error of $\|Ax_k - b\|_2$ for the increasing of iteration numbers. We can see that the underrelaxation version of the randomized Kaczmarz algorithm performs better than the other three cases.

4 Conclusion

In this paper, we prove that the randomized Kaczmarz algorithm with relaxation converges with expected exponential rate for the system of $Ax = b$ and $Ax \approx b+r$.

Experimentally, we conclude that the randomized Kaczmarz algorithm with relaxation $\lambda = 1 + n/m$ converges faster than others in the consistent case of the equation (1) linear system. For the inconsistent of (1) and the noisy version of (5), the randomized Kaczmarz algorithm with underrelaxation performs the best. The relaxation parameter is chosen from 0.25 to 0.75 in general. In our future work, we will study the randomized Kaczmarz algorithm for the real image reconstruction problems.

Acknowledgment. The authors would like to thank the Natural Science Foundations of Jiangxi Province $(2009GZS0021, 2007GQS2063)$ for financial support.

References

1. Björck, A., Elfving, T.: Accelerared projection methods for computing pseudoinverse solutions of systems of linear equations. BIT 19(2), 145–163 (1979)
2. Censor, Y., Gordan, D., Gordan, R.: Component averaging: An efficient iterative parallel algorithm for large sparse unstructured problems. Parallel Comput. 27, 777–808 (2001)
3. Censor, Y., Herman, G.T., Jiang, M.: A note on the behavior of the randomized Kaczmarz algorithm of Strohmer and Vershynin. J. Fourier Anal. Appl. 15, 431–436 (2009)
4. Cimmino, G.: Calcolo approssimato per le soluzioni dei sistemi di equazioni lineari, La Ricerca Scientifica, XVI, Series II, Anno IX, 326–333 (1938)
5. Feichtinger, H.G., Cenker, C., Mayer, M., Steier, H., Strohmer, T.: New variants of the POCS method using affine subspaces of finite codimension, with applications to irregular sampling. In: Proc. SPIE: Visual Communications and Image Processing, pp. 299–310 (1992)
6. Gordon, R., Bender, R., Herman, G.T.: Algebraic reconstruction techniques (ART) for the three-dimensional electron microscopy and X-ray photography. J. Theor. Biol. 29, 471–481 (1970)
7. Herman, G.T., Meyer, L.B.: Algebraic reconstruction techniques can be made computationally efficient. IEEE Transactions on Medical Imaging 12(3), 600–609 (1993)
8. Kaczmarz, S.: Angenäherte auflösung von systemen linearer gleichungen. Bull. Internat. Acad. Polon. Sci. Lettres A, 335–357 (1937)
9. Needell, D.: Randomized kaczmarz solver for noisty linear systems. BIT Numer. Math. 50(2), 395–403 (2010)
10. Strohmer, T., Vershynin, R.: A Randomized Solver for Linear Systems with Exponential Convergence. In: Díaz, J., Jansen, K., Rolim, J.D.P., Zwick, U. (eds.) APPROX 2006 and RANDOM 2006. LNCS, vol. 4110, pp. 499–507. Springer, Heidelberg (2006)
11. Strohmer, T., Vershynin, R.: A randomized Kaczmarz algorithm with exponential convergence. J. Fourier Anal. Appl. 15, 262–278 (2009)
12. Strohmer, T., Vershynin, R.: Comments on the randomized Kaczmarz method. J. Fourier Anal. Appl. 15, 437–440 (2009)

The Experiment and Simulation Study of Respiration on the Dose Distribution in Radiotherapy[*]

Xiao Xu and Keqiang Wang

Abstract. Objective: To evaluate the influence of respiration on the radiation dose distribution in radiotherapy with matlab simulation and film dosimetry. Methods: Radiation of 50MU was delivered in a square, round, ellipse, dumb bell, or female shaped field to the films within a moving or static Respiration Motion Phantom respectively, the dose distributions for the two motion status were measured and compared. In order to further verify the impact of amplitude of respiration movement, the matlab simulation with movement amplitude of 0 cm, 0.5 cm, 1.0 cm, 1.5 cm, 2.0 cm, and 2.5cm were done respectively. The dose distributions in different status were measured and compared with film dosimetry. Iso-dose line comparison, NAT (Normalized Agreement Tests) and γ comparison were used for the comparison of dose distributions. F_s was used as an index to evaluate the differences of the areas that surrounded by iso-dose lines in different situations (F_{S90}, F_{S50}, F_{S25} delegates the ratio of the areas that surrounded by 90%, 50%, 25% iso-dose line in different situation respectively). Results: (1) For round field, the matlab simulation showed that S_{90} decreased as the increase of the movement. S_{90} was almost 0 when the amplitude became to half of the diameter of the field. S_{25} varied inversely. (2) The experiment showed that in horizontal movement situation compared with in static situation, the F_{S90} became smaller and the F_{S25} became larger. The more the displacement became larger, the more the F_{S90} and the F_{S25} deviate remarkable. In vertical movement situation, F_s changed significantly in square field and dumb bell shaped field while changed a little in the others. (3) γ and NAT comparison: In the horizontal movement situation, compared with in the static phantom, the P_γ was $<60\%$ and the P_{NAT} was $<75\%$ in every radiation field. In vertical movement situation, the P_γ was less than 85% for all the square, round, dumb bell and female shaped fields. Conclusions: The respiration can impact on the dose distribution within the target volume in radiotherapy, leading to a smaller area of higher dose level and an expanded area of lower dose level. The influence will become more significant with larger movement of the target.

Xiao Xu · Keqiang Wang
The General Hospital of Tianjin Medical University Tianjin, China
e-mail: xxucn@yahoo.com.cn

[*] Supported by National Natural Science Foundation of China (No.50937005).

F.L. Gaol et al. (Eds.): Proc. of the 2011 2nd International Congress CACS, AISC 145, pp. 475–482.
springerlink.com   © Springer-Verlag Berlin Heidelberg 2012

1   Introduction

Respiratory motion affects normal tissue and tumors in thoracic and abdominal tumors. Radiation delivery in the presence of motion causes an averaging or blurring of the static dose distribution over the path of the motion. The displacement results in a deviation between the intended and the delivered dose distribution. At present, studies about impact of respiratory movement on radiation therapy were mainly on its impact on target volume. Few studies estimated its impact on radiation dose distribution[1]-[3]. In order to evaluate the impact of respiration on the dose distribution, we measured and compared the radiation dose distribution in the static and moving (horizontal and vertical) Respiration Motion Phantom respectively. Further more, we simulated the respiration with matlab in order to study the relationship between the magnitude of respiration and the dose distribution.

2   Materials and Methods

2.1   Materials

American Kodak X-omat V low-speed film, the saturation dose is 100cGy, the field size dependence of response <2% for field size of 2cm×2cm∼10cm×10cm. Microtek 8700 laser scanner (Taiwan). JYT-2000 X-ray automatic film processor (Tianjin Jieyate Medical Company Limited). Doselab 4.0 dose comparison software. American Capintec 292 dosimeter and PR-06C 0.65cm^3 ion chamber for absolute dose measurement. Phantom and accelerator :   QUASAR Respiration Motion Phantom from Modus Company of Canada. Siemens Primus-M Medical Linear Accelerator. Matlab software.

2.2   Methods

2.2.1   Matlab Simulation of Respiration

In order to research the relationship between the magnitude of respiration and the dose distribution, matlab simulations of round field with movement amplitude of 0cm, 0.5cm, 1.0cm, 1.5cm and 2.0cm were done respectively. Normal breathing involved rhythmic inhaling and exhaling. Therefore, we could quantify breathing using trigonometric equations. In general, we described the movement magnitude by a sine wave:

$$M = A\sin(2\pi t / p) \qquad (1)$$

where M was the magnitude of movement, A was the amplitude, p was the period, t was the time. In this simulation, we supposed p=3s, A=0cm, 0.5cm, 1.0cm, 1.5cm, 2.0cm respectively and the diameter of the round field was 5cm. In order to improve the precision, the time interval was set to 0.1s and size interval was set to 0.01cm.

2.2.2 Film Calibration

In order to convert a map of optical densities into a map of absorbed radiation doses, a calibration curve is necessary. Measurements were performed at water equivalent depth of 1.5cm in a slab phantom with field size of 3cm×3cm. At least 2cm distance between fields was needed to prevent effects from each other. Doses of 15, 25, 35, 45, 55, 65, 75, 85 Mu were used. Radiation was performed with 6MV X-ray and SSD (Source Surface Distance) = 100cm and so 1Mu = 1cGy. The calibration curve was obtained by the measurement data.

2.2.3 Radiation to the Respiration Motion Phantom

The magnitude of tumor motion may reach 3cm in lung cancer[4]-[6]. We use Respiration Motion Phantom to simulate the status with magnitude of 2.5cm along AP (anterior-posterior) axis and 1cm along SI (superior-inferior) axis respectively. The respiratory frequency was given 20 breaths per minute. Square, round, ellipse, dumb bell, or female shaped field with maximum diameter of 5.5cm were used (The central width of the female shaped field was 1.5 cm). The film was located at a depth of 8cm below the surface of the phantom and SAD (Source Axis Distance) = 100cm. Radiations of 50 Mu were delivered in static, horizontal (AP) and vertical (SI) movement status with each field. Then we got 15 films. Besides, the absolute dose was measured with ion chamber in static status for every radiation field.

2.2.4 Analysis and Comparison of Relative Dose

We compared the dose distribution between different statuses from the obtained films. Methods used included **A**): Iso-dose line comparison, 90%, 50%, 25% iso-dose lines were chosen; **B**): γ value pixel to pixel test: γ value was the commonly used measure for relative dose evaluation. γ value indicates that the dose deviation is acceptable in that pixel, otherwise unacceptable. Pγ indicates the percentage of pixels that the γ value ≤ 1. The NAT(Normalized Agreement Tests) index was used to represent the agreement between the moving and static dose distributions. The NAT value was set to 0 when the dose criteria were fulfilled. P_{NAT} was defined as the percentage of pixels with NAT value = 0. Smaller value of NAT index represented better agreement. Due to the influence of signal noise, we chosen the acceptable criteria as the percentage deviation $\leq 5\%$ and the equivalence distance deviation ≤ 3 mm[7]; **C**):Iso-dose line deviation index F_S: F_{S90}, F_{S50}, F_{S25} represent the ratio of the areas that surrounded by 90%, 50%, 25% iso-dose line in movement situation compared to the areas in static situation, respectively.

3 Results

3.1 Dose Distribution in Matlab Simulation

Dose distribution in different amplitude of movements was shown in Fig.1, the red line was 90% iso-dose line and the blue line was 50% iso-dose line. It was obvious that with the increase of the movement, high dose region decreased and low dose region increased. The homogeneity of dose distribution reduced. When the amplitude of movement was 1.5cm(about a quarter of the diameter of the round field), S_{90} was 6.15cm^2 ,about a third of the field areas. When the amplitude of movement was 2.5cm(half of the diameter of the round field), S_{90} was almost 0, S_{25} was 36.4cm^2,about twice of the field areas. S_{50} changed a little in all situations.

3.2   Respiration Motion Phantom

3.2.1   Comparison between Horizontal Movement and Static Situation

For every field situation, we got $P_\gamma < 60\%$, $P_{NAT} < 75\%$, and NAT index > 30. The results suggested that the dose distributions were obviously different between horizontal movement and static situation. As shown in Table 1 and Fig.2, all the F_{S90} were < 1 and all the F_{S25} were > 1, which suggested that high dose region decreased and low dose region increased in movement situation. For circle shaped field, the areas surrounded by 90%, 50%, and 25% iso-dose lines were 15.6, 17.9, 20.3cm^2 in static situation and 6.4, 17.9, 26.4cm^2 in movement situation. S_{90} decreased 59%, S_{25} increased 31% and S_{50} changed a little. Because of the respiration, the accuracy of the dose decreased significantly.   Fig.3 showed the dose distribution in the two statuses for female shaped field. Two lightened female shape appeared, showing high dose region separated and low dose region increased. As shown in Table 2, the radiation dose to the female shape was 11.8-12.8cGy in static situation, and it was 3.8-5.4 cGy in horizontal movement situation.

3.2.2   Comparison between Vertical Movement and Static Situation

As shown in Table 3, both the P_γ and the P_{NAT} were greater than 90% for ellipse shaped field. The P_γ were 70%~90% and the P_{NAT} were 70-95% for the other 4 shaped fields. As shown in Table 1, F_{S90} and F_{S25} were 0.9~1.1 for square and dumb bell shaped field. F_{S90}, F_{S50} and F_{S25} were all almost equal to 1 for other fields. The results suggested that the effect of vertical movemnt on dose distribution is smaller than that of horizontal movement.

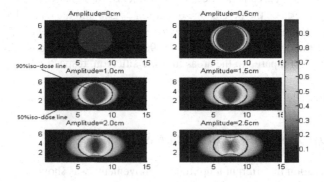

Fig. 1 The dose distribution map with different amplitude of movemnt in matlab simulation

Fig. 2 The iso-dose line of circle shaped field in static and horizontal movement situation

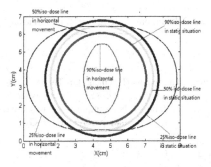

Fig. 3 Comparison of dose distribution between horizontal movement and static situation for female shaped field

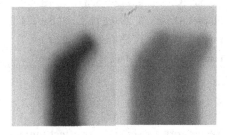

Table 1 The areas surrounded by 90%, 50%, 25% iso-dose line with every shaped field in different situations

Fields and Situation	S90 cm^2	F$_{S90}$	S50 cm^2	F$_{S50}$	S25 cm^2	F$_{S25}$
Square and Static	19.4	1.00	22.5	1.00	24.8	1.00
Square and Horizontal	17.1	0.88	22.5	1.00	27.9	1.13
Square and Vertical	18.5	0.95	22.5	1.00	25.5	1.03
Round and Static	15.6	1.00	17.9	1.00	20.3	1.00
Round and Horizontal	6.4	0.41	17.9	1.00	26.4	1.32
Ellipse and Static	16.4	1.00	18.8	1.00	19.3	1.00
Ellipse and Horizontal	15.6	0.95	31.1	1.65	35.7	1.85
Dumb bell and Static	10.7	1.00	12.7	1.00	15.6	1.00
Dumb bell and Horizontal	7.9	0.74	17.9	1.41	21.9	1.40
Dumb bell and Vertical	9.8	0.92	12.7	1.00	14.5	0.93
Female and Static	1.3	1.00	4.1	1.00	7.1	1.00
Female and Horizontal	0.8	0.62	12.4	3.02	14.6	2.06

Table 2 The mean dose in different situations

Fields	The mean target dose in static situation (cGy)	The mean target dose in horizontal situation·cGy·	The mean target dose in vertical situation·cGy·
Square	34.6±1.6	30.4±4.7	31.5±1.7
Round	32.8±0.9	21.0±3.3	28.7±1.0
Ellipse	33.1±1.3	29.9±11.1	32.5±1.4
Dumb	23.3±0	13.4±3.3	22.4±0.7
Female	12.3±0.5	4.6±0.8	14.0±0.9

Table 3 Comparison between vertical movement and static situation in respiration motion phantom

Fields	P (%)	P_{NAT} (%)	NAT factor
Square	75.3	72.0	60.7
Round	71.3	73.6	72.3
Ellipse	93.3	93.7	2.3
Dumb bell	72.8	91.1	7.1
Female	82.9	86.1	127.1

4   Discussion

Respiratory movement can affect tumors in the thorax and abdomen significantly. As described by Balter[8] that respiratory movement can impact on the three-dimensional conformal radiotherapy in liver cancer. The precision of the dose delivered to target volume reduces by the movement of the tumors. The present experiment showed that there was a significant difference in dose distribution between the horizontal moving and static situation. High dose region decreased, low dose region increased and the precision and homogeneity of dose distribution reduced. Part of clinical tumor volume (CTV) will not receive adequate dose and the likelihood of treatment-related complications will increase. The same conclusion can be included in the matlab simulation. The more the displacement is, the more remarkable the F_{S90} and the F_{S25} deviate. At the same time, the shape of the isodose line varies according to the motion of the tumor. This result can and should be used in the CTV-to-PTV margin used for treatment planning. The entire range of motion should be considered when establishing the internal margin. Differences in high dose region, low dose region and dose distribution between the vertical moving and static situation also exist.

To cover the respiratory motion by adding treatment margins is suboptimal, because this increases the radiation field size and consequently the volume of healthy tissues exposed to irradiation. This increased treatment volume increases the likelihood of treatment-related complications in traditional radiotherapy. So it is recommended that the margin be tailored to the individual patient by measuring the magnitude of the respiration in radiotherapy. With the development of precise radiation therapy, some methods are used in the management of respiratory movement, including respiratory gated techniques, breath-hold techniques, forced shallow-breathing methods and respiratory-synchronized techniques[9]-[13]. Besides, Image guided radiotherapy (IGRT) may be a good method to track the motion[14].

Film dosimetry is used in this experiment due to its wide accessibility and the flexibility to place the film in phantoms. It is advantageous because of good llinearity of dose, high spatial resolution and saturation dose. The main problem in using film is the dependence of optical density on washing condition, film plane orientation, films of different types or batches and strong energy dependence. The same condition should be used to reduce the errors in film calibration and film measurement.

The present experiment estimated the change of dose distribution in single field irradiation. Further study needs to be done to evaluate the changes of dose distribution in multiple fields' irradiation situations. Simulation study of the dose distribution of respiration with matlab is an effective method. It's convenient to get dose distribution map and to do a quantitative study.

References

1. Shirato, H., Shimizu, S., Kunieda, T.: Physical aspects of a real-time tumor-tracking system for gated radiotherapy. Int. J. Radiat. Oncol. Biol. Phys. 48, 1187–1195 (2000)
2. Shimizu, S., Shirato, H., Kagei, K.: Impact of respiratory movement on the computed tomographic images of small lung tumors in three-dimensional (3D) radiotherapy. Int. J. Radiat. Oncol. Biol. Phys. 46, 1127–1133 (2000)
3. Maxim, P.G., Loo Jr., B.W.: Quantification of motion of different thoracic locations using four-dimensional computed tomography: implications for radiotherapy plannings. Int. J. Radiat. Oncol. Biol. Phys. 69, 1395–1401 (2007)
4. Liu, H.H., Balter, P., Tutt, T.: Assessing respiration-induced tumor motion and internal target volume using four-dimensional computed tomography for radiotherapy of lung cancer. Int. J. Radiat. Oncol. Biol. Phys. 68, 531–540 (2007)
5. van Sornsen de Koste, J.R., Lagerwaard, F.J., Nijssen-Visser, M.R.: Tumor location cannot predict the mobility of lung tumors: A 3D analysis of data generated from multiple CT scans. Int. J. Radiat. Oncol. Biol. Phys. 56, 348–354 (2003)
6. Seppenwoolde, Y., Shirato, H., Kitamura, K.: Precise and real-time measurement of 3D tumor notion in lung due to breathing and heartbeat, measured during radiotherapy. Int. J. Radiat. Oncol. Biol. Phys. 53, 822–834 (2002)
7. Low, D.A., Harms, W.B., Mutic, S.: A technique for the quantitative evaluation of dose distributions. Med. Phys. 25, 656–661 (1998)
8. Balter, J.M., Ten-haken, R.H., Lawarence, T.S.: Uncertainties in CT-based radiation therapy treatment planning association with patient breathing. Int. J. Radiat. Oncol. Biol. Phys. 36, 167–174 (1996)
9. Rosenzweig, K.E., Hanley, J., Mah, D.: The deep inspiration breath-hold technique in the treatment of inoperable non-small-cell lung cancer. Int. J. Radiat. Oncol. Biol. Phys. 48, 81–87 (2000)
10. Shirato, H., Shimzu, S., Kunieda, T.: Physical aspects of a real-time tumor-tracking system for gated radiotherapy. Int. J. Radiat. Oncol. Biol. Phys. 48, 1187–1195 (2000)
11. Shirato, H., Shimizu, S., Kitamura, K.: Four-dimensional treatment planning and fluoroscopic real-time tumor tracking radiotherapy for moving tumor. Int. J. Radiat. Oncol. Biol. Phys. 48, 435–442 (2000)

12. Kim, D.J.W., Murray, B.R., Halperin, R.: Held-breath self-gating technique for radiotherapy of non-small-cell lung cancer: a feasibility study. Int. J. Radiat. Oncol. Biol. Phys. 49, 43–49 (2001)

13. Sixel, K.E., Aznar, M.C., Ung, Y.C.: Deep inspiration breath hold to reduce irradiated heart volume in breast cancer patients. Int. J. Radiat. Oncol. Biol. Phys. 49, 199–204 (2001)

14. Rijkhorst, E.J., van Herk, M., Lebesque, J.V.: Strategy for online correction of rotational organ motion for intensity-modulated radiotherapy of prostate cancer. Int. J. Radiat. Oncol. Biol. Phys. 69, 1608–1617 (2007)

Level Based Flooding for Node Search
in Wireless Sensor Network

Yanhong Ding, Tie Qiu*, Honglian Ma, and Naigao Jin

Abstract. In this paper, we pay attention to the problem of node search in Wireless Sensor Networks (WSNs). The main problem of basic flooding is that the cost is too high. In order to solve above problem, we propose a variant of Flooding called Level Based Flooding (LBF). In LBF, the whole network is divided into several layers according to the distance (hops) between the nodes and the sink node. The sink node knows the level information of each node. The search packet is broadcast in the network according to levels of nodes and its *TTL* is set to the level of the target node. When the target node receives the packet, it sends its data back to the sink node in random walk within *level* hops. We show by extensive simulations that the energy consumption of LBF is much better than that of basic flooding.

1 Introduction

Wireless Sensor Networks (WSNs) are consisted of tiny resource-impoverished battery powered devices which have sensing, computation, and communication capabilities [1]. As sensor nodes get energy from a battery source, their energy should be used judiciously. When designing proposals for WSNs, energy efficiency is one of the important considerations. In recent years, WSNs are used in varied areas such as wild life monitoring, target tracking, home and industry automation, and military systems.

Yanhong Ding · Honglian Ma · Naigao Jin
School of Software, Dalian University of Technology, Dalian, China
e-mail: dyh2010@gmail.com, {qiutie,mhl,ngjin}@dlut.edu.cn

Tie Qiu
School of Software, Dalian University of Technology,
Dalian, China

* Corresponding author.

F.L. Gaol et al. (Eds.): Proc. of the 2011 2nd International Congress CACS, AISC 145, pp. 483–488.
springerlink.com   © Springer-Verlag Berlin Heidelberg 2012

Node discovery in WSNs is a widely study problem [2, 3, 4, 5, 6, 7]. It is a key problem especially in PULL based UNSTRUCTURED WSNs [8, 9]. In PULL WSNs, the sink node searches for the nodes that send emergency data. And in UN-STRUCTURED WSNs, the search initiator has no clue about any locations of target node. One important strategy is flooding. This searching mechanism has been extensively studied in the context of WSNs [10, 11, 12, 13, 14]. The basic flooding strategy makes the whole network with heavy loads and nodes' energy are used up quickly. The aim of this paper is to propose a searching strategy to reduce the energy consumption of basic flooding strategy and the load of the network.

In this paper, we propose a new search mechanism, Level Based Flooding (LBF), which reduces the energy consumption of searching for targets compared to basic flooding mechanism. In LBF, when we deploy the network, the sink node broadcasts the level building packet to the network first. After this, all nodes in the network are divided into layers according to the hops between the nodes and the sink node, and the sink node also gets each node's level information. The level of nodes stands for the minimum hops to the sink node. When the sink node searches for the emergency node, it broadcasts the search packet and sets the *TTL* to the level of the target. When the target receives the search packet, it sends its emergency data back to the sink node in random walk within *level* hops. Through this, energy consumption is reduced.

2   Level Based Flooding Strategy

2.1   Level Building Process

After the deployment of the WSN, the sink node broadcasts a *LevelBuilding* packet by adding its level information to it. The level of the sink node is considered to be zero. The level of sensor nodes is initialized to be an integer that is larger than the maximize level of the network. When a sensor node receives this packet, if the packet's *level* is smaller than the node's *level*, the node updates its level as the packet's level value plus one. Then the sensor node rebroadcasts the *LevelBuilding* packet by updating the packet's level with its level. And the nodes also sends its level information to the sink node. How the information packet goes back to the sink node will be introduced next. If the packet's *level* is larger than or equal to the node's *level*, the node drops the packet directly. In this process, a sensor node could receive more than one *LevelBuilding* packets, but it sets its level value to the minimum level of the received *LevelBuilding* packets. So we can know that each node's level stands for minimum hops between the sensor node and the sink node.

In this process, each node also collects its neighbors' level information. When a node receives a *LevelBuilding* packet from its neighbor, if the neighbor's information is already in the node's cache and the information is the same as that in the cache, the node records noting except deciding whether to broadcast the packet; if the node has nothing about the neighbor, the node caches the neighbor's level information. After collecting all neighbors' information, nodes can segregate their neighbors into the following three sets: *LowLevelNeighbors*: Set of all neighbors whose level is smaller than that of the node; *HighLevelNeighbors*: Set of all neighbors whose level is larger than that of the node and *EqualLevelNeighbors*: Set of all neighbors whose level is equal to that of the node. When the node broadcasts the query packet, it only sends the packet to the neighbors in *HighLevelNeighbors* set. When the node receives the back packet, it chooses a node from the *LowLevelNeighbors* set as the next hop.

When the sensor's level is updated in this process, the node sends a reply packet to inform the sink node. When a sensor node receives the reply packet, it transmits the packet and chooses a node from the *LowLevelNeighbors* set as the next hop of the packet. So the reply packet gets closer to the sink node at each hop, and through *level* hops, the reply packet gets to the sink node. When the sink node receives reply packets, if it doesn't have the node's level information, it records it; or it updates the node's level information. The sink node records the level information of all sensor nodes. The back process is shown in Fig. 1.

The level building process is only executed one-time and it is part of the initial setup of WSN. Since we assume a static network, there will be no sensor nodes added to the network. When a node's energy is used up, it broadcasts a message to its neighbors to notice that it dies, and then its neighbors remove its information from the cache. The lifetime of the *LevelBuilding* packet is not determined by its *TTL*. When all nodes' level value gets to the minimum value, this process will finish.

2.2 Query Process

When the sink node senses emergency sensor nodes, for example the sink node receives emergency data from a node, broadcasts a query packet to the network. The sink node sets the packet's *TTL* according to the target node's level information in its cache. Only the sensor nodes in levels between zero and *TTL* will process the query packet. So only when the node is far away from the sink node, the query packet would be broadcast through the whole network. If the node is very close to the sink node, the energy consumption will be reduced a lot. So the average situation will be much better than the basic flooding.

Fig. 1 The sensor node sends reply to the sink node in Random Walk. Node n selects a neighbor from its *LowLevelNeighbor* set as the next hop of reply packet. All the nodes receiving the reply packet do the same thing as *Node n* does until the reply packet reaches to the sink node.

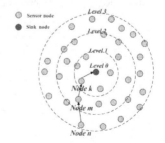

When a node receives the query packet, if the packet's *hopCount* (number of hops traveled by the current query packet) is equal to the packet's *TTL* or the node has processed the packet, it drops the packet directly; otherwise it caches the query packet's basic information which is used to avoid reprocessing the query packet. The query packet is marked by the sender's ID (the ID of the node that starts the query) and the sequel number (*seqNum*) uniquely. If the packet's *hopCount* is larger than the node's level, the node also drops the packet directly. Each node only processes each query packet at most once.

When the target receives the query packet, it sends its emergency data to the sink node. The process is a little like the process introduced in Sect. 2.1. The emergency data packets get to the sink node in random walk within *level* hops. Every time the node that transmits the packets chooses node from its neighbors with lower level as next hop of packets. When the hop number equals to *TTL*, the data packet gets to the sink node.

3 Simulation

3.1 Simulation Setup

We evaluate performance of LBF and basic flooding protocols using NS-2, a popular discrete event network simulator. We compare the cost and latency of LBF and Basic Flood mechanism. In order to get the compare results of different scale WSNs, we have designed five simulate scenarios. Sensor nodes are uniformly deployed in the monitored area and the sink node is nearly placed at the center of terrains. The number of the nodes chosen is to make sure that the sink node can get to the each node.

3.2 Simulation Results

Table 1 shows the results of level building process. With the increase of monitored area's square and the number of sensor nodes, the max level is increased. The max level number means that if the query packet wants to go through the whole network,

Table 1 Results of level building process

Scene No.	Max level	Average level
1	3	1.92
2	5	2.90
3	15	7.78
4	32	16.85
5	49	24.27

Fig. 2 This is the simulation results of average load of each node to find all targets. The query packet is broadcast from the nodes near the sink node to the nodes far away from the sink node. If the nodes are closer to the sink node, the average load will be heavier, because they have to transmit more packets.

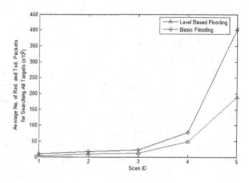

Fig. 3 This shows the comparison simulation results. We can know that the energy consumption is increased with the improvement of the network scale for each searching strategy. But the increase of LBF is much slower than that of basic flooding. The energy consumption is about 54.4% of that of basic flooding. With the increase of the area square and number of nodes, the effect is more obviously.

the max hops that is needed. The average hop number means that the average hops that the sink node wants to find the target node at the best situation.

Fig. 2 shows the average load of each node to find all targets. And Fig. 3 shows the performance of LBF and the basic flooding strategy.

4 Conclusions

In this paper, we proposed LBF, an energy efficient query resolution protocol. The basic principle of the protocol is to divide the whole network into different layers according to the hops between the sink node and the sensor nodes. The sink node knows the level information of each sensor node. When the sink node wants to find the target node, it broadcasts the query packet to the network and sets the packet's *TTL* according to the level information of the target. And the packet is broadcast according to levels of nodes. When the target receives the packet, it sends the data back in random walk. The simulations show that LBF incurs less energy consumption compared to that of basic flooding.

Acknowledgements. This work is partially supported by Natural Science Foundation of China under Grant No. 60773213, Program for New Century Excellent Talents in University (NCET-09-0251), the Fundamental Research Funds for the Central Universities.

References

1. Di Francesco, M., Shah, K., Kumar, M., Anastasi, G.: An Adaptive Strategy for Energy-Efficient Data Collection in Sparse Wireless Sensor Networks. In: Silva, J.S., Krishnamachari, B., Boavida, F. (eds.) EWSN 2010. LNCS, vol. 5970, pp. 322–337. Springer, Heidelberg (2010)
2. Sadagopan, N., Krishnamachari, B., Helmy, A.: The ACQUIRE mechanism for efficient querying in sensor networks. In: Proceedings of the 1st IEEE International Workshop on Sensor Network Protocols and Applications, pp. 149–155 (May 2003)
3. Chang, N.B., Liu, M.: Controlled flooding search in a large network. IEEE/ACM Transactions on Networking 15(2), 436–449 (2007)
4. Intanagonwiwat, C., Govindan, R., Estrin, D., Heidemann, J., Silva, F.: Directed diffusion for wireless sensor networking. IEEE/ACM Transactions on Networking 11(1), 2–16 (2003)
5. Gehrke, J., Madden, S.: Query processing in sensor networks. IEEE Pervasive Computing 3(1), 46–55 (2004)
6. Zuniga, M., Avin, C., Hauswirth, M.: Querying Dynamic Wireless Sensor Networks with Non-Revisiting Random Walks. In: Silva, J.S., Krishnamachari, B., Boavida, F. (eds.) EWSN 2010. LNCS, vol. 5970, pp. 49–64. Springer, Heidelberg (2010)
7. Nascimento, M.A., Alencar, R.A.E., Brayner, A.: Optimizing Query Processing in Cache-Aware Wireless Sensor Networks. In: Gertz, M., Ludäscher, B. (eds.) SSDBM 2010. LNCS, vol. 6187, pp. 60–77. Springer, Heidelberg (2010)
8. Huang, H., Hartman, J., Hurst, T.: Data-centric routing in sensor networks using biased walk. In: Proceedings of the 3rd IEEE Communications Society Conference on Sensor, Mesh and Ad Hoc Communications and Networks, pp. 1–9 (September 2006)
9. Ahn, J., Kapadia, S., Pattem, S., Sridharan, A., et al.: Empirical evaluation of querying mechanisms for unstructured wireless sensor networks. In: ACM SIGCOMM Computer Communication Review, vol. 38(3), pp. 17–26 (July 2008)
10. Cheng, Z., Heinzelman, W.: Flooding Strategy for Target Discovery in Wireless Networks. Wireless Networks 11(5), 607–618 (2005)
11. Avin, C., Brito, C.: Efficient and robust query processing in dynamic environments using random walk techniques. In: Proceedings of the 3rd International Symposium on Information Processing in Sensor Networks, pp. 277–286 (April 2004)
12. Braginsky, D., Estrin, D.: "Rumor Routing Algorithm For Sensor Networks. In: Proceedings of the 1st ACM International Workshop on Wireless Sensor Networks and Applications, pp. 22–30 (September 2002)
13. Khan, M., Gabor, A.: An Effective Compiler Design for Efficient Query Processing in Wireless Sensor Networks. In: Internal Conference on Circuit and Signal Processing, pp. 157–159 (November 2010)
14. Chakroaborty, A., Lahiri, K., Mandal, S., Patra, D., et al.: Optimization of Service Discovery in Wireless Sensor Networks. Wired/Wireless Internet Communications, 351–362 (June 2010)

Numerical and Experimental Study of Hydrogen Release from a High-Pressure Vessel

Sang Heon Han and Daejun Chang

Abstract. The dispersion characteristics of hydrogen leaking through a small hole from a high-pressure reservoir are investigated numerically and experimentally to provide a guideline in determining the safety distances for hydrogen stations. The studies were carried out for the leaking holes with diameters of 0.5, 0.7 and 1.0mm and for the release pressures at 100, 200, 300 and 400 bar. Because Froude numbers in these realistic hydrogen leaking conditions are very large, it is reasonable to employ a 2-dimensional axisymmetric approach in the numerical simulations as well as in hydrogen samplings.

1 Introduction

Hydrogen is one of the most reactive compounds. On the other hand, hydrogen has a very low volumetric energy density and a great tendancy to buoyantly disperse away from the leaking source. Consequently, the hydrogen safety in filling stations can be greatly improved by promoting the hydrogen dispersion before being exposed to the ignition sources. When developing the hydrogen station safety code that will guarantee the safety for the station operators as well as for the public, the better quantified hydrogen jet characteristics will enable us to maximize the safety potential by specifying the guideline to promote the dispersion and dilution of highly reactive hydrogen. There have been lots of studies to achieve this kind of purpose [1-6].

It is the objective of the present study to quantify the characteristics of hydrogen leaking from a high pressure system through a small rupture hole. In this study, a particular characteristics of hydrogen jets leaking from a high pressure storage vessel is

Sang Heon Han · Daejun Chang
Division of Ocean System Engineering KAIST 373-1 Guseong-dong
Yuseong-gu Daejeon 305-701 Korea
e-mail: freezia@kaist.ac.kr, djchang@kaist.edu

F.L. Gaol et al. (Eds.): Proc. of the 2011 2nd International Congress CACS, AISC 145, pp. 489–494.
springerlink.com

utilized in order to facilitate the experiments and numerical calculations. The typical velocity and length scales for high pressure hydrogen leakage are assumed to be u~1500m/s, corresponding to the hydrogen sound speed, and d~1mm respectively. The resulting Froude number, defined to be u2/(g•d), is of order of 108, so that the buoyancy effect is negligible and two dimensional axisymmetric condition can be assumed to significantly reduce the experimental and numerical efforts.

2 Experimental and Numerical Approach

The experimental set up to investigate the characteristics of hydrogen jets is schematically described in Fig. 1. Hydrogen is released from a cylindrical chamber representing the vessel containing high pressure hydrogen. The chamber is 140mm and 80mm in the internal length and diameter, thereby having an internal volume of 0.7L, and has a release hole at the center of its cap. The release hole diameters are 0.5mm, 0.7mm and 1.0mm. Five gas samplers are employed to measure the hydrogen concentration profiles along the jet centerline. The sampling probes are located at 1m, 3m, 5m, 7m, and 9m from the jet exit.

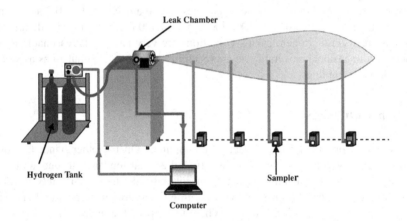

Fig. 1 Schematics of experimental set up

Thanks to the large Froude number limit, the buoyancy is neglected and the entire system involving hydrogen jet is assumed to be 2D axisymmetric as shown in Fig. 2.

The conservation equations for mass, momentum, energy, turbulent kinetic energy, eddy dissipation rate, and species, i.e. hydrogen, can be written in the following generalized form

$$\frac{\partial}{\partial x}(\rho u \phi) + \frac{1}{r}\frac{\partial}{\partial r}(\rho v r \phi) = \frac{\partial}{\partial x}\left(\Gamma_\phi \frac{\partial \phi}{\partial x}\right) + \frac{1}{r}\frac{\partial}{\partial r}\left(\Gamma_\phi r \frac{\partial \phi}{\partial r}\right) + S_\phi \tag{1}$$

where ϕ denotes any one of the dependent variables, and ρ, u, v, Γ_ϕ and S_ϕ represent the density, axial velocity, radial velocity, source term and the exchange coefficient, respectively. The exchange coefficient and source term corresponding to each conservation equation are presented in Table 1. The computation has been done by using the commercial code FLUENT.

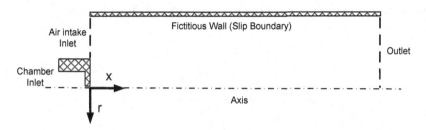

Fig. 2 Geometrical model for computation

Table 1 Variables and source terms of generalized equation

Equation	ϕ	Γ_ϕ	S_ϕ
Continuity	1	0	0
u-momentum	u	μ_{eff}	$-\dfrac{\partial p}{\partial x}+\dfrac{\partial}{\partial x}\left(\mu_{eff}\dfrac{\partial u}{\partial x}\right)+\dfrac{1}{r}\dfrac{\partial}{\partial r}\left(\mu_{eff}\dfrac{\partial v}{\partial x}\right)$
v-momentum	v	μ_{eff}	$-\dfrac{1}{r}\dfrac{\partial p}{\partial x}+\dfrac{\partial}{\partial x}\left(\mu_{eff}\dfrac{\partial u}{\partial r}\right)+\dfrac{\partial}{r\partial r}\left(\mu_{eff}r\dfrac{\partial v}{\partial x}\right)-\dfrac{2\mu v}{r^2}$
k-equation	k	μ_{eff}/σ	$G-\rho\varepsilon$
ε-equation	ε	μ_{eff}/σ	$\dfrac{\varepsilon}{k}(C_1 G-\rho C_2\varepsilon)$
Energy	h	$\dfrac{\mu}{Pr}+\dfrac{\mu_t}{\sigma}$	0
Species	Y_i	Sc_i	0

$$\mu_t=C_\mu\rho\frac{k^2}{\varepsilon},\ \mu_{eff}=\mu+\mu_t,\ h=\sum_k Y_k\int_{T_{ref}}^{T}C_k dT,\ T_{ref}=298$$

$$G=\mu_{eff}\left\{2\left[\left(\frac{\partial u}{\partial x}\right)^2+\left(\frac{\partial v}{\partial r}\right)^2+\left(\frac{v}{r}\right)^2\right]+\left(\frac{\partial v}{\partial r}+\frac{\partial u}{\partial x}\right)^2\right\}$$

3 Results and Discussion

In this study, the characteristics of hydrogen jets are investigated with three different hole diameters at d=0.5mm, d=0.7mm, and d=1.0mm. For each hole diameter, the experiments are carried out at four different leak chamber pressures (Pc) at 100bar, 200bar, 300bar, 400bar. The hole diameters are chosen in such a way that the hole area increases approximately twice for each diameter increase.

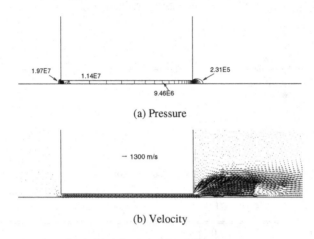

(a) Pressure

(b) Velocity

Fig. 3 Computational results for d=1.0mm and Pc=200bar

Fig. 3 shows (a) pressure distribution and (b) velocity vector for the hydrogen jet with d=1.0mm and Pc=200bar. The pressure of hydrogen flow drops through three stages. The first pressure drop is associated with the acceleration of the flow inside the pressure chamber, where the flow is relatively quiescent except the entrance to the hole. Entering the hole, the pressure is found to be close to 120 bar. The second stage is inside where the pressure drops by 50bar due to both acceleration and the viscous dissipation associated with high flow speed. The final stage is expansion area formed just in front of the exit of the hole. The pressure of hydrogen flow drops to the ambient pressure at the last stage.

The velocity profile in Fig. 3-(b) also reveals that the flow develops quite rapidly in the inlet region of hole. The relatively quiescent flow in the chamber, where the velocity is in the range of 0.05m/s, is rapidly accelerated up to 200m/s at the inlet of the hole. The velocity magnitude is reaching 1300m/s, sonic speed of hydrogen, just before exiting the hole, then the flow is choked. The hydrogen jet is further accelerated by the rapid expansion. Eventually the maximum flow velocity occurs at 10D distance from the hole exit, with a Mach number corresponding to 5.7.

Exiting the hole, the flow characteristics become much more complicated to adjust to the ambient condition. The flow immediately outside of the hole shows a character similar to that of supersonic nozzle, in that the flow is accelerated to a supersonic speed. Since the pressure possesses the shortest response time, it drops to the ambient pressure level very rapidly as seen from Fig. 3-(a). When the hydrogen passing through the hole still with a high pressure is exposed to the ambient condition, the flow begins to accelerate because of the gas expansion caused by sudden pressure drop.

The dilution lengths for LFL (Lean Flammability Limit,4% H_2), 1/2LFL (2% H_2), and 1/4LFL (1% H_2), denoted by L_{1LFL}, $L_{1/2LFL}$ and $L_{1/4LFL}$, are plotted in Fig. 4. In case of d=1.0mm, the data for 300bar and 400bar are not available because the jet lengths for 1/4LFL exceed the   domain of experiment and numerical simulation. The experimental and numerical   data are well match for L_{1LFL} and $L_{1/2LFL}$. However, the predicted $L_{1/4LFL}$ has some discrepancy. The discrepancy was particularly large for $L_{1/4LFL}$, reaching almost 35%.

Without any doubt, such differences arise from the inaccurate design of the experiment, including the numerical experiment. Among a number of causes contributing to the experimental errors, the following two causes appear to be outstanding. First, the virtual wall introduced for the numerical stability perhaps contributed over-prediction of the hydrogen concentrations from the numerical experiments. As the jet being developed, the jet diameter becomes wide enough to be affected by the virtual wall, which prevents the jet from further expanding. Consequently, the numerical experiments would over-predict the hydrogen concentration profiles particularly in the downstream region. In addition, there is another cause of the measurement error. The concentration of hydrogen is very sensitive to the wind in the far downstream of jet, even though the wind speed is very small.

However, this large discrepancy may not be important to determine a safety distance. It was reported that the hydrogen jet cannot be ignited even at the LFL location This is also confirmed from our own simple ignition test, in which the ignition source had to be placed much closer to the jet exit than the LFL location in order to ignite the jet. Considering the facts that $L_{1LFL \rightarrow 1/2LFL}$, the distance between 4% and 2% locations, is almost comparable to L_{1LFL} and that hydrogen jet is extremely unlikely to be ignited below 4% hydrogen volume fraction (1LFL), it would be sufficient that $L_{1/2LFL}$ be used as a guideline to determine the safety distance.

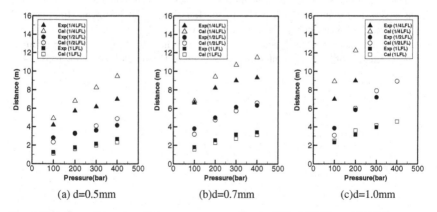

(a) d=0.5mm   (b)d=0.7mm   (c)d=1.0mm

Fig. 4 Hydrogen dispersion distances, corresponding to LFL, 1/2LFL and 1/4LFL

4 Conclusion

The characteristics of hydrogen jet, formed by high pressure release through a small hole, are investigated experimentally and numerically in order to provide the physical basis to determine the safety distance in hydrogen filling stations.

The Froude numbers, corresponding to the accidental releases of high pressure hydrogen, are estimated to be of order of 10^8, so that the effect of buoyancy is negligible in determining the trajectory of jet movement. The 2-D axisymmetric nature is verified by the comparison between experimental and numerical data of hydrogen concentration. The two results are well matched with each other in the meaningful region.

From the centerline hydrogen mole fraction measurements and numerical simulations, the hydrogen dispersion distances, corresponding to 4%, 2% and 1% hydrogen volume fractions, are obtained for several hole diameters and release pressures. Considering the facts that , the distance between 4% and 2% locations, is almost comparable to and that hydrogen jet is extremely unlikely to be ignited below 4% hydrogen volume fraction (1LFL), it would be sufficient that be used as a guideline to determine the safety distance.

Acknowledgments. This work was conducted on behalf of the Korean Ministry of Land, Transport and Maritime Affairs under a program called "Development of Technology for CO2 Marine Geological Storage".

References

[1] Ruffin, E., Mouilleau, Y., Chaineaux, J.: Large scale characterization of the concentration field of supercritical jets of hydrogen and methane. J. of Loss Prev. Process Ind. 9, 279–284 (1996)
[2] Tanaka, T., Azuma, T., Evans, J.A., Cronin, P.M., Johnson, D.M., Cleaver, R.P.: Experimental study on hydrogen explosions in a full-scale hydrogen filling station model. International Journal of Hydrogen Energy 32, 2162–2170 (2007)
[3] Takeno, K., Okabayashia, K., Kouchia, A., Nonaka, T., Hashiguchia, K., Chitose, K.: Dispersion and explosion field tests for 40MPa pressurized hydrogen. International Journal of Hydrogen Energy 32, 2144–2153 (2007)
[4] Houf, W., Schefer, R.: Analytical and experimental investigation of small-scale unintended releases of hydrogen. International Journal of Hydrogen Energy 33, 1435–1444 (2008)

SIMD Computer Using 16 Processing Elements for Multi-Access Memory System

Jea-Hee Kim, Kyung-Sik Ko, and Jong Won Park

Abstract. Improving the speed of image processing is in great demand according to spread of high definition visual media or massive images applications. A Single Instruction Multiple Data (SIMD) computer attached to a host computer can accelerate various image processing and massive data operations. Especially, MAMS (Multi-access Memory System) can not only reduce the access time of the memory but also improve the cost and complexity for the large volume of data demanded in visual media applications. This paper presents implementation of simulator for image processing algorithms and speed comparison of an SIMD architectural multi-access memory system, MAMS-PP16 which consists of MAMS and 16 processing elements (PEs), executed with MAMS-PP4 which consists of MAMS and 4 processing elements(PEs). The newly designed MAMS-PP16 has a 64 bits instruction format and application specific instruction sets. The result of performance analysis verifies performance improvement and the consistent response of MAMS-PP16 through the morphology filter operation in image processing algorithms, comparing with a CISC-based(Intel Processor) Serial Processor.

Keywords: Instruction Format, Multi-Access Memory System(MAMS), Single Instruction Multiple Data (SIMD)Computer.

1   Introduction

Expanding such high definition visual media applications as medical services demands to improve the processing speed of the visual media, which has usually been investigated with a parallel processing system like a single instruction multiple data computer. An SIMD computer attached to a host computer can accelerate various image processing and massive data operations [1]. Multi-Access Memory System

Jea-Hee Kim · Kyung-Sik Ko · Jong Won Park
Information Communications Engineering Dept. Chungnam National
University Daejeon, Korea
e-mail : jwpark@cnu.ac.kr

F.L. Gaol et al. (Eds.): Proc. of the 2011 2nd International Congress CACS, AISC 145, pp. 495–501.
springerlink.com   © Springer-Verlag Berlin Heidelberg 2012

(MAMS), originally proposed by Park, was a sound architecture to reduce the access time of the memory system [1]-[4]. MAMS was especially suitable for the SIMD computer. MAMS also improved the cost and complexity that involved in controlling the large volume of data demanded in visual media applications. MAMS is thus appropriate for the high-speed processing of visual media [4]. The MAMS support simultaneous access to various Subarray types of data elements that are related by a constant interval, where the number of memory modules (MMs) of the MAMS is greater than or equal to the number of Processing Elements (PEs) [1]. A MAMS-PP16 System consists of 16 PEs and 17 MMs in a single chip. This machine aims to achieve high performance of processing high definition class images in real time. This paper presents implementation of image processing algorithms and performance analysis for a SIMD architectural multi-access memory system, MAMS-PP16. MAMS-PP16 which consists of 16 PEs with 17 MMs is an extension of the prior work, MAMS-PP4 [4], which consists of 4 PEs with 5 MMs to processor. In order to implement the algorithms we designed a 64-bit long instruction format and developed instructions specialized for processing various images sizes in the MAMS-PP16 System. We additionally developed a simulator of the MAMS-PP16 System, which implemented algorithm can be executed on speed comparison has been done with this simulator executing implemented algorithms of processing images. Section 2 briefs the MAMS architecture and the MAMS-PP16 platform as a target machine of designing an instruction format and instructions. Section 3 discusses implementation of the MAMS-PP16 Simulator, and then Section 4 describes speed comparison of the MAMS-PP16 System.

2   MAMS (Multi-Access Memory System)

MAMS is a multi-access memory system which is, along with multiple PEs, adequate for establishing a high performance SIMD computer. MAMS supports simultaneous access to pq data elements within a horizontal (1 X pq), a vertical (pq X 1) or a block (p X q) subarray with a constant interval in an arbitrary position in an M X N array of data elements, where the number of MMs, m, is a prime number greater than pq.

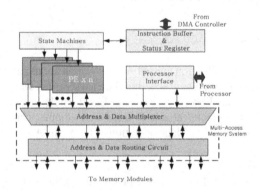

Fig. 1 MAMS based SIMD computer.

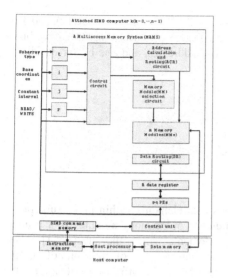

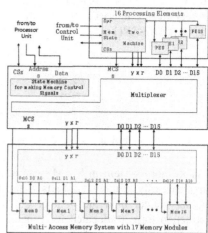

Fig. 2 General design of MAMS.   **Fig. 3** Functional block of MAMS-PP16.

MAMS reduces the memory access time for an SIMD computer and also improves the cost and complexity for the large volume of data demanded in visual media applications. Fig. 1 shows a MAMS based SIMD computer applicable to high speed images or visual data processing, and Fig. 2 shows a general design of MAMS. MAMS-PP4, the first implementation of a MAMS-based parallel processing architecture, consists of 4 PEs and 5 MMs in a single chip [4].

By analyzing operations of the Morphological filter operations, 16 instructions are designated for the MAMS-PP4 System. MAMS-PP16 System is an extended version up MAMS-PP4 System in order to achieve higher performance of processing high definition class images in real time, which includes 16 PEs and 17 MMs in a single chip. MAMS-PP16 System is an extended version up MAMS-PP4 System in order to achieve higher performance of processing high definition class images in real time, which includes 16 PEs and 17 MMs in a single chip. A functional block diagram of the MAMS-PP16 machine is shown in Fig. 3. Opr and Mem are referred to general instruction and memory instruction, respectively. CSs are for control signals and MCSs for memory control signals. D and A are initial letters of data and address, respectively

3   Desingn of Simulator

3.1   Instruction Fetch Module

The instruction format for the MAMS-PP16 System is shown in Fig. 4. The instruction format consists of totally 64 bits, where bits 0 through 31 are allocated for general operations and bits 32 through 63 are allocated for memory access operations. Memory access operations include data location coordinates. The instruction format including data location coordinates consists of both 12 bits for x-axis, X(12), and 12 bits for y-axis, Y(12), four bits subarray types, Type(3), for accessing data, three bits memory storing interval, INTV(4), to calculate addresses, and one bit for deciding read or write, R/W(1).

63	62	59 58	56 55	44 43	32
RW	INTV	Type	X(12)	Y(12)	

31	24 23	16 15	0
OpC	Offset	Op	

Fig. 4 Instruction format.

A memory system in MAMS can access data elements within given subarrays simultaneously without conflicts if the number of memory modules is a prime number greater than the number of data elements within the subarray[1]. The instruction format allocated for general operations consists of eight bits of an op code, OpC(8), for classifying operations, eight bits of an offset, Offset(8), and 16 bits of an operand, Op(16), for operation parameters. Instruction Fetch Module receives the instruction whose is shown in Fig. 4 from Host Computer, and drives the State Machine Module and Processing Elements Module.

3.2 State Machin Module

The MAMS system has prime number of memory modules and provides 8-SubArray Type which is simultaneous access method as shown Fig. 5[1].

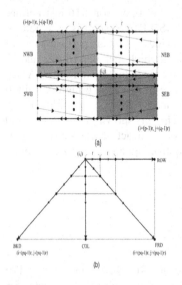

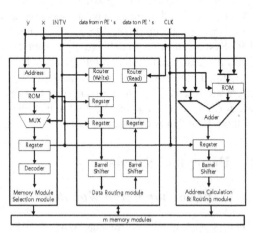

Fig. 5 Subarray Types.

Fig. 6 Functional block diagram of State Machine Module.

In order to apply the provided method to MAMS-PP16, the state machine module is configured with RW, INTV, Type, X and Y provided from the Instruction Fetch Module as shown Fig. 6.

3.3   Processing Elements Module

Processing Elements Module is provided with OpC, Op and Offset from the Instruction Fetch Module, thereby accomplishes with arithmetic, logical, and conditional search operation same general instruction, In order to approach the memory it access a memory address from State Machine Module.

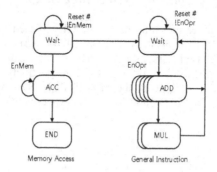

Fig. 7 Flow chart of the state machine of Pocessing Element Module.

Processing Elements Module executes Erosion, Dilation, Opening and Closing of Morphological filter operations. Fig. 7 shows the state transition machine easy to perform general instruction and memory access of PE.

4   Speed Comparison of MAMS-PP16 System
Has Been Compared with CISC-Based Serial Processor

In this paper speed of the MAMS-PP16 System has been compared with CISC-based serial processor for the Erosion of Morphological filtering, with image sizes of 64X64, 128X128, 256X256 and 512X512. After simulating born systems by Visual C++.The serial processor uses only one PE and MAMS-PP16 System uses with image sizes of 64X64, 128X128, 256X256 and 512X512, each of which uses the Erosion of Morphological filtering in testing the performance.

4.1   Comparison of Image Results

In regard of 128X128 pixel projection to test the performance Fig. 8(a) shows the result image by MAMS-PP16 System and Fig. 9(a) shows the result image by CISC-based Serial Processor. Each of Fig. 8(b) and Fig. 9(b) shows the same pixel value of the result image of Fig. 8(a), respectively and Fig. 9(a).

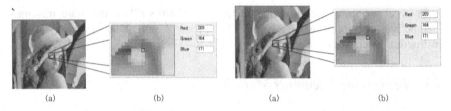

(a)　　　　　　　　　　(b)　　　　　　　　　(a)　　　　　　　　　(b)

Fig. 8 Pixel value for MAMS-PP16.　　**Fig. 9** Pixel value for CISC-based Serial Processor.

4.2　Speed Comparison for Each Image Sizes

Fig.10 shows the number of sampling of the CISC-based Serial Processor(Intel Processor) and MAMS-PP16 System. The Fig.10 (a) ~ (d) represent the number of samples of the CISC-based Serial Processor(Intel Processor) for the image sizes of 64X64, 128X128, 256 X256, and 512 X512, respectively. The Fig.10 (e) ~ (h) represent the number of samples of the MAMS-PP16 System for the image sizes of 64X64, 128X128, 256 X256, and 512 X512, respectively.

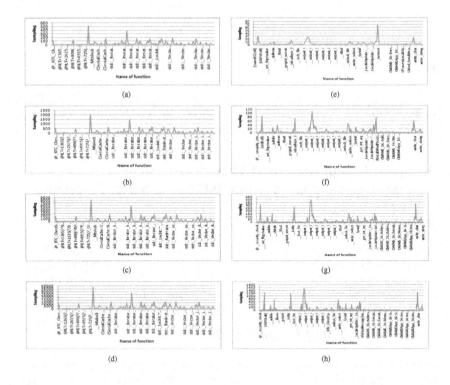

(a)　　　　　　　　　　　　　　　　　(e)

(b)　　　　　　　　　　　　　　　　　(f)

(c)　　　　　　　　　　　　　　　　　(g)

(d)　　　　　　　　　　　　　　　　　(h)

Fig.10 Number of Samples.

The Profiler collects one sample every 10 million CPU cycles, and Ratio of Intel Processor to MAMS-PP16 for each image size is shown in Table 1.The speed improvement of MAMS-PP16 System to Intel Processor is 9.3 times, 10.55 times, 11.1 times, and 12.8 times for image size of 64X64, 128X128, 256X256, and 512 X512, respectively.

Table 1 Experimental results

IMAGE SIZE	Processor		
	MAMS-PP16 (number of samples**)	Intel Processor (number of samples**)	Ratio of Intel Processor to MAMS-PP16 (times)
64x64	491	4,613	9.3
128x128	1,692	17,866	10.55
256x256	6,548	72,688	11.1
512x512	26,104	336,741	12.8

** The sampling period in profiler is 10^7 CPU cycles.

5   Conclusions

In this paper, MAMS-PP16 System which was an extension of MAMS-PP4 [2] was simulated by using Visual C++, and compared with CISC-based Serial Process by Visual C++. MAMS-PP16 System consists of MAMS with 17 MMs and 16 PEs, where MAMS-PP4 System with 5 MMs and 4 PEs. In addition, the Instruction Format was expanded from the previous 34 bits to 64 bits. Also interval of 1, 2, 3, and 4   and 8-SubArray Type were implemented in the MAMS-PP16 System. MAMS-PP16 System was compared with CISC-based Serial Process for Erosion Morphology filter. The application of MAMS-PP16 System can be general SIMD Computer algorithms as well as image processing algorithms. A hardware FPGA Platform is now being designed for MAMS-PP16 System.

References

[1] Park, J.W.: Multiaccess Memory System for Attached SIMD Computer. IEEE Trans. on Computers 53(3), 353–359 (2004)
[2] Lee, H., Park, J.W.: Parallel processing for multi-access memory system. In: World Multi-Conference on Systems, Cybernetics, and Informatics (SCI 2000), vol. VII, part I, pp. 561–565 (March 2001)
[3] Park, J.W.: An efficient buffer memory system for subarray system for subarray access. IEEE Trans. Parallel and Distributed Systems 12(3), 316–335 (2001)
[4] Lee, H., Cho, H.K., You, D.S., Park, J.W.: An MAMS-PP4: Multi-Access Memory System used to improve the processing speed of visual media applications in a parallel processing system. IEICE Trans. Fundamentals E87-A (11), 2852–2858 (2004)

Detection of Broken Rotor Bars in Induction Motors Using Unscented Kalman Filters

Damian Mazur

Abstract. Determining the number of broken bars due to the failure or the number of ring segments of the rotor cage of the asynchronous motor is the important problem during the operation of such machine. UKF filter, used to achieve this goal, gave exact information: numbers of broken bars and ring segments. For receiving such exact information the filter needed the measurements of the stator currents and currents of the additional one-winding coils placed at stator teeth and loaded by big resistance (~1 kΩ). The measurement of the rotation speed of the rotor and connected with this the angle of the machine shaft rotation was needed also. The method of determining the numbers of broken bars and ring segments of the rotor and the method of detecting only the broken bars of the rotor cage were tested. The presented method can be used for protecting big and expensive asynchronous drives.

1 Introduction

There are similarities between EKF and UKF filters found. A rule of work of EKF filter applied by authors in [5] is shown for nonlinear system:

$$\begin{cases} x_k = f(x_{k-1}, v_{k-1}, u_{k-1}) \\ y_k = h(x_k, \eta_k, u_k) \end{cases} \tag{1}$$

Where x_k are state variables for time sample k -tej, v is process noise, η is observation noise, u - input voltage, y - observed quantities. For the step $k = 0$ comes initialization

Damian Mazur
Rzeszow University of Technology, The Faculty of Electrical
and Computer Engineering, ul. W. Pola 2, 35-959 Rzeszow, Poland
e-mail: mazur@prz.edu.pl

F.L. Gaol et al. (Eds.): Proc. of the 2011 2nd International Congress CACS, AISC 145, pp. 503–511.
springerlink.com

$$\hat{x}_0 = E[x_0]$$
$$P_{x_0} = E[(x_0 - \hat{x}_0)(x_0 - \hat{x}_0)^T]$$
$$P_V = E[(v - \overline{v})(v - \overline{v})^T]$$
$$P_\eta = E[(\eta - \overline{\eta})(\eta - \overline{\eta})^T]$$
$$(2)$$

For the next time steps, k we perform:
a) prediction step, i.e.
a1) process Jacobian model is calculated:

$$F_{x_k} = \nabla_x f(x, \overline{v}, u_{k-1})|_{x=\hat{x}_{k-1}}$$
$$G_V = \nabla_V f(\hat{x}_{k-1}, v, u_k)|_{v=\overline{v}}$$
$$(3)$$

a2) time value of mean state variables and their covariance are calculated:

$$\hat{X}_k^- = f(\hat{x}_{k-1}, \overline{v}, u_k)$$
$$P_{x_k}^- = F_{x_k} \cdot P_{x_k} \cdot F_{x_k}^T + G_V \cdot P_V \cdot G_V^T$$
$$(4)$$

b) correction step, including:
b1) observation process Jacobian model is calculated:

$$H_{x_k} = \nabla_x h(x, \overline{n}, u_k)|_{x=\hat{x}_k^-}$$
$$D_\eta = \nabla_\eta h(\hat{x}_k^-, \eta, u_k)|_{\eta=\overline{\eta}}$$
$$(5)$$

b2) estimation is adjusted with the aid of the last observation

$$K_k = P_{x_k}^- H_{x_k}^T (H_{x_k} \cdot P_{x_k}^- H_{x_k}^T + D_\eta \cdot P_\eta \cdot D_\eta^T)^{-1}$$
$$\hat{X}_k = \hat{X}_k^- + K_k(y_k - h(\hat{x}_k^-, \overline{\eta}))$$
$$P_{x_k} = (I - K_k \cdot H_{x_k})P_{x_k}^-$$
$$(6)$$

In order to obtain bigger accuracy, however UKF filter was used for diagnostics testing of asynchronous machine rotor [1]. For the sake of the fact that a character of process v and measurement η noises is not known (they are not most probably of addictive character) an extended model of UKF filter was used. State variables of the extended model are variables of process state x, process noises v and measurement noises η combined in one vector x^a [1].
For the step $k = 0$ comes initialization:

$$\hat{x}_0 = E[x_0]$$

$$P_0 = E[(x - \hat{x}_0)(x - \hat{x}_0)^T]$$

$$\hat{x}_0^a = E\left[x^a\right] = \left[\hat{x}_0^T, [0]^T, [0]^T\right]^T$$

$$P_0^a = E\left[(x_0^a - \hat{x}_0^a)(x_0^a - \hat{x}_0^a)^T\right] = \begin{bmatrix} P_0 & [0] & [0] \\ [0] & R^V & [0] \\ [0] & [0] & R^\eta \end{bmatrix} \tag{7}$$

where (0) denotes a matrix in zero elements.

Next, for the next time step k UT transformation is calculated, i.e. sigma points χ:

$$\chi_{k-1}^a = \left[\hat{x}_{k-1}^a, \ \hat{x}_{k-1}^a + \gamma \cdot \sqrt{P_{k-1}^a}, \ \hat{x}_{k-1}^a - \gamma \cdot \sqrt{P_{k-1}^a}\right] \tag{8}$$

If dimension of the extended vector of state variable χ^a is L, then $\gamma = \sqrt{L + \lambda}$ where scaling parameter λ [1]:

$$\lambda = \alpha^2 \cdot (L + \kappa) - L \tag{9}$$

$\alpha = 0.001$, $k = 3 - L$ where assumed in the calculations which were presented here. A parameter β was also introduced which informed about predicted character of distribution of random state variables x. The Gauss distribution i.e. $\beta = 2$ was assumed. Instead of performing a big quantity of process samples for the k-step like in Monte Carlo method, UKF filter method under consideration uses UT transformation and examines distribution of samples χ_{k-1}^a (8), by quantity $2L + 1$.

The following steps of UKF filtration are performed, including:
a) correcting step:

$$\chi_{k|k-1}^x = f\left(\chi_{k-1}^x, \chi_{k-1}^v, u_{k-1}\right) \tag{10}$$

$$\hat{x}_k^- = \sum_{i=0}^{2L} W_i^{(m)} \cdot \chi_{i, k|k-1}^x \tag{11}$$

$$P_k^- = \sum_{i=0}^{2L} W_i^{(c)} \cdot \left[\chi_{i, k|k-1}^x - \hat{x}_k^-\right]\left[\chi_{i, k|k-1}^x - \hat{x}_k^-\right]^T \tag{12}$$

$$\Upsilon_{k|k-1} = h\left(\chi_{k|k-1}^x, \chi_{k|k-1}^\eta, u_k\right) \tag{13}$$

$$\hat{y}_k^- = \sum_{i=0}^{2L} W_i^{(m)} \cdot \Upsilon_{i, k|k-1} \tag{14}$$

b) prediction step:

$$P_{\bar{y}_k \, \bar{y}_k} = \sum_{i=0}^{2L} W_i^{(c)} \cdot \left[\Upsilon_{i,\,k\,|\,k-1} - \hat{\bar{y}_k} \right] \left[\Upsilon_{i,\,k\,|\,k-1} - \hat{\bar{y}_k} \right]^T \qquad (15)$$

$$P_{\bar{x}_k \, \bar{y}_k} = \sum_{i=0}^{2L} W_i^{(c)} \cdot \left[\mathcal{X}_{i,\,k\,|\,k-1}^x - \hat{\bar{x}_k} \right] \left[\Upsilon_{i,\,k\,|\,k-1} - \hat{\bar{y}_k} \right]^T \qquad (16)$$

reinforcement:

$$K_k = P_{\bar{x}_k \, \bar{y}_k} \cdot P_{\bar{y}_k \, \bar{y}_k}^{-1} \qquad (17)$$

$$\hat{x}_k = \hat{x}_k^- + K_k \cdot (y_k - \hat{y}_k^-) \qquad (18)$$

where y_k is measurements

$$P_k = P_k^- - K_k \cdot P_{\bar{y}_k \, \bar{y}_k} \cdot K_k^T \qquad (19)$$

$W_i^{(m)}$ and $W_i^{(c)}$ coefficients were used to calculate a mean value and covariance in the presented equations, which are equal [1]:

$$\begin{cases} W_0^{(m)} = \dfrac{\lambda}{L+\lambda} \\ W_0^{(c)} = \dfrac{\lambda}{L+\lambda} + (1-\alpha^2+\beta) \\ W_i^{(m)} = W_i^{(c)} = \dfrac{1}{2(L+\lambda)} \end{cases} \qquad (20)$$

When calculating sigma points $\mathcal{X}_{k-1}^a$ from the formula (8) $\sqrt{P^a}$ designation determines a lower triangle of Choleski factorization matrix P^a.

2 Electric Machine and Its Model

It means extension of the set of variables of a machine conditions by estimated parameters i.e. bar and ring segment resistance. Equations of the system (1), which correspond to these new estimated. Variables are their quality in k-1 and k-time step [9, 10]. The time step dt = 1e-4 s was assumed. Simultaneously, this is a period of time to which inverter control from an object was set. A model used by UKF filter had averaging supply voltage fed to division time dt. In the dy4 program, which estimated only bar resistance, state vector variables x were:

$x(1:13)$ - bar resistances

$$x(14:27) - \int_0^t [R_{\text{rotor eyelets}}][i_{\text{rotor eyelets}}]dt$$

$$x(28:39) - \int_0^t [i_{\text{stator teeth coil}}]dt$$

(21)

$$x(40:42) - \int_0^t [i_{st}]dt \quad - \quad \text{current charge of the stator winding}$$

$$x(43:56) - \int_0^t [i_{\text{rotor eyelets}}]dt$$

$nw = 13$ eyelets formed by adjacent bars and ring segments to match them were isolated in the rotor cage. An additional eyelet was a circuit composed only of ring segment on the rotor left side. Estimation ability of ring segment resistance was not forecast in the dy4 program. The other dy2 program, that was worked out, forecast except for bar resistance estimation also ring segment resistance estimation on the both rotor sides. Therefore, state variables x for the dy2 program represented:

$x(1:13)$ - bar resistance
$x(14:26)$ - ring segment resistance on the rotor left side
$x(27:39)$ - ring segment resistance on the rotor right side   (22)
$x(40:82)$ - like $x(14:56)$ in the dy4 program,
deseribed by the formula (21).

3   Simulations in Calculation

Estimated bar and ring segment resistance should be pluses. This dependence should be included in calculations in the form of constrains. If limitations are exceeded according to [1], estimated quantities should be viewed on limitations and views of these quantities should be taken for further calculations. However, temporary minus values of estimated resistance were allowed for calculation performed. A function suitably shaped was used which diminished possible minus values of resistance which could appear in calculations. It is shown in the figure 1.

P_k covariance matrix monitoring from the formula (19) turned out to be equally important for stability of calculations. Too big values of this matrix elements can lead to big changes of estimated resistances. Calculations can then reach local minimum for estimated resistances and indications of numbers of broken bars or ring segments may turn out to be incorrect. On the other hand, it is easily possible to predict maximum values for groups of state variables e.g. shown in the formula (22).

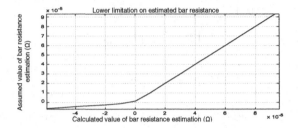

Fig. 1 Dependence of bar resistance (axis y) assumed for further calculations on this quantity calculated by UKF-a similar function was used for ring segment resistance

Also it is naturally essential here to predict increased values $x(1{:}39)$, for the sake of assumed break of bar or ring segment. It is assumed that covariance diagonal value P_k from the formula (19) should not be larger than square of these correct maximum values of variables, multiplied by number 0.0001 selected here for resistance and by 0.01 for the other variables. Let's mark this number as α. If not, this quantity is brought on i-diagonal $P_k\,(i,i)$ to assumed value a, in the way given below through:

$$
\begin{cases}
wsp = \sqrt{a/P_k(i,i)} \\
P_k(:,i) = P_k(:,i) \cdot wsp \\
P_k(i,:) = P_k(i,:) \cdot wsp
\end{cases}
\tag{23}
$$

According to the notation of the MATLAB program used here for calculations, the marking $P_k\,(:,i)$ stands for i-matrix column P_k (similarly $P_k(I,:)$ stands for i-row of matrix P_k). Owning to this procedure, local minima trap was avoided.

4 Results of Calculations

Calculations were performed al bevel of rotor bars which made estimation of numbers of broken bars or ring segments difficult. The dy4 program estimated broken bars assuming that ring segments are free from damage. At speed about 300 rad/s the program started to point at the broken bar in number 8, though it was free from damage. Maybe the cause was its proximity to damaged bars (from 1 to 7) and bevel of machine bars.

Estimation of broken bars 1, 4, 7, 10, 13 at rings free from damage was shown in the figure 2.

The dy4 program assumed to calculate that all ring segment are free from damage which, however, did not prevent from correct estimation.

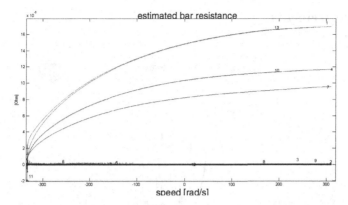

Fig. 2 Estimation of broken bars 1, 4, 7, 10, 13 by the dy4 program (rings free from damage)

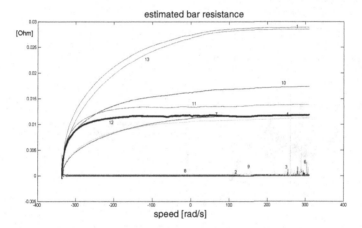

Fig. 3 Estimations of broken bars 1, 4, 7, 10, 13 by the dy4 program at additionally broken 9÷13 ring segments, on the left side which were not estimated by the program (in rotational speed function)

During estimation of broken bars 1, 4, 7, 10, 13 by the dy4 program at additionally broken segments 9÷13 of the left ring, shown in the figure 3, the dyh4 program additionally pointed at a break of bars 11 and 12. The dy4 program found abnormality within these ring segments but their estimation was not performed and as a substitution it pointed at break of bars falling under the failure, thought they were free from damage. Whereas the dy2 program estimated resistance to bars and ring segments on the rotor both sides.

5 Measurements

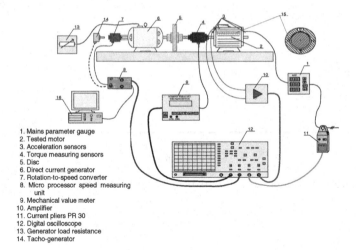

1. Mains parameter gauge
2. Tested motor
3. Acceleration sensors
4. Torque measuring sensors
5. Disc
6. Direct current generator
7. Rotation-to-speed converter
8. Micro processor speed measuring
   unit
9. Mechanical value meter
10. Amplifier
11. Current pliers PR 30
12. Digital oscilloscope
13. Generator load resistance
14. Tacho-generator

Fig. 4 The measurement verification circuit

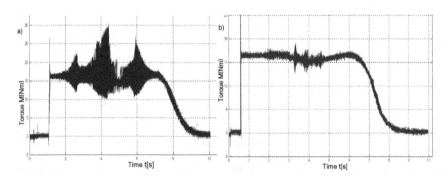

Fig. 5 Shaft torque measurements a) in the motor with dynamic eccentricity and 28 straight rotor cage bars; b) in the motor free from damage with 26 straight rotor cage bars

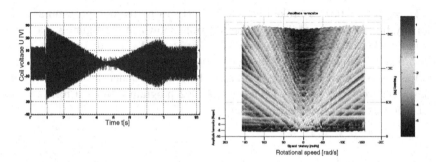

Fig. 6 Measuring of induced voltage in the coil in the motor free from damage and with 28 straight cage rotor bars

6 Conclusions

Considering the fact that a break of ring segments is a rare occurrence, the dy4 program that includes fewer variables for estimation, seems sufficient in practice. A relatively good estimation is owed to the exact machine model [2,5], which gives consideration to higher harmonics magnetic field in air-gap. Also, inverter feed which includes higher harmonics voltages, activates correct currents in windings furnishing more complete information for estimation process.

The dy2 and dy4 programs presented here represented a part of software of the dy programs that are used for diagnosing of the asynchronous machine.

References

1. Wan, E.A., van der Merwe, R., Nelson, A.T.: Dual Estimation and the Unscented Transformation. In: Solla, S.A., Leen, T.K., Müller, K.-R. (eds.) Advances in Neural Information Processing Systems, pp. 666–672. MIT Press (2000)
2. Gołębiowski, L.: Analiza falkowa przebiegów maszyny asynchronicznej. In: XLI International Symposium on Electrical Machines, SME 2005, Opole, pp. 525–530 (2005)
3. Gołębiowski, L.: Analiza falkowa przebiegów maszyny asynchronicznej. Przegląd Elektrotechniczny (10), 30–34 (2005)
4. Noga, M., Gołębiowski, L., Gołębiowski, M., Mazur, D.: Sterowanie silnikiem synchronicznym z wewnętrznymi magnesami stałymi IPMS uwzględniające ograniczenia. Maszyny Elektryczne: zeszyty problemowe 82, 55–62 (2009)
5. Gołębiowski, L., Gołębiowski, M., Noga, M., Skwarczyński, J.: Strumień osiowy w modelu 3D maszyny indukcyjnej. Elektrotechnika i Elektronika, Akademia Górniczo-Hutnicza w Krakowie 25, 2, 147–152 (2006)
6. Gołębiowski, L., Gołębiowski, M., Mazur, D.: Estymacja Kalmana parametrów[R], [L_{rozpr}] trójkolumnowego, 15-uzwojeniowego autotransformatora w układzie prostownika 18-pulsowego clean power. In: VIII Seminarium Naukowe WZEE 2008, Białystok-Białowieża, September 22-24 (2008)

Author Index